Multiple Criteria Decision Making

Series editor

Constantin Zopounidis
School of Production Engineering and Management
Technical University of Crete
Chania
Greece

Multiple Criteria Decision Making

This book series focuses on the publication of monographs and edited volumes of wide interest for researchers and practitioners interested in the theory of multicriteria analysis and its applications in management and engineering. The book series publishes novel works related to the foundations and the methodological aspects of multicriteria analysis, its applications in different areas in management and engineering, as well as its connections with other quantitative and analytic disciplines.

In recent years, multicriteria analysis has been widely used for decision making purposes by institutions and enterprises. Research is also very active in the field, with numerous publications in a wide range of publication outlets and different domains such as operations management, environmental and energy planning, finance and economics, marketing, engineering, and healthcare.

More information about this series at http://www.springer.com/series/13834

Julio Berbel • Thomas Bournaris • Basil Manos •
Nikolaos Matsatsinis • Davide Viaggi
Editors

Multicriteria Analysis in Agriculture

Current Trends and Recent Applications

Springer

Editors
Julio Berbel
Department of Agricultural Economics
University of Cordoba
Cordoba, Spain

Thomas Bournaris
Department of Agricultural Economics
Aristotle University of Thessaloniki
Thessaloniki, Greece

Basil Manos
Department of Agricultural Economics
Aristotle University of Thessaloniki
Thessaloniki, Greece

Nikolaos Matsatsinis
Decision Support Systems Laboratory
Technical University of Crete
Chania, Greece

Davide Viaggi
Department of Agricultural and Food Sciences
University of Bologna
Bologna, Italy

ISSN 2366-0023 ISSN 2366-0031 (electronic)
Multiple Criteria Decision Making
ISBN 978-3-030-08327-4 ISBN 978-3-319-76929-5 (eBook)
https://doi.org/10.1007/978-3-319-76929-5

Printed on acid-free paper

This Springer imprint is published by the registered company Springer International Publishing AG part of Springer Nature.
The registered company address is: Gewerbestrasse 11, 6330 Cham, Switzerland

Preface

Economics is the science that tries to allocate scarce resources for the satisfaction of multiple and frequently conflicting objectives. Decision-making in agriculture is always the process that typically involves multiple criteria, especially when new policies are applied in agriculture and the environment. Multicriteria analysis is a methodology that considers multiple criteria simultaneously in a wide range of concerns in complex decision-making processes by taking into account multiple conflicting criteria.

This book integrates 12 chapters and outlines the current trends in the use of multicriteria analysis in agriculture by providing recent applications for modelling agricultural decision-making. Specific case studies using multicriteria analysis as a method for selecting multi-attribute discrete alternatives or solving multi-objective planning problems are also considered. The volume intends to target agricultural and environmental economists, engineers and all scientists concerned with the management of agricultural resources and decision-making in agriculture.

In chapter "Multi-scaling Agroclimatic Classification for Decision Support Towards Sustainable Production" a decision support system by using multicriteria analysis combining different criteria to a utility function under a set of constraints concerning different categories of agroclimatic, social, cultural and economic conditions and to achieve an optimum agricultural production plan is developed by conducting contemporary agroclimatic classification based on remote sensing and geographic information system.

Chapter "Review of Multicriteria Methodologies and Tools for the Evaluation of the Provision of Ecosystem Services" focuses on studies that used multicriteria decision analysis for an ecosystem service assessment, attempting to cover a representative sample of case studies of ecosystem service assessments through multicriteria decision analysis. It also discusses advantages and disadvantages of different methodological choices in ecosystem service evaluation. For this reason, a literature review has been performed that covers an overview of various methodologies that seek to improve the knowledge base of existing tools and methodologies in ecosystem services assessment.

Chapter "Integrating AHP and GIS Techniques for Rural Landscape and Agricultural Activities Planning" aims at providing some insights into the usefulness of the analytic hierarchy process in the context of geographic multicriteria analysis, applied to geographic information system techniques for rural landscape and agricultural activities planning. This chapter illustrates four case studies from Tuscany Region (Italy) where this approach has been applied.

Chapter "The Use of the Analytic Network Process for the Analysis of Public Goods Supply from Agricultural Systems: Advances and Challenges Ahead" aims to review and summarize the potentials and limitations of the use of the analytic network process in the context of the evaluation of public goods provision from agricultural systems. The chapter provides a description and a step-by-step explanation of the method and presents insights from three selected recent papers using the analytic network process to analyse public goods supply from agricultural systems.

Chapter "Allocating Shadow Prices in a Multi-objective Chance Constrained Problem of Biodiesel Blending" tries to allocate shadow prices in a multi-objective chance-constrained problem of biodiesel blending. The chapter presents a model that determines the optimal blend that minimizes production costs and GHG emissions and assesses the influence of technical constraints on the decision objectives. For this purpose, an algorithm for the allocation of shadow prices to the constituent parts of the composite objective function was implemented.

Chapter "Promoting Small-Scale Biofuel Production: A Qualitative GIS-OWA Methodology for Land Suitability Analysis of Winter Rapeseed" focuses on promoting small-scale biofuel production as biofuels could be a possible solution to promote agricultural development in rural areas by increasing farm income. As land planning issues are complex problems with multiple decision-makers and criteria, a spatial multicriteria analysis model for supporting decision-makers in the site selection process for winter rapeseed production is proposed.

Chapter "Multi-criteria Decision Analysis: Linear and Non-linear Optimization of Aqueous Herbal Extracts" aims to present a multicriteria decision analysis with linear and non-linear optimization of aqueous herbal extracts. Modelling is an indispensable part of food production, from "farm to fork", where it is used to optimize the initial production of food and feed as well as in the food and feed processing. Different particle sizes of olive leaves were used in extraction of biologically active components using water as a solvent.

Chapter "Methodology and Criteria for the Allocation of Reused Water in Agriculture" provides a methodology to support the water allocation decision, applied to reused water. The tool has been developed for the Guadalquivir River Basin Authority, allowing decision-makers to rank the water users asking for an entitlement to urban water for agriculture when demand is higher than supply.

The objective of chapter "Simulating Farmers' Decision-Making with a Cobb-Douglass MAUF: An Application for an *Ex-Ante* Policy Analysis of Water Pricing" is to provide more in-depth knowledge about simulating farmers' behaviour by using non-linear multi-attribute utility functions, developing a new non-interactive method to elicit Cobb–Douglas multi-attribute utility functions based on farmers' actual behaviour that overcomes some shortcomings of traditional additive multi-attribute utility functions.

Chapter "Perceptions and Attitudes of Greek Farmers Towards Adopting Precision Agriculture: Case Study Region of Central Greece" aims to examine the concepts of crop producers regarding the prospects that arise through the adoption of Precision Agriculture in Greece, a country with problematic primary sector, with particular climatic conditions and varied microclimates while competing with countries of low labour costs.

Chapter "Multi-criteria Optimization Methods Applied in Agricultural Touring" is a comprehensive review of the multi-objective nature of agricultural touring and focuses on multi-objective formulations that arose in the literature so far in touring, especially concerning tourism, and with regard to them being applied under agritourism scenarios.

Finally, chapter "Life Cycle Assessment and Multi-criteria Analysis in Agriculture: Synergies and Insights" outlines the integration of life cycle assessment and multicriteria analysis methodologies and develops a complete literature review regarding the sustainability of the agricultural sector. In this review, scientific papers integrating life cycle assessment and different multicriteria methodologies in agriculture are analysed.

The main lesson learned from the papers in this book is in the fact that there are a number of emerging areas of application of multicriteria analysis, and this requires an effort for methodological adaptation. Some of these areas actually bring more and more complex issues either due to the comprehensiveness or width of the topic (ecosystem services, bioeconomy issues) or to the "fuzziness" of impacts (e.g. information tools). In addition, multicriteria analysis is part of an important process of integration of different tools, e.g. with LCA or models, as well as with participatory instruments.

This seems to lead to a number of new challenges for researchers to meet the future needs of practitioners and decision-makers.

Cordoba, Spain — Julio Berbel
Thessaloniki, Greece — Thomas Bournaris
Thessaloniki, Greece — Basil Manos
Chania, Greece — Nikolaos Matsatsinis
Bologna, Italy — Davide Viaggi

Contents

Multi-scaling Agroclimatic Classification for Decision Support Towards Sustainable Production

Nicolas R. Dalezios, Kostas Mitrakopoulos, and Basil Manos

Abstract Agriculture is highly affected by environmental conditions and the assessment of the agroclimatic potential is necessary for sustainability and productivity. The climate is among the most important factors that determine the agricultural potentialities of a region and the suitability of a region for a specific crop, whereas the yield is determined by weather conditions. In this chapter the first objective is to identify sustainable production zones in Thessaly by conducting contemporary agroclimatic classification based on remote sensing and GIS. The agroclimatic conditions of agricultural areas have to be assessed in order to achieve sustainable and efficient use of natural resources in combination with production optimization. Thus, a quantitative understanding of the climate of a region is essential for developing improved farming systems. The second objective derives from the first; it develops a decision support system (DSS) by using multi-criteria analysis combining different criteria to a utility function under a set of constraints concerning different categories of agroclimatic, social, cultural and economic conditions and so we can achieve an optimum agricultural production plan. In order to support the realization of the proposed production zoning and DSS in real-time, a Sensor Web service platform is proposed to be implemented based on the Sensor Web technologies, which extracts Real-time environmental and agronomic data.

Keywords Agroclimatic classification · Production zones · Agroclimatic indices · Decision support system · Multi-criteria analysis · Web production platforms

N. R. Dalezios (✉) · K. Mitrakopoulos
Department of Civil Engineering, University of Thessaly, Volos, Greece

B. Manos
Department of Agricultural Economics, Aristotle University of Thessaloniki, Thessaloniki, Greece

J. Berbel et al. (eds.), *Multicriteria Analysis in Agriculture*, Multiple Criteria Decision Making, https://doi.org/10.1007/978-3-319-76929-5_1

1 Introduction

Agriculture is a primary productivity sector, which is highly dependent on environmental conditions. The agroclimatic potential of agricultural areas has to be assessed in order to achieve sustainable and efficient use of natural resources in combination with optimal production (Tsiros et al. 2009). Temperature and rainfall, in terms of quantity and spatiotemporal variability, are variables which determine the type of crops suitable to a given location. These variables, in combination with soil type and geomorphology, also determine areas, where high levels of production are appropriate, avoiding the threat of degrading the natural resources. Extreme climate variables and climate extremes, such as droughts, are projected to experience significant changes over the twenty-first century, just as they have during the past century, in many areas, including the Mediterranean basin, among others (IPCC 2012, 2013). Thus, climate change has to be considered in all the aspects of agroclimatic analysis.

Agroclimatic classification is very useful in identifying sustainable production zones within a climatic region. Recently, a methodology has been developed based on remote sensing data and methods and applied in semi-arid region (Tsiros et al. 2009). Specifically, two drought indices have been computed, namely aridity index (AI) and vegetation health index (VHI), in order to define zones adequate for sustainable farming according to water limitations, called water limited growth environment (WLGE) zones. At the second stage, WLGE zones are combined with soil maps and a digital elevation model (DEM) of the region and two more indices are computed, namely the Growing Degree Days (GDD) and Net Radiation (Rn), in order to define sustainable production zones. In addition, viticultural zoning has been developed at a European scale based on seven bioclimatic indices. Thus, climatic classification at meso-scale level has been achieved leading to zones with different viticultural suitability and their possible geographical shift under future climate change scenarios using regional climate model simulations (Malheiro et al. 2010). As a result, optimal production can be computed for each sustainable production zone, at a multi-scale level, within a major climatic region, which is a research need that has not been investigated yet.

Moreover, during the last decade many Decision Support Systems (DSS) have been developed to supply decision makers with tools on several aspects of climate change and extreme events, as well as mitigation and preparedness planning. In particular, Spatial Multi-Criteria Evaluation (SMCE) approach has been widely used in the development of Decision Support System (DSS) due to its ability to take into account simultaneously all the basic aspects of the problem including heterogeneous sets of many different factors and constraint (Steinemann et al. 2005). Although many applications of DSS already exist, there are just a few recent developments in climate change and vulnerability assessment and in hydrometeorological hazard risk management (e.g. EU project CHANGES) (Van Westen 2013). However, DSSs for assessing adaptation in agroecosystems or drought preparedness and mitigation plans are still not available, although they remain a current research need.

There are several recent applications of web environmental platforms, such as satellite derived online web-maps (Gong et al. 2009). For example, satellite imagery can monitor crop phenology, measure and estimate meteorological parameters, measure the effectiveness of irrigation systems and evaluate the risk of droughts and floods, among others. Similarly, high frequency multispectral imagery, or automated change detection, is used for developing insurance products for small farmers or trends in agricultural production. Immediate information on crop health is critical to enable early warning systems for potential drops in crop production or soil moisture monitoring for agricultural drought. Remote sensing data is increasingly used for famine early warning systems. Indeed, before disasters occur, satellite images can provide means to reduce vulnerabilities and prepare for different scenarios, such as high resolution digital elevation maps for assessing the risk of floods or droughts. Remote sensing information is immensely useful to gain oversight of the current situation after an event and to identify areas that are most affected. All these are achieved through web platforms. This summarizes the major research question: to develop a holistic, hierarchical, scientifically objective and multi-scaling agroclimatic classification approach for optimal production in a web platform, which is still not available.

The objective of this chapter is twofold: first, to identify sustainable production zones by conducting contemporary agroclimatic classification based on remote sensing and GIS and second to achieve an optimum agricultural production plan. The first objective is achieved in four steps: in the first step hydroclimatic zones are developed leading to water limited growth environment (WLGE) based on drought indices; the second step further develops the non-crop specific agroclimatic zones; in the third step crop-specific agroclimatic zones are produced, which identify sustainable production classes; finally, in the fourth step multi-scaling agroclimatic zones are developed. Specifically, agroclimatic classification at meso-scale level based on additional seven bioclimatic indices can be achieved leading to zones with different suitability and their possible geographical shift under future climate change scenarios using regional climate model simulations. The second objective is based on the outcome of the first stage and develops a decision support system (DSS) by using multi-criteria analysis in each production class combining different criteria to a utility function under a set of constraints concerning different categories of agroclimatic, social, cultural and economic conditions. Moreover, to support the realization of the proposed production zoning and DSS in real-time, a Sensor Web service platform is proposed to be implemented based on the Sensor Web technologies. Real-time environmental and agronomic data, e.g. soil moisture and temperature data, meteorological data, production data for each zone or remotely sensed data, are managed in the Sensor Web service platform. It is stated that the method integrating real-time GIS data model and Sensor Web Service Platform is an effective way to manage production and environmental data under the Geospatial Service Web framework.

2 Background

Common Agricultural Policy (CAP) continues to play a major role in affecting agriculture and agricultural production, as well as the farming population. The focus on non-protected national markets and the enlargement of the European Union are creating a new reality for agriculture and rural areas, in general. The overall economic contribution of farm-households in rural areas depends on the weight of agriculture in each area. The policies of the European Union (EU) highlight the multi-functional role of the rural areas, which extends beyond the role of agriculture to also include other activities. Additionally, the issue of maintaining economically vital rural communities, particularly in disadvantaged regions, where alternative income opportunities have been limited, it is a traditional argument connected to Common Agricultural Policy. However, in the last decades, a full range of new issues has emerged (Manos et al. 2011). A Communication about the CAP post 2013 (EC COM 672 2010) suggests three main objectives for the future CAP. The third objective "Balanced territorial development" aims to support rural employment and maintaining the social fabric of rural areas, to improve the rural economy and promote diversification to enable local actors to unlock their potential and to optimize the use of additional local resources and finally to allow for structural diversity in the farming systems, as well as improve the conditions for small farms and develop local markets (European Commission 2009). When modelling the dynamics of agricultural systems, economists recognized that farm households vary and that this variation is important, but rather than attribute this variation to different factors, they concentrated on defining farm types by structural variables, such as farm size and enterprise mix. However, common sense suggests that not all farmers within any given farm type are similar, and it is becoming increasingly apparent that few individuals maximize financial gain. Given this situation, one alternative approach would be to develop a methodological procedure including a model, which assumes some degree of commonality in the behavior of individuals, but also recognizes that the characteristics of the individuals may influence the specifics of any generalized response (Manos et al. 2010b).

Extensive research has been conducted on climate variability and change involving, among others, General Circulation Models (GCMs), Regional Climatic Models (RCMs), downscaling, as well as assessment of impacts, vulnerabilities, mitigation and recently adaptation of different sectors of the economy (Eitzinger et al. 2007; Olesen and Bindi 2002) (e.g. EU targeted projects: CECILIA, ADAGIO, PRUDENCE, ENSEMBLE, also IRENA, ARIDE, MEDALUS, ENVASSO, RAMSOIL, PRODIM, CLIVAGRI). There are several recent research efforts on climate change impacts in the Mediterranean region (Politi et al. 2016; Kioutsioukis et al. 2010). It is clear that results of climate change impacts should be considered in several aspects of agroclimatic zoning, such as preparedness planning, computation of indices and future scenarios. Moreover, climate change research on long term observations of past decades suggests that characteristic recurrence frequencies, intensities and durations of certain extreme events have already increased (IPCC 2013; Bruce

1994; Dilley and Heyman 1995). There is medium confidence that since the 1950s some regions of the world have experienced more intense and longer droughts (IPCC 2012). Land use changes have potential impacts on environmental hazards (Arneth et al. 2014), the anthropogenic forcing has contributed to the global trend towards increased extreme events, such as droughts, in the second half of the twentieth century. Extreme climate variables and climate extremes, such as droughts, are projected to experience significant changes over the twenty-first century, just as they have during the past century, in many areas, including Southern Europe, among others (IPCC 2012; Nastos et al. 2016; Tarquis et al. 2013). There is also medium confidence that the duration and intensity of hydrological droughts will increase in the twenty-first century in some seasons and areas, due to reduced precipitation and/or increased evapotranspiration (ET), although other factors leading to a reduction in river flows or groundwater recharge are changes in agricultural land cover and upstream interventions.

Besides, the climate is among the most important factors that determine the agricultural potentialities of a region and the suitability of a region for a specific crop, whereas the yield is determined by weather conditions. Since agriculture is highly dependent on environmental conditions, a quantitative understanding of the climate of a region is essential for developing improved farming systems (Pereira 2017). Even though crop production depends on environmental conditions, almost all agroclimatic classifications take into account temperature and rainfall. These climatic parameters in combination with soil type and geomorphology can determine areas, where high levels of production are appropriate, avoiding the threat of degrading the natural resources (Mavi and Tupper 2004). It is understood that vulnerable agriculture and its related impacts operate on a variety of time scales. There is, thus, a research need for agroclimatic classification at different spatial scales and for an objective procedure for the assessment of production within each agroclimatic zone, which still has not been attempted (Tsiros et al. 2008, 2009; Dalezios et al. 2014). The international research community continues working toward newer and potentially better monitoring of agricultural production and environment that can also account for a changing climate, where there may be a shift in both temperature and precipitation regimes (Waseem et al. 2015; Zargar et al. 2011).

Agroclimatic classification is very useful in identifying sustainable production zones within a climatic region. There are many climatic and agroclimatic classifications seeking to describe the moisture conditions of crops (Thornwaite 1948). These classifications vary in complexity, ranging from the use of one parameter to methods incorporating a number of parameters. Most of these agroclimatic classifications use rainfall and potential evapotranspiration in order to delimit the growth environment of crops (Badini et al. 1997). Specifically, an investigation of the Water Limited Growth Environment (WLGE) has been conducted for millet cultivation in Burkina Faso, where rainfed production is a major source of food and income. In this study, Aridity Index (AI) and Crop Water Stress Index (CWSI) have been used for defining such environments. Kogan (2001) proposed the Vegetation Health Index (VHI) for monitoring the impact of weather to vegetation, and used it for agricultural drought and agricultural production monitoring.

Recently, a methodology has been developed based on remote sensing data and methods and applied in a semi-arid region (Tsiros et al. 2009). Specifically, two drought indices have been computed, namely aridity index (AI) and vegetation health index (VHI), in order to define zones adequate for sustainable farming according to water limitations, namely water limited growth environment (WLGE) zones. Moreover, sustainable production zones are identified in terms of water efficiency, fertility (appropriate or not for agricultural use), desertification vulnerability and altitude restrictions. Thus, WLGE zones are combined with soil maps and a Digital Elevation Model (DEM) of the semi-arid region and two more indices are computed, namely the Growing Degree Days (GDD) and Net Radiation (Rn), in order to define sustainable production zones (Tsiros et al. 2009). In order to apply new management techniques, transfer new technologies and plan alternative crops according to the bio-physical characteristics of each region, a quantitative understanding of the relationships among crop, climate and soil are needed (Badini et al. 1997). Defining areas of sustainable crop production is a major step for identifying agroclimatic zones, considering environmental limitations and the sustainable use of natural resources. In addition, multi-scaling agroclimatic classification at meso-scale level based on additional seven bioclimatic indices can be achieved leading to zones with different suitability and their possible geographical shift under future climate change scenarios using regional climate model simulations (Malheiro et al. 2010). These indices are: length of the growing season (LGS), growing season precipitation (GSP), Huglin heliothermal index (HI), cool night index (CI), hydrothermal index (HyI), dryness index (DI) and composite index (CompI). The proposed combination of all 11 indices can achieve finer scale agroclimatic classification even in hilly terrain. As a result, optimal production can be computed for each sustainable production zone, at a multi-scale level, within major climatic regions.

The main driving force behind the use of satellite products as inputs to agroclimatic zoning indices is the lack of long records from weather stations in many developing areas, as well as lack of available data in remote areas (Thenkabail et al. 2004). It is also recognized that remote sensing has gradually become an important tool for the detection of the spatial and temporal distribution and characteristics of drought at different scales. Moreover, there is a gradually increasing reliability and accuracy in remote sensing data and methods (Dalezios et al. 2017). In addition, the new satellite systems have better spatial resolution, more bands and new sensors for environmental parameters and vegetation (Niemeyer 2008). Needless to say, the continuous technological advancements also offer additional computational capabilities. The tendency is to extract data from gridded satellite datasets and biophysical data. This is particularly useful and necessary when dealing with combined use of several satellite systems, where the selected cell resolution is that of the coarsest input dataset (McVicar and Jupp 1998; Brown et al. 2008).

Moreover, during the last decade many Decision Support Systems (DSS) have been developed to supply decision makers with tools on several aspects of climate change, extreme events and agroenvironment, as well as mitigation and preparedness planning (Steinemann et al. 2005). In particular, Spatial Multi-Criteria Evaluation (SMCE) approach has been widely used in the development of DSSs due to its ability to take into account simultaneously all the basic aspects of the problem including heterogeneous sets of many different factors and constraints. Although several applications of DSS already exist, there are just a few recent developments in climate change and vulnerability assessment (Sivakumar et al. 2005) and in hydro-meteorological hazard risk management (e.g. EU project CHANGES) (Van Westen 2013). For this purpose, an extension of existing methodologies (Sumpsi et al. 1997; Amador et al. 1998) is conducted for the analysis and simulation of agricultural systems based upon multi-criteria techniques. In these approaches, weighted goal programming has been proposed as a methodology for the analysis of decision making. Specifically, a Multi-Criteria Decision Making (MCDM) model is used in order to achieve better policy-making procedures and the simulation of the most realistic decision process. The utility of MCDM approach in comparison with other approaches, such as linear programming, or cost benefit analysis, can achieve optimum farm resource allocations (land, labour, capital, water, etc.) that imply the simultaneous optimization of several conflicting criteria, such as the maximization of gross margin, the minimization of fertilizers, the minimization of labour used and similar aspects.

Furthermore, web content management systems (also known as WCM or CMS platforms) have been recently used to design, implement and administer digital content, including satellite imagery, on the web. Most are bundled software packages that deliver a platform for developers to build and customize web templates and institute a publishing workflow. Specifically, with the development of information technologies, such as Web services and interoperable services, a Geospatial Service Web (GSW) has been recently proposed in the geospatial community (Gong et al. 2009). GSW is a virtual geospatial infrastructure based on the Internet, and it integrates various geospatial-related resources, such as sensor resources, data resources, processing resources, information resources, knowledge resources, computing resources, network resources, and storage resources to manage data, extract information, and obtain knowledge in the geospatial community domain (Gong et al. 2009). GSW unifies the functions of a geospatial acquisition system, data transformation system, distributed spatial data collection, high-capability server system, large volume storage system, remote sensing, and a geographic information system (GIS), where the functions are implemented by Web services and communicated through the standardized protocols of the Internet. The mission of GSW includes the following: (1) acquire global spatial data for all seasons, all days, and all directions using all kinds of sensors on satellite, aircraft, and ground surface; (2) chain the whole process seamlessly from sensors to application services using unified information networks, including satellite communication, data relay network, and wired or wireless computer communication networks; (3) register sensors, computing resources, storage resources, internet resources, manipulate software and spatial data on the internet, and process spatial data online quantitatively, automatically, intelligently, and in real time; and (4) provide geospatial services, compose virtual service chains and transmit user-required information in the most effective and efficient ways. Using GSW for real-time environmental data management will help describe, organize, manage, manipulate, interchange, search, and release environmental data in a unified framework (Gong et al. 2015).

There are several recent applications of web environmental platforms, such as satellite derived online web maps. For example, satellite imagery can monitor crop phenology, measure and estimate meteorological parameters, measure the effectiveness of irrigation systems and evaluate the risk of droughts and floods, among others. Similarly, high frequency multispectral imagery, or automated change detection, is used for developing insurance products for small farmers or trends in agricultural production. Immediate information on crop health is critical to enable early warning systems for potential drops in crop production or soil moisture monitoring for agricultural drought. Remote sensing data is increasingly used for famine early warning systems. Indeed, before disasters occur, satellite images can provide means to reduce vulnerabilities and prepare for different scenarios, such as high resolution digital elevation maps for assessing the risk of floods or droughts. Remote sensing information is immensely useful to gain oversight of the current situation after an event and to identify areas that are most affected. All these are achieved through web platforms. This summarizes the major research question: to develop a holistic, hierarchical, scientifically objective and multi-scaling methodology for agroclimatic zoning in a web production platform.

New satellite systems offer online open information for web platforms, such as the European Copernicus system with six Sentinel satellites (2014–2021) to monitor land, ocean, emergency response, atmosphere, security and climate change (ESA 2014; Dalezios et al. 2017), or NASA's new online satellites for climate change, Global Precipitation Measurement Core Observatory, Orbiting Carbon Observatory-2, and active-passive Soil Moisture. Moreover, massive cloud computing resources and analytical tools for working with big datasets make it possible to extract new information from environmental satellites' imagery with varying spatial resolution, such as Landsat-8 imagery (15 m), RapidEye (5 m), Worldview-3 (.31 m) or Pleiades (.5 m). Thus, digital data processing and analysis for agroecosystems, including satellite imagery, monitoring and preparedness planning, including DSS, could be incorporated into a dynamic web production platform.

3 Methodology

The objective of the proposed methodology is to identify sustainable production zones by conducting contemporary agroclimatic classification based on remote sensing and GIS and considering also climate change and then to achieve an optimum agricultural production plan.

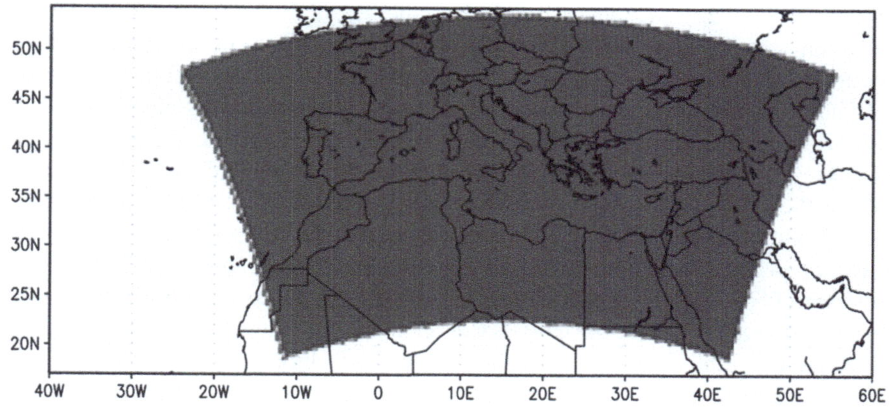

Fig. 1 WRF used coupled model domain (from Dalezios 2015)

3.1 Climate Change and Agroclimatic Zoning

3.1.1 Climate Variability and Change

This is a significant step involving the Weather Research and Forecasting (WRF) model for downscaling and assessment of impacts on drought at different spatial and temporal scales. The WRF is a non-hydrostatic model, with several available dynamic cores, as well as many different choices for physical parameterizations suitable for a broad spectrum of applications across scales ranging from meters to thousands of kilometers. The physics package includes microphysics, cumulus parameterization, planetary boundary layer (PBL), land surface models (LSM), long-wave (LW) and short-wave (SW) radiation. The WRF is used in its version 3.6.1 (Skamarock et al. 2006). WRF is setup with two nested grids, one at 20 km (Domain 1—Europe) and a second at 5 km (Domain 2—Greece and Cyprus) horizontal grid spacing, 40 vertical levels, using one-way nesting (Politi et al. 2016). The outer grid is centered in the Mediterranean basin, while the second in Greece and have 265 × 200 and 185 × 185 grid points (for Greece), respectively (Fig. 1). The high resolution downscaling improves the spatial and temporal variability of climate parameters due to better representation of topography (Soares et al. 2012).

The ERA-Interim reanalysis dataset is used to provide initial and boundary conditions. The lateral boundary conditions and sea surface temperature are both updated every 6 h, from ERA-Interim. Two different schemes are tested for the boundary layer, the first (WRF_pp1) with the Yonsei University (YSU) (Hong and Lim 2006) along with the corresponding surface layer of the revised version of MM5, the Kain-Fitch convective parameterization (Kain 2004) only for the domain 1, none for domain 2. The second (WRF_pp2) is the Mellor-Yamada-Janjic (MYJ) (Janjic 2001) PBL, the Monin-Obukhov similarity theory for the surface layer. Finally, WSM 6 class single-moment and Thomson represent the microphysics schemes, respectively. The common schemes in both cases, are the Noah land surface (LSM) and the newer version of the Rapid Radiative Transfer Model radiation scheme. To evaluate the model simulations, surface observations of daily minimum and maximum temperatures are used, and daily precipitation values available for Greece-Cyprus from the European Climate Assessment and Data (ECA&D) project (http://www.ecad.eu/) station dataset. To assess the performance of the model, four statistical indices were estimated (Soares et al. 2012; Zittis et al. 2014), the root mean square error (RMSE), the BIAS, the mean absolute error (MAE) and the Pearson's correlation coefficient (COR). Before the statistical analysis, an elevation correction of constant lapse rate 6 °C/km should be applied, to both minimum (TN) and maximum temperatures (TX), due to different horizontal position and smooth model topography (Soares et al. 2012). The work includes mapping of simulated and already available results of climate change impacts on agroenvironment assessment including climate stresses, water availability and land degradation at different spatial and temporal scales including agroclimatic zoning, which is innovative and new. The results of WRF are used in agroclimatic zoning at different scales for future scenarios, as well as in DSS.

3.2 Agroclimatic Zoning

3.2.1 Hydroclimatic Zones

The first step is to identify zones adequate for sustainable farming according to water limitations using GIS and remote sensing. In order to determine such zones and classify the Water Limited Growth Environment (WLGE), satellite derived Vegetation Health Index (VHI) and Aridity Index (AI) are used (Tsiros 2009; Tsiros et al. 2008, 2009). The WLGE zones are identified through superposition of the two indices' images over an area and describe the hydroclimatic component of the agroclimatic zoning.

Water limited growth environment The first index used to identify WLGE is VHI. VHI represents overall vegetation health (moisture and thermal conditions) and is used for identification of vegetative stress and drought affected areas (Kogan 1995, 2001). VHI is a combination of VCI and TCI derived by a long term NDVI and channel 4 images from NOAA/AVHRR satellite. NDVI, is obtained by combining the channels 1 and 2, the visible and near infrared, respectively, of NOAA/AVHRR. NDVI is a quick and efficient way for the estimation of vivid vegetation. NDVI is indicative of the level of photosynthetic activity in the vegetation monitored, reflecting whether the vegetation is stressed or not. After stressed conditions, significant reduction in NDVI of the field is expected. VCI and TCI characterize the moisture and thermal conditions of vegetation, respectively (Bhuiyan et al. 2006; Kogan 1995, 2001) and are given by the equations:

$$VCI = 100^* \frac{NDVI - NDVI_{\min}}{NDVI_{\max} - NDVI_{\min}} \quad (1)$$

$$TCI = 100^* \frac{BT_{\max} - BT}{BT_{\max} - BT_{\min}} \quad (2)$$

where NDVI, $NDVI_{max}$ and $NDVI_{min}$ are the smoothed 10-day normalized difference vegetation index, is multi-year maximum and minimum, respectively; BT, BT_{max} and BT_{min} are the smoothed 10-day radiant temperature, its multi-year maximum and minimum, respectively, for each pixel, in a given area. Thermal conditions are especially important when moisture shortage is accompanied by high temperature, increasing severity of agricultural drought, having direct impact to vegetation's health. VCI and TCI vary from zero, for extremely unfavorable conditions, to 100, for optimal conditions. Thus, higher VCI and TCI values represent healthy and unstressed vegetation.

Both indices are based on the same concept. Maximum amount of vegetation is developed in years with optimal weather conditions, whereas minimum vegetation amount develops in years with extremely unfavorable weather (mostly dry and hot). Therefore, the absolute maximum and minimum values of NDVI and BT, calculated from several years, contain the extreme weather events (drought and no drought conditions). The resulted maximum and minimum values can be used as criteria for quantifying the environmental potential of a region (Kogan 1995). VHI is expressed by the following equation:

$$VHI = 0.5^*(VCI) + 0.5^*(TCI) \quad (3)$$

In VHI computation, an equal weight has been assumed for both VCI and TCI, since moisture and temperature contribution during the vegetation cycle is currently not known. The five classes of VHI that represent agricultural drought are illustrated in Table 1 (Bhuiyan et al. 2006; Kogan 2001).

The other index used to identify WLGE zones is AI. AI represents climatic aridity and is used to determine the adequacy of rainfall in satisfying the water needs of crops. AI is a function of the ratio of precipitation to potential evapotranspiration. The categories as they are defined by the values of AI are illustrated in Table 2 (UNESCO 1979). The index is calculated on multiyear basis, using monthly values. The potential evapotranspiration is calculated with the use of Blaney-Criddle method (Tsiros 2009; Tsiros et al. 2009; Blaney and Criddle 1950). The method estimates potential evapotranspiration (ETp) using monthly air temperature data, the ratio of

Table 1 VHI drought classification scheme (Kogan 2001)

VHI values	Vegetative drought classes
<10	Extreme drought
<20	Severe drought
<30	Moderate drought
<40	Mild drought
>40	No drought

Table 2 Dryland categories (UNESCO 1979)

Aridity Index: P/PET	Rainfall (mm)	Classification
PET>P		Desert climate
<0.03	<200	Hyper-arid
0.03 to <0.20	<200 (winter)	Arid
	<400 (summer)	
0.20 to <0.50	200–500 (winter)	Semi-arid
	400–600 (summer)	
0.50 to <0.65	500–700 (winter)	Dry sub-humid
	600–800 (summer)	
>0.65		No desertification

daytime hours (month/year), and a weighted crop coefficient (K_c). Regarding the weighted crop coefficient, 12 maps with grid cell size of 100 × 100 m (one for each month) have been utilized. K_c values are defined according to land use provided by CORINE 2001 database.

In ETp calculations, land surface temperature (LST) is used instead of air temperature. The generation of LST maps is based on the "split-window" algorithm from Becker and Li (1990), which uses the differential absorption effects in channels 4 and 5 of NOAA/AVHRR for correcting atmospheric attenuation mainly caused by water vapour absorption. For estimating surface emissivity, the relationship given by Van de Griend and Owe (1993) is applied. In order to avoid over-estimating ETp, LST is converted to air temperature using a linear empirical relationship. The relationship has been derived by applying a regression analysis to the LST and air temperature data of the time series fpr the station of Larisa (R2 = 0.84). Results are depicted in Fig. 2.

Since both indices have been computed, two maps are created. From the VHI images a final map is obtained using the frequency of occurrence of agricultural drought events. The derived map is combined with the climatic aridity map and leads to the definition of WLGE zones. The generalized thematic classification scheme is shown in Table 3.

3.2.2 General Agroclimatic Zoning

The second step is to identify sustainable production zones to characterize the general agroclimatic zones in terms of water efficiency, fertility (appropriate or not for agricultural use), desertification vulnerability and altitude restrictions. Thus, WLGE zones are combined with Soil Maps and a DEM (Tsiros 2009; Tsiros et al. 2009).

Soil map and DEM Overlapping WLGE zones, a soil map and a DEM of the study area has led to the definition of regions, where crop production is sustainable and agriculture is the best suited agronomic use. Soil types are digitized according to fertility (appropriate or not for sustainable agricultural use) and desertification vulnerability. The sustainable agronomic use and the desertification risk according to soil

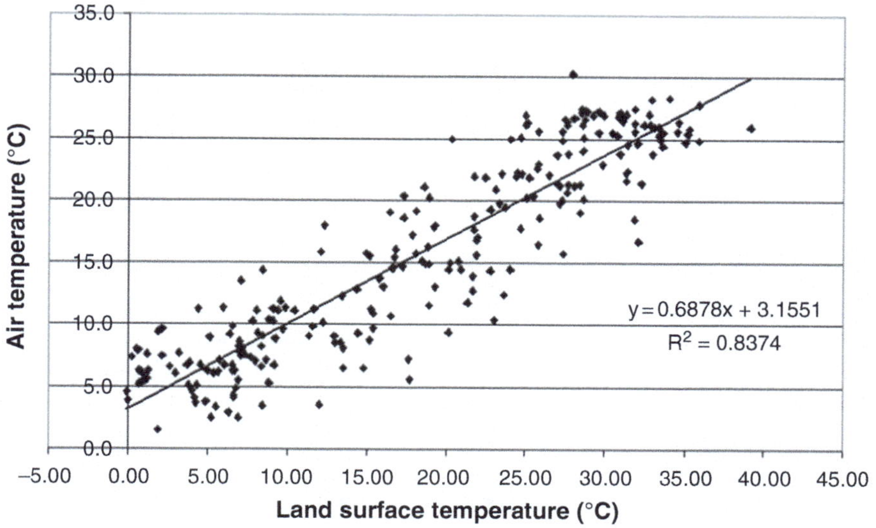

Fig. 2 Application of linear regression analysis to land surface temperature and air temperature for Larisa (from Tsiros et al. 2009)

Table 3 WLGE generalized classification scheme (Tsiros et al. 2009)

Agricultural drought classes	Aridity classes	WLGE classes
Extreme drought	Extremely dry	Limited environment
Severe drought	Dry	
Moderate drought	Semi-dry	Partially limited environment
Mild drought	Semi-wet	
No drought	Wet	No limitations

category are adopted (Yassoglou 2004). Soil types are grouped into three classes during the digitization, namely soils appropriate for agricultural use, controlled agricultural use, and no agricultural use. The classification pattern is illustrated in Table 4. Finally, the digitized vector map is converted to raster (grid) with cell size of 100 m × 100 m. Three major crop growth zones are selected according to altitude limitations (Dalezios 2015; Danalatos 2007). The first, ranging from 0 to 600 m, is appropriate for most of the crops. The second, ranging from 600 m to 900 m is appropriate for non-tropic crops and fruit trees (maize, winter wheat, apple trees, chestnuts, etc.). The last one, having altitudes higher than 900 m is not appropriate for crops.

Supervised classification During the supervised classification, the parallelepiped technique is used in order to combine the WLGE zones, the soil map and the DEM and define the sustainable production zones. During the classification, the following rule pattern is used. Crop production is:

Table 4 Classification scheme of soil types for sustainable use and desertification vulnerability (from Tsiros et al. 2009)

Class name	Sustainable agronomic uses	Desertification vulnerability	Soil types category
No agricultural use	Wild nature,	Very high	Rock outcrops Leptosols
	Forest		Regosols (low quality)
	Controlled pasture	High	Cambisols (medium-low quality)
Controlled agricultural use	Controlled agriculture	Medium	Regosols (medium quality)
	Pasture		Cambisols (medium-high, high quality)
	Forest		Luvisols (medium quality)
Agricultural use	Agriculture	Low	Fluvisols, Vertisols, Luvisols (high quality)

- "Unsustainable" in areas characterized by any of the "limiting" classes.
- "Sustainable under restrictions" when "partial limitations" regarding to WLGE or soil map or DEM (intermediate classes) exist.
- "Sustainable for non-tropic crops" in regions with "no limitations" and 600–900 m altitude range.
- "Sustainable" in areas with "no limitations" and altitude lower than 600 m.

3.2.3 Specific Agroclimatic Zoning

Two indices are used to identify areas suitable for cultivation, namely Growing Degree Days (GDD) (°C d) and Net Radiation (Rn) (Tsiros 2009; Tsiros et al. 2009). Rn is used to define areas, where crop growth is not restricted due to limitations related to the radiation component. This leads to production classification zones, i.e. high, medium and low productivity zones, where, for instance, in low productivity zones energy crops could be considered. Figure 3 summarizes the flow chart of the various steps of the agroclimatic zoning methodology.

3.2.4 Multi-scaling Agroclimatic Zones

Agroclimatic classification at meso-scale level based on seven bioclimatic indices has been achieved leading to zones with different suitability and their possible geographical shift under future climate change scenarios using regional climate model simulations (Malheiro et al. 2010; Kogan 2001). These indices are: length of the growing season (LGS), growing season precipitation (GSP), Huglin heliothermal index (HI), cool night index (CI), hydrothermal index (HyI), dryness index (DI) and composite index (CompI). The proposed combination of all 11 indices for finer scale agroclimatic classification even in hilly terrain is new. As a result, optimal production can be computed for each sustainable production zone, at a

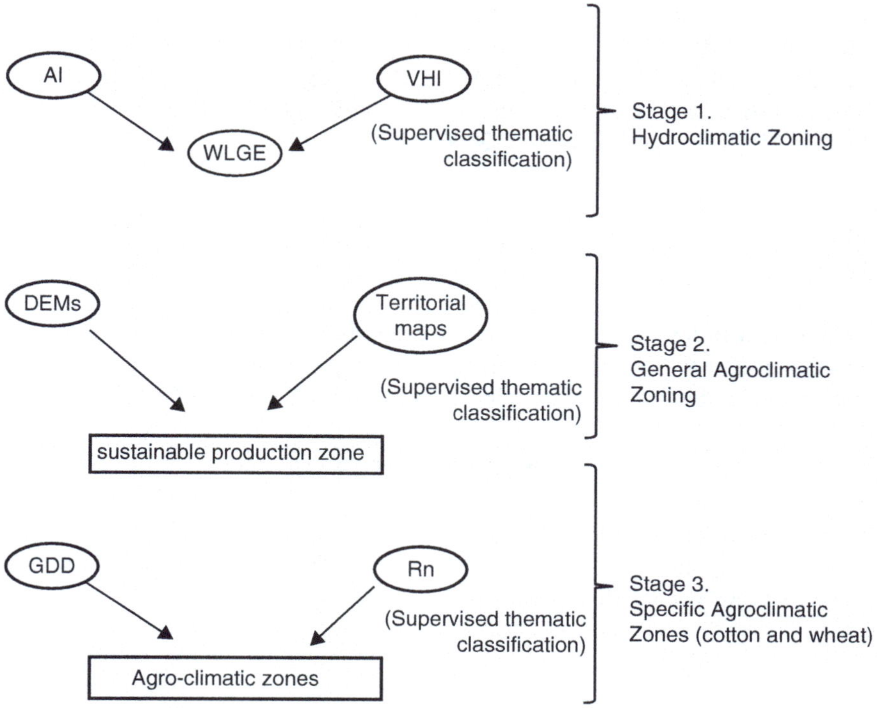

Fig. 3 Flow chart of the agroclimatic zoning methodology (from Tsiros 2009)

multi-scale level, within major climatic regions. Malheiro et al. (2010) in their paper considered climate change scenarios applied to viticultural zoning in Europe. In their study, the above seven indices were described as follows.

1. **Length of the growing season (LGS)** is calculated as the number of days with mean temperatures above 10 °C (growing degree days). Although depending on the different varieties, a region is commonly considered appropriate for vine growing for LGS higher than 182 days.

$$\text{LGS} = \text{Number of days with } \text{T}_{avg} > 10\,^{\circ}\text{C} \tag{4}$$

2. **Growing season precipitation (GSP):** Precipitation from April to September was found to be one of the most discriminating climatic variables in northwestern Spain for current conditions.

$$\text{GSP} = \sum_{\text{Apr}}^{\text{Sep}} (P) \tag{5}$$

3. **Hulgin Heliothermal Index (HI)** is a degree-day formulation that weights maximum temperatures above daily mean temperatures and applies a latitude-

varying day-length adjustment. The day length coefficient was linearly interpolated from 1.02 to 1.06 within latitude belt 40–50 °N. Southwards of 40 °N the coefficient takes a value of 1.00. For latitudes 50–60 °N was considered a linear extrapolation. HI is grouped into six categories of climate from very cool (HI < 1500) to very warm (HI > 3000).

$$\mathrm{HI} = \sum_{\mathrm{Apr}}^{\mathrm{Sep}} \frac{(T - 10) + (T_{max} - 10)}{2} d \tag{6}$$

4. **Cool night Index (CI)** provides a relative measure of ripening potential with four categories, namely very cool nights (CI ≤ 12°C), cool nights (12 < CI ≤ 14 °C), temperature nights (14 < CI ≤ 18 °C) and warm nights (CI > 18 °C).

$$\mathrm{CI} = \text{September average } T_{min} \tag{7}$$

5. **The Hydrothermic index (HyI)** considers precipitation and temperature regimes for estimating the risk of downy mildew disease. HyI < 2500 °C mm, Low risk; HyI > 5100 °C mm, High risk.

$$HyI = \sum_{\mathrm{Apr}}^{\mathrm{Aug}} (T\, x\, P) \tag{8}$$

6. **Dryness Index (DI)** defines four viticulture climates ranging from very dry (DI ≤ −100 mm) to humid (DI > 150 mm). DI defines the soil water component of the climate, assessing the level of dryness relevant for wine production in a specific region.

$$\mathrm{DI} = \sum_{\mathrm{Apr}}^{\mathrm{Sep}} (W_O + P - T_V - E_S) \tag{9}$$

7. **The Composite index (CompI)** summarises the main results obtained from the previous indices. It is computed for each year separately and extends between 0 and 1, depending on four conditions HI ≥ 1400; DI ≥ −100; HyI ≤ 5100 and daily minimum temperatures never below −17 °C are simultaneously accomplished.

$$\mathrm{CompI} = \begin{cases} \mathrm{HI} \geq 1400; \\ \mathrm{DI} \geq -100; \\ \mathrm{HyI} \leq 5100; \\ \mathrm{T}_{min}\mathrm{always} \geq -17\ ^{\circ}\mathrm{C} \end{cases} \tag{10}$$

3.3 *Multi-criteria Modelling*

The selected model is a Multi-criteria Programming (MCDM) model for planning the agricultural production. In order to analyze how Common Agriculture Policy may influence farm production decisions, the methodologies for the analysis and simulation of agricultural systems based upon multi-criteria techniques are extended (Sumpsi et al. 1997; Amador et al. 1998). Indeed, a weighted goal programming is proposed as a methodology for the analysis of decision making. This methodology has been successfully implemented on real agricultural systems (Gomez-Limon and Sanchez-Fernandez 2010; Gomez-Limon and Berbel 2000; Berbel and Rodriguez 1998; Bartolini et al. 2007a, b; Gomez-Limon et al. 2002; Gomez-Limon and Riesgo 2004; Manos et al. 2006, 2007, 2008, 2010a, b). Specifically, an MCDM model is used in order to achieve better policy-making procedures and the simulation of the most realistic decision process. The MCDM model is selected due to the variety of criteria taken into account by farmers in crop planning, broadening in this way the traditional assumption of profit maximization. It also assembles the multi-functionality of agriculture involving variables related to economic, social and environmental aspects. The use of the MCDM approach in comparison to other approaches, such as linear programming, or cost benefit analysis, can achieve optimum farm resource allocations (land, labour, capital, water, etc.) that imply the simultaneous optimization of several conflicting criteria, such as the maximization of gross margin, the minimization of fertilizers, or the minimization of labour used, among others.

Weighting goal programming for policy analysis This methodology is employed to estimate a surrogate utility function in order to simulate farmers' decision-making processes, broadening in this way the traditional profit-maximizing assumption. This surrogate utility function is then used to estimate the value of decoupled payments in crop production. Briefly, the methodology can be summarized as follows:

1. Establish a tentative set of objectives that may be supposed to be most important for farmers. Questionnaires and descriptive research are sufficient for this purpose.
2. Determine the pay-off matrix of the above set of objectives. Using this matrix, estimate a set of weights that optimally reflect farmers' preferences.

3.3.1 Components of the Multi-criteria Objective Function

Multi-criteria model definition A system is defined via a mathematical simplification of the relevant variables and their relationships in order to understand the effect of any modifications of the initial conditions that characterize the system. Every system has variables that control the processes involved and that belong to the decision-making process as 'decision variables'. The selected crop plan determines changes in certain attributes of the system. Attributes are relevant functions deduced from the decision variables, but there are also attributes that are not relevant to the decision makers. Attributes to which decision makers assign a desired direction of improvement are considered objective functions. In this analysis not only the farmers' objectives are considered, but also attributes that are relevant to policy makers.

Variables Each farmer has a set of variables Xi (crops), as described in the previous section. These are the decision variables that can assume any value belonging to the feasible set.

Objectives This model optimises at the same time different criteria as profit maximisation, fertilizer minimisation etc. At the preliminary stage, three objectives must be regarded as belonging to the farmer's decision-making process, namely profit maximization, fertilizer minimisation and minimisation of the labour inputs.

Profit maximization Farmers wish to maximise profits, but calculation of profit requires the computation of some relatively difficult factors, such as depreciation. Therefore, for convenience it is assumed that gross margin (GM) is a good estimator of profit, and maximisation of profit is equivalent in the short run to maximisation of gross margin. The objective function included in the model is defined as follows:

$$MaxGM = \Sigma GM_i \times X_i \tag{11}$$

where GM is the total gross margin, X_i is crop i and GMi is the gross margin of crop i.

Fertilizer minimization Fertilizer minimization is a public objective. For this reason it is not considered in the decision process by farmers. The most obvious indicators are those related to the consumption of water and use of pesticides that are directly related to the pollution of water resources and appear more directly quantifiable at farm level. They are, nevertheless, not obviously subject to aggregation at higher level and their effects on the environment can be evaluated only after some elaboration of prediction models based on diffusion functions.

Fertilizer minimization is the main form for calculating the surpluses of nitrogen, which are potentially dangerous for the environment. It would also be the main indicator of the impact of farming on the environment as far as groundwater quality. In this way, all nitrogen reaching the cultivated soil is included as input. Similar indicators can be designed for other nutrients, such as phosphorus and potassium. For this reason, fertilizer is computed as the sum of fertilizers used for all crops (TF), and its objective function is as follows:

$$Min\,TF = \Sigma F_i \times X_i \tag{12}$$

Minimization of labour inputs The minimization of labor implies not only a reduction of input cost, but also an increase of leisure time and reduction of administration and management processes. The farmers usually show an aversion to hiring labor. An explanation of this behavior is that this parameter is connected with the complexity of crops, because the hired labor adds a degree of complexity to family farming. For this reason, labor is calculated as the sum of labor for all farm activities (TL), therefore the objective function becomes:

$$Min\,TL = \Sigma TL_i \times X_i \tag{13}$$

No other objectives are proposed in advance. It is assumed that at the preliminary stage that the three objectives mentioned above are sufficient to explain farmers' behaviour.

3.3.2 Constraints of the Multi-criteria Modelling

In order to analyze CAP's impacts several constraints are used as a result of the implementation of the new CAP. The selected constraints are the following:

Total cultivation area The sum of all crops (X_i) must be equal to 100. This constraint is only introduced in order to obtain the results of the model in percentages. The sum of total available land for all crops is a second constraint. Finally, another constrain is the sum of irrigable available land for irrigated and non-irrigated crops.

CAP A large proportion of agricultural income depends upon CAP subsidies, and farmers cannot afford to ignore CAP regulations that affect most of the crops available for cultivation. For this reason, in accordance with RDP measures, a set-aside activity (SA) is included related to the subsidized crops (which are the majority):

$$\Sigma \mathrm{X}_i + \mathrm{SA} = 100 \tag{14}$$

This SA, as a CAP requirement, must be at least the 25% of the irrigated land according to the RDP measure "protection of nitrate sensitive areas". A second CAP constraint "Production Rights" has been also included in the model. The sum of production rights (PR) according to CAP for crops (Xi) following CAP regulations has the following objective function:

$$\mathrm{PR} = \Sigma \mathrm{PR_i} \times \mathrm{X_i} \tag{15}$$

Market and other constraints Such payments, by being up to 100% decoupled from current production, allow farmers to make production decisions based more on market signals than on policy interventions. Some other crops are not subject to CAP rules, but marketing channels put an upper limit on short-term variations. This is the case for alfalfa. This crop needs to be produced in quantities that processing facilities, the marketing system or livestock in the vicinity of the production area is expected to demand without price distortions. For this circumstance, a 'greater-than' constraint has been included in the model. This upper limit has been fixed on the basis of the maximum historical cultivation.

Rotational and agronomic considerations It is regarded as agronomic sound policy not to cultivate a crop, such as a cereal, if, during the previous year, the same plot has grown another cereal. This is called a rotational constraint. A rotational constraint limits the cultivated area for a crop to a maximum number of the total available area, and it is applied to all cereals. All this information has been included in the model that forms the basis for the MCDM simulation.

3.4 Web Agricultural Production Platform

With the development of information technology, the purpose is to determine a method to realize agricultural production or environmental data management under the Geospatial Service Web (GSW) framework, which is innovative and new. The aim for GSW is to develop a real-time GIS data model to manage real-time data. Indeed, GIS data models have evolved from static GIS data models, to temporal GIS data models, and then to real-time GIS data models (Gong et al. 2015). The real-time GIS data model is developed from the temporal GIS data model and emphasizes the time efficiency of data management (Hatcher and Maher 2000) although the real-time GIS data model is still in an immature stage and needs further study. This method is based on a novel real-time GIS data model and the model's implementation, namely the Sensor Web Service Platform with Sensor Web technologies (Bröring et al. 2011). Real-time GIS is an important new research domain, transforming the study of historical changed data to real-time data in GIS. The real-time GIS data model represents further progress for static and temporal GIS data models.

The data model is the core of GIS, since an appropriate data model plays a decisive role in constructing a GIS application. The primary task of a spatiotemporal data model is the organization and management of spatiotemporal data, as well as analysis and expression of the content and relationships of spatiotemporal change. The data model should have five characteristics: (1) the model takes into account both traditional GIS and real-time GIS; (2) it can express the dynamic data from moving object; (3) it is highly effective for storing and retrieving real-time data from various sensors; (4) it can support the dynamic simulation of spatiotemporal processes from real-time GIS data; and (5) the model can represent the relationships among its factors, including geographical objects, states, events, processes, sensors and observations.

The real-time GIS data model is shown in Fig. 4. Figure 4a is the Entity-Relationship diagram (Chen 1976) and Fig. 4b is the conception diagram. Some relevant elements of the conceptual model are described as follows: (1) Sensor (Sensor): Various sensors containing space-borne, air-borne, and ground sensors; (2) Observation (Observation): The behavior of observable attributes from various

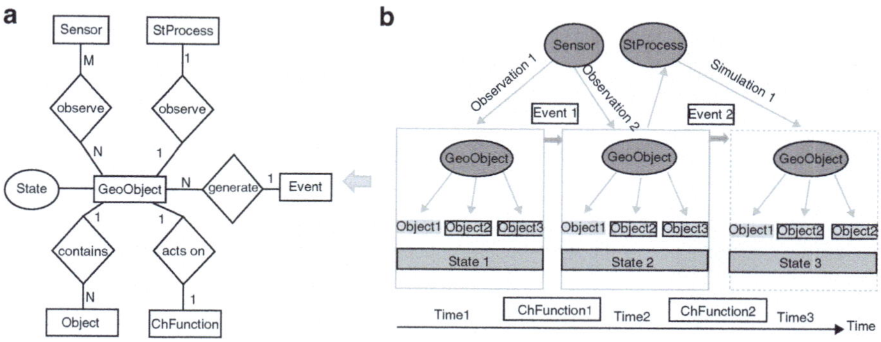

Fig. 4 Real-time GIS data model: (**a**) entity-relationship diagram; (**b**) conception diagram (from Gong et al. 2015)

sensors provides observational data for the model; (3) Geographical Object (Geo-Object): Either physical entities or social phenomenon formed naturally or artificially, expressed with clear boundaries or not, as the objects of GIS research in the real world; (4) Object (Object): Single entity in the real world; a Geo-Object can contain one or multiple objects; (5) Spatiotemporal Process (StProcess): The Spatiotemporal Process is a periodized change process of a complex geographic phenomenon in a timeline, and the processes refer to a series of Geo-Objects and their interactions; (6) Simulation (Simulation): Simulation is the imitation of the operation of a real-world process or system over time; (7) Event (Event): An event is an occurrence of the Geo-Object change, and is the reason for the change of Geo-Objects; (8) State (State): A snapshot of a geographic object at a point of time in the change process; and (9) Change Function (ChFunction): In the time of research, the correspondence between an instant and the values of geospatial and thematic properties. This function can be derived from industry, scientific computing, and relevant experience.

A geographical object consists of three basic indivisible features: time, space, and thematic attributes (Gong et al. 2009, 2015). A geographical object contains both unchangeable attributes and time-varying attributes. Time-varying attributes are associated with state sequences. The time-varying attributes may be different at different states. A sensor is a special geo-object that contains self-parameters and observations. The sensor, described by its metadata, is a tool to observe the spatial attributes and the thematic attributes of geographical objects. As a result, a sensor is the primary means of obtaining the changed information of a geographical object. In fact, one sensor may observe many geo-objects and a geo-object can be observed by many sensors. The wide use of sensors has brought revolutionary changes to data acquisition by improving the accuracy, speed, timely perception, and timely transmission of spatiotemporal data. This change has resulted in the generation of a large volume of data, such as spatiotemporal data, thematic attribute data, image data, and video stream data. This information, which may be remote sensing image collected by a remote sensor, environmental parameters collected by in-situ sensors, or only

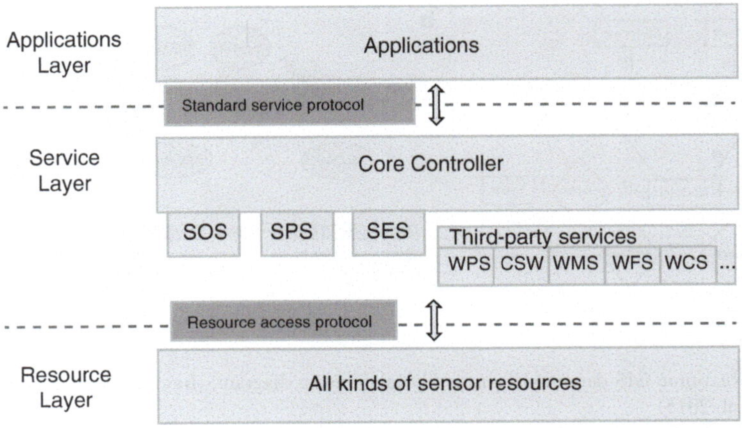

Fig. 5 Sensor Web service platform (from Gong et al. 2015)

position information acquired by a Global Navigation Satellite System, is recorded in a series of observations along with the time.

Complex spatiotemporal changes in geographical phenomena refer to three core components: spatiotemporal processes, geographical objects, and events. There is interoperable usage of sensor resources by enabling their discovery, access, and tasking, as well as eventing and alerting in a standardized way. The Open Geospatial Consortium (OGC) Sensor Web Enablement defines the Sensor Web information model and interface model. The information model defines the encoding standards of sensor observations and sensor metadata, such as the Observations & Measurement (Cox) and the Sensor Model Languages (SensorML) (Botts 2012). The interface model specifies the interfaces of the different Sensor Web services, such as the Sensor Observation Service (SOS) (Bröring et al. 2012) the Sensor Planning Service (SPS) (Simonis and Echterhoff 2009) and the Sensor Event Service (SES) (Echterhoff and Everding 2008). The SOS provides a standardized interface to manage and retrieve metadata and observations from heterogeneous sensor systems. The SPS defines interfaces for queries that provide information about the capabilities of a sensor and how to task the sensor. The SES is an enhancement of the OGC Sensor Alert Service, and it provides operations to register sensors at the service application and let clients subscribe to observations available at the service. Indeed, the Sensor Web Service Platform (Fig. 5) follows a layer-based framework with three layers: a resource layer, service layer, and application layer. The service layer not only provides the standard Sensor Web services (SOS, SPS, and SES), but also enables integrating third-party services, such as the commonly used OGC services Web Processing Service (WPS) (Schut) Catalog Service for the Web (CSW), Web Map Service (WMS), Web Feature Service (WFS), and Web Coverage Service (WCS), using a core controller component. The service layer interacts with the resource layer and the application layer using a resource access protocol and standard service protocol, respectively.

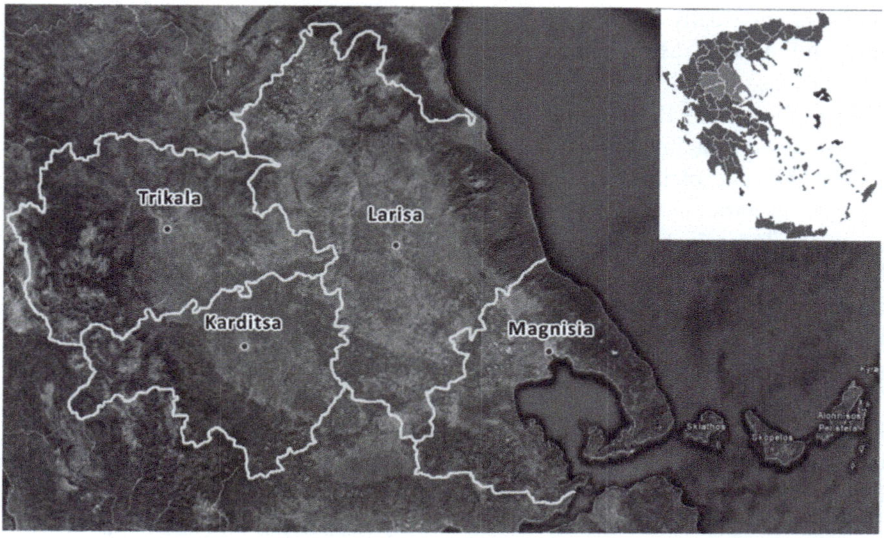

Fig. 6 Map of the region of Thessaly (from Dalezios et al. 2014)

In summary, to support the realization of the proposed real-time GIS data model, a Sensor Web service platform is implemented based on the Sensor Web technologies. Real-time environmental data, e.g. soil moisture and temperature data, meteorological data or remotely sensed data, are managed in the Sensor Web service platform. It is stated that the method integrating real-time GIS data model and Sensor Web Service Platform is an effective way to manage drought data under the Geospatial Service Web framework.

4 Study Area and Database

4.1 Study Area

The study area is the climatological region of Thessaly (Fig. 6). Thessaly is a hydrological district located in the central part of the country with a total area of 14,036 km^2, which roughly represents 10.6% of the whole country. Moreover, in Thessaly 36% of the land is flat, 17.1% is semi-mountainous and the remaining 44.9% is mountainous. The region of Thessaly is characterized by a highly variable landscape and the terrain is such that high mountains surround the plain, which is the largest in the country. It comprises the prefectures of Larisa, Magnisia, Trikala and Karditsa, together with the Northern Sporades group of islands, the largest of which are Skiathos, Skopelos and Alonisos (Fig. 6). The climate of Thessaly is continental in the west part with cold winters, hot summers and large seasonal temperature range. In the east part of Thessaly the climate is typical Mediterranean. In Thessaly, summers are usually hot and dry with temperatures occasionally reaching 40 °C in July and August. Mean annual precipitation over Thessaly is about 700 mm, unevenly distributed in space and time, varying from about 400 mm at the central plain to more than 1850 mm at the western mountain peaks. The mountain areas receive significant amounts of snow during winter months.

The thessalic plain constitutes the main agricultural area of the country. The increase in agricultural activities and the intensive type of agricultural practices applied in Thessaly, has resulted in insufficient use of the available natural resources. Moreover, since rainfall is, in general, rare from June to August, the resulted water deficit is replaced by irrigation in order to satisfy agricultural water needs. The irrigated areas are expected to further increase in Thessaly, thus, the future water needs are also expected to increase, despite scheduled crop restructuring programs. Thessaly is characterized by vulnerable agriculture, since extreme hydrometeorological events, such as floods, hail and droughts are quite common in the catchment, but also due to the existing water deficit for agriculture. Having the higher percent of flatlands than any other district in Greece, 38.7% of the population is occupied in the primary productivity sector and thus, Thessaly is a major supplier of agricultural products. Furthermore, the labour force is underemployed, agricultural machinery underused and land is expensive to buy or lease. As a result, costs in the region are relatively high. The utilized agricultural area (UAA) in Thessaly covers an area of 432,059 ha. Arable crops are the main cultivation for the majority of the agricultural holdings. In arable crops are included cereals, cotton, maize, alfalfa, sugar beets and vineyards.

4.2 Data Base and Preprocessing

Data base The research carried out is based on the available statistical data. The modelling approach suggested requires data collection from the specific regions. The data base consists of NOAA/AVHRR satellite data and conventional data for 20 hydrological years, from October 1981 to September 2001. In specific:

- Normalized Difference Vegetation Index (NDVI), and CH4 and CH5 Brightness Temperature (BT) 10-day composite images (8×8 km^2 spatial resolution).
- Monthly rainfall maps with grid cell size 50×50 km^2.
- Mean monthly air temperature measurements from Larissa meteorological station (National Meteorological Service, NMS).
- Soil map of the study area (Yassoglou 2004).
- Digital Elevation Model derived from 100 m contours.

All satellite data are obtained on-line by NASA archives. NDVI maps are 10-day Maximum Value Composite (MVC) images. CH4 and CH5 images are converted to BT's using the equation provided by the info file of the data set. Using the 10-day images, NDVI and BT images are composed over a monthly period using the MVC and mean pixel value, respectively. Missing data due to cloud cover or sensor's technical problems are completed using monthly climatic values derived from the images of the time series which presented no blunders. The rainfall maps are produced using the data of ISPRA European database. The subset satellite images and the rainfall maps cover the entire Greece. After all computations have been carried out, the area under investigation is isolated.

Preprocessing Before using NDVI and BT images, fluctuations induced by noise must be removed. The combination of the filtering and the MVC can significantly reduce the noise from residual clouds, fluctuating transparency of the atmosphere, target/sensor geometry, and satellite orbital drift (Goward et al. 1991). Other noise can be related to processing, data errors, or simple random noise (Kogan 1995). In the current study, a "4253 compound twice" filter (Van Dijk et al. 1987) is applied to the NDVI images, whereas a "conditional" statistical mean spatial filter (window size ranging from 3×3 to 7×7, according to image needs) has been used for smoothing the BT series (Tsiros et al. 2008). The BT series presented continuous spatial fluctuations and thus a spatial filter (statistical mean) has been preferred for smoothing Channel 4 and Channel 5 BTs. "Conditional" means that the filter is applied only to the pixels that presented errors.

5 Application of the Two-Stage Methodological Approach

The proposed methodology consists of two distinct stages. The first stage comprises of the development of agroclimatic zones based on remote sensing and GIS leading to optimum production classification and the second stage develops a DSS based on multi-criteria analysis for production planning within each class.

5.1 Agroclimatic Zoning

Agroclimatic zoning is an innovative approach based on GIS and time series of satellite data. Remote sensing is already a useful and reliable tool to analyze the vegetation dynamics and there are several studies showing the inter-annual differences in vegetation parameters mainly due to water availability (Al-Bakri and Taylor 2003; Weiss et al. 2004). Agroclimatic zoning is developed in three sequential steps, namely hydroclimatic zoning, general agroclimatic zoning and specific agroclimatic zoning. A description follows.

5.1.1 Hydroclimatic Zoning

This first step aims to define zones adequate for sustainable farming according to water limitations (Tsiros et al. 2009). As crop growth is affected by water supply, these zones are named Water Limited Growth Environment (WLGE) zones. The WLGE zones describe the hydroclimatic component of the agroclimatic zoning, characterizing the moisture conditions during crop growth. In order to determine such zones and classify the WLGE, two satellite derived indices are jointly used, namely Vegetation Health Index (VHI) and Aridity Index (AI).

The computation of the two indices leads into the development of two maps for the period under consideration, where one characterizes areas according to the frequency of agricultural drought episodes, i.e. VHI, and the other represents climatic aridity, i.e. AI. The analysis results of the two maps indicate that there is no area in Thessaly water district, where the climate is "dry" or "extremely dry" based on AI, and "severe" and "extreme" drought events are frequent based on VHI. The WLGE zones result from the combination of these two maps. The thematic classification procedure follows certain steps. At first, a number is assigned to every class of the two indices (five classes each). Specifically, number one corresponds to "wet" and "no drought" classes, grading the sequence up to five, which corresponds to "extremely dry" and "extreme drought" classes. Then, by adding those numbers, three classes are produced to delimit WLGE zones: (1) "limited" (values from 7 to 10); (2) "partially limited" (values from 3 to 6); and (3) "no limitation" exists according to water availability (values equal to 2). The map of WLGE zones is presented in Fig. 7.

Figure 7 indicates that there is no area in Thessaly water district, where plant growth is prohibited by water availability. The identification of "limited" growth environment indicates areas, where moisture and rainfall cannot satisfy crop water requirements or even a portion of them. In order to satisfy crop water needs in those areas, large quantities of water supply from irrigation are required, leading to unsustainable use of water resources and increase the cost of the final product. Areas of "partially limited" growth environment due to water availability require smaller amount for irrigation, whereas areas with "no limitation" even smaller. In such areas, an effective use of water resources could be achieved, since a major part of crop water requirements is supplied by rainfall and existing moisture conditions.

5.1.2 General Agroclimatic Zoning

The combination of the WLGE zones with soil maps and DEM results into the development sustainable production zones, or general agroclimatic zones or non-crop specific agroclimatic zones. This is achieved through image processing, namely parallelepiped supervised classification. The zones of sustainable use according to soil characteristics and altitude based on crop growth zones in Thessaly lead to the development of the sustainable production zones map of Fig. 8.

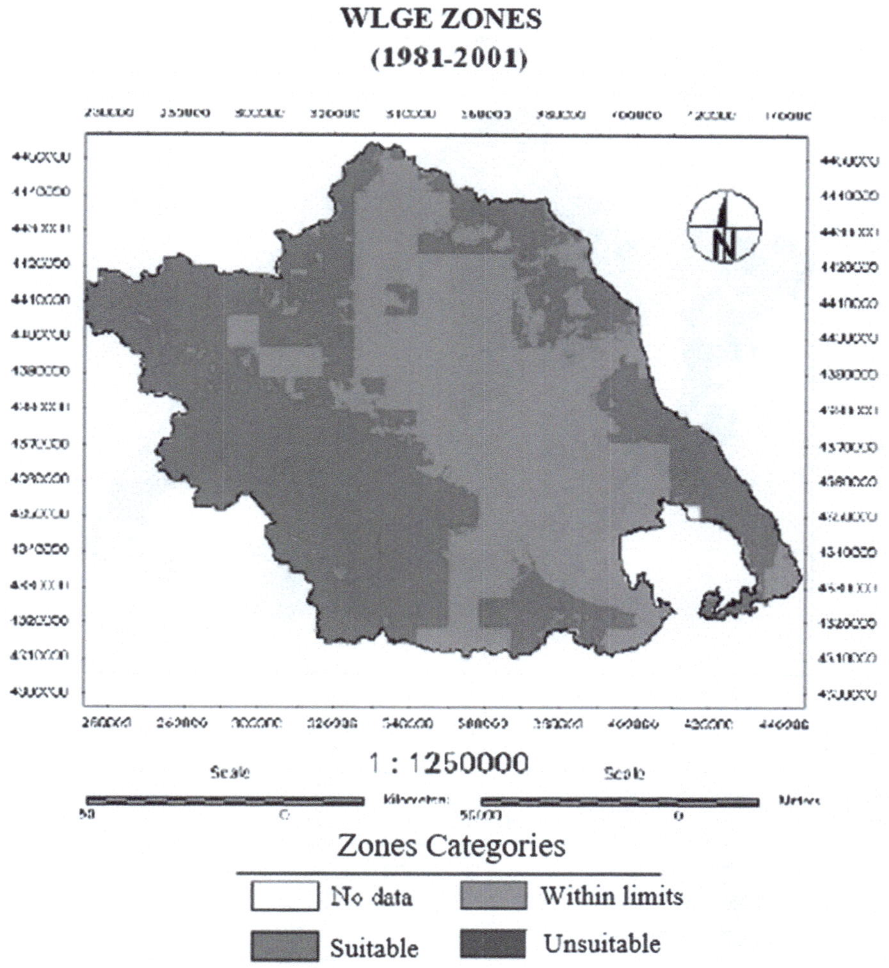

Fig. 7 Map of Thessaly showing WLGE zones (first stage: hydroclimatic zoning) (from Tsiros et al. 2009)

Specifically, Fig. 8 shows that in the 35% of Thessaly water district agriculture is not a sustainable due to water, altitude, or soil limitations. The term "sustainable under restrictions" refers to the cultivation of crops that do not need large quantities as "input" regarding irrigation and fertilizers. Also, "sustainable under restrictions" indicates that the type of cultivation preferred to those areas is extensive and not intensive. Further work has to be done in order to define the type of crops and cultivation techniques applied to those areas. The sustainable production areas for non-tropic crops have small spatial coverage, because they are delimited by the relatively high altitudes. Lastly, sustainable production zones cover about 25% of Thessaly indicating that those areas of the water district are suitable for any

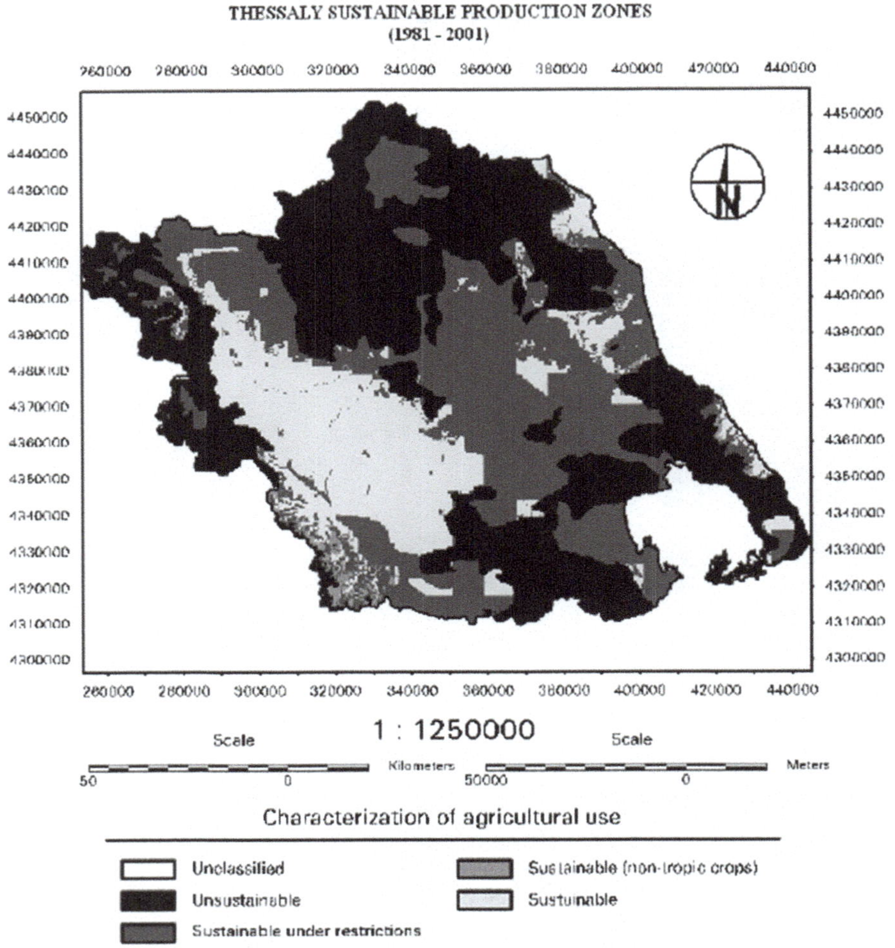

Fig. 8 Map of Thessaly showing sustainable production zones (second stage: general agroclimatic zoning, non-crop specific) (from Tsiros et al. 2009)

agricultural use. But, in order to obtain sustainability, farming management practices such as crop rotation, use of crop cover, and combination with livestock grazing out of the growing season are essential. Most of the times, monocultures are not sustainable systems.

5.1.3 Specific Agroclimatic Zoning

In this application, two indices are used to characterize areas suitable for cultivation, namely Growing Degree Days (GDD) (°C d) and Net Radiation (Rn) (Tsiros et al. 2009). This leads to specific agroclimatic zones, i.e. high, medium and low

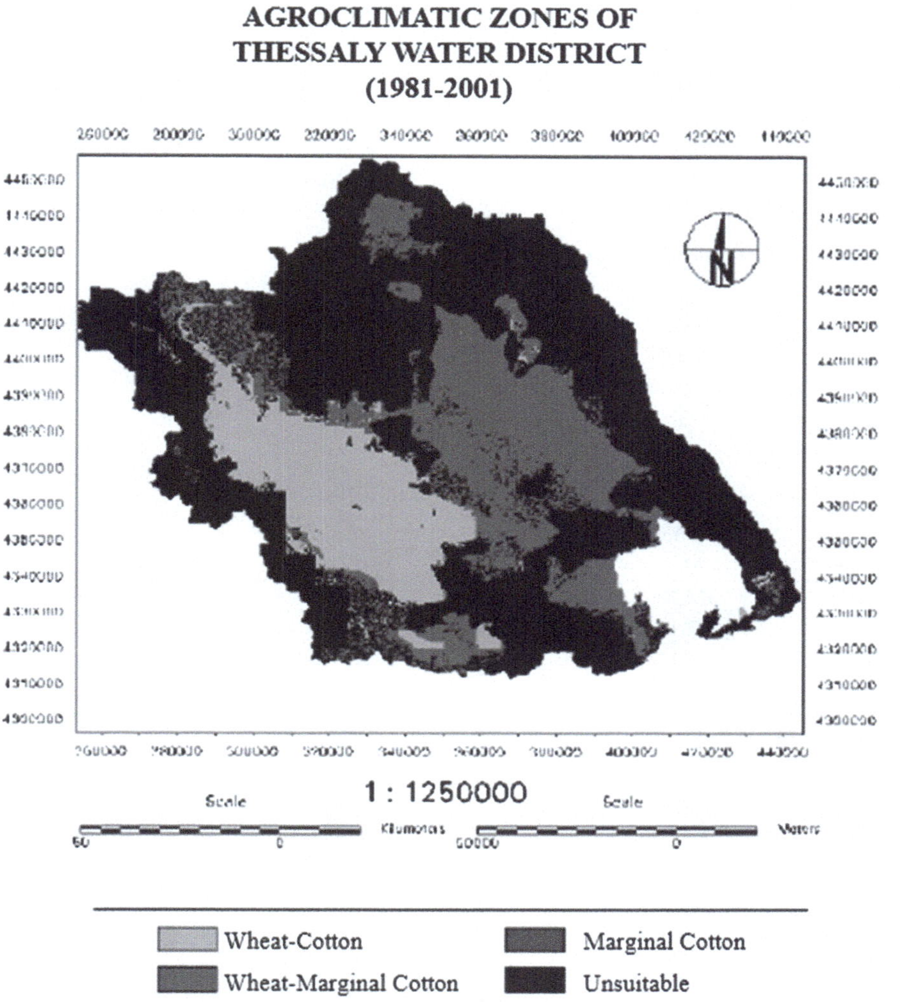

Fig. 9 Map of Thessaly showing specific agroclimatic zones (third stage) (from Tsiros et al. 2009)

productivity zones. Figure 9 shows the map of specific agroclimatic zones of Thessaly based on major existing crops, such as cotton and wheat.

5.2 *Application of the Multi-criteria Methodology*

5.2.1 Pay-Off Matrix

The weighted goal programming technique is applied to the model. The three objectives are: (1) Maximization of Total Gross Margin (GM); (2) Minimization of

Table 5 Pay of matrix in Karditsa (from Manos et al. 2010a)

	Optimum			
Values	GM	TF	TL	Real (observed values)
GM	**52,652**	48,207	47,484	51,867
TF	37,412	**35,801**	36,603	39,038
TL	14,656	13,599	**13,581**	14,996

Table 6 Pay of matrix in Larisa (from Manos et al. 2010a)

	Optimum			
Values	GM	TF	TL	Real (observed values)
GM	**76,696**	61,353	49,491	75,861
TF	30,808	**29,948**	29,951	32,565
TL	12,615	11,362	**11,176**	13,410

Table 7 Pay of matrix in Magnisia (from Manos et al. 2010a)

	Optimum			
Values	GM	TF	TL	Real (observed values)
GM	**324,706**	292,555	255,286	319,597
TF	37,115	**36,869**	36,879	37,123
TL	22,151	19,505	**18,920**	22,154

Table 8 Pay of matrix in Trikala (from Manos et al. 2010a)

	Optimum			
VALUES	GM	TF	TL	Real (observed values)
GM	**77,962**	69,805	69,443	76,685
TF	33,596	**32,348**	32,760	35,628
TL	15,875	15,029	**15,021**	16,511

Fertilizer use (TF); and (3) Minimization of Total Labor (TL). The pay-off matrices for the four Prefectures of Thessaly are shown in the next tables (Tables 5, 6, 7, and 8), where the diagonal values of these tables are bold to signify the optimal values.

The last column shows real data (observed) for each analyzed study area. These values show for each Prefecture the actual crop distribution (considering a theoretical 100 ha farm) and the relation among different crops and the objectives considered [gross margin (GM), fertilizers (TF) and labour (TL)]. This is an indication on how far the real situation (2009) is from any single optimum (column). This may lead to an attempt to combine objectives as a better simulation of farmers' behavior. Besides, this is the basis for the multi-criteria theory and for the methodology described. With the values of Tables 5–8 for each Prefecture the set of weights is obtained that best reflects farmers' preferences. These are shown in Table 9.

These weights show a type of farmers' behavior that combines profit maximization and total labor minimization. The minimization of total labor is an important

Table 9 Weights that best reflects farmers' preferences (from Manos et al. 2010a)

	Karditsa	Larisa	Magnisia	Trikala
W1 = (maximize GM)	0.9998	0.9693	0.9264	0.8501
W2 = (minimize TF)	0.00	0.00	0.00	0.00
W3 = (minimize TL)	0.0002	0.0307	0.0736	0.1499

criterion, since it has a weight of 14.99% in Trikala, 7.36% in Magnisia and 3.07% in Larisa agricultural area. This is combined with the criterion of profit maximization that has a large weight (85.01% in Trikala, 92.64% in Magnisia, 96.93% in Larisa and 99.98% in Karditsa). On the contrary, fertilizers minimization is not considered as a relevant criterion in these particular agricultural systems.

The estimation of these weights has been based on the current situation (2009). In this sense, it is important to note that the set of weights can be considered as a structural factor. As these weights correspond to the psychological attitudes of the producers, it is reasonable to assume that they will be kept at the same level in the short and the medium run, and this is actually an important pre-assumption in the simulation. The weightings given above are used in order to represent the farmers' utility function. For each Prefecture the utility function can be as follows:

- $\text{Karditsa} : \mathbf{U = 99.98\%GM - 0.02TL}$ (16)
- $\text{Larisa} : \mathbf{U = 96.93\%GM - 3.07TL}$ (17)
- $\text{Magnisia} : \mathbf{U = 92.64\%GM - 7.36TL}$ (18)
- $\text{Trikala} : \mathbf{U = 85.01\%GM - 14.99TL}$ (19)

5.2.2 Comparison of the Observed and Simulated Situations

It is essential to compare the real (observed) situation with the situation predicted with the help of the estimated utility function (Eqs. 16, 17, 18 and 19). Tables 10, 11, 12 and 13 show that the adopted methodology produces a better approximation to observed values. The results of the MCDM model for Karditsa suggest the abandonment of soft wheat, oat and sugar beets cultivations (Table 10). There is a decrease of 36.1% in the cultivated area of maize, and 9.07% in the cultivated area of hard wheat. In addition, there is an increase of 50% in the cultivated area of vines, 27.66% in the area of tomatoes, 20% in the area of barley, 18.75% of vetch, 0.51% of cotton and an increase 9.83% in the cultivated area of alfalfa. The participation of set aside in the optimal production plan increases, as compared with the existent production plan, by 13.3% of the total cultivated area of Karditsa. From the comparison of the existent and optimal production plans, it is observed that gross margin is increased by 1.51% (Fig. 10). In addition, a reduction is observed of fertilizers' use by 4.17%. Regarding labor use, a reduction is observed of 2.27%, due to increased set aside and finally water demand decreased by 3.02%.

Table 10 Comparison between observed values and MCDM model for Karditsa (from Manos et al. 2010a)

		MCDM model	
	Existent plan	Model values	% deviation
Gross margin (€)	48,987,305	49,728,414	+1.51
Fertilizer use (kg)	36,870,717	35,334,832	−4.17
Total labour (h)	14,163,554	13,841,931	−2.27
Water demand (m^3)	455,526,255	441,753,606	−3.02
Soft wheat	2.29%	0.00%	−100
Hard wheat	32.74%	29.77%	−9.07
Barley	1.10%	1.32%	+20.00
Oat	0.51%	0.00%	−100
Maize	5.79%	3.70%	−36.10
Sugarbeets	0.26%	0.00%	−100
Cotton	42.98%	43.20%	+0.51
Alfalfa	5.19%	5.70%	+9.83
Vetch	0.64%	0.76%	+18.75
Tomatoes	0.47%	0.60%	+27.66
Vines	1.10%	1.65%	+50.00
Set aside	6.94%	13.30%	+91.64
Total	100%	100%	

The results of the MCDM model for Larisa suggest the abandonment of rye, and sugar beets cultivations (Table 11). There is a decrease of 40% in the cultivated area of soft wheat, and 86% in the cultivated area of maize. In addition, there is an increase of 17% in the cultivated area of tomatoes, 9.88% in the area of vines, 9.21% of oat, 5% of hard wheat, 4.90% of barley, 4.96% of cotton, 4.35% of vetch, 4.75% of olives, 2.04% of apples and an increase of 4.91% in the cultivated area of alfalfa. The participation of set aside in the optimal production plan increases, as compared with the existent production plan, by 7.9% of the total cultivated area of Larisa. From the comparison of the existent and optimal production plans, it is observed that gross margin is increased by 1.10% (Fig. 11). In addition, a reduction is observed of fertilizers' use by 5.39%. Regarding labor use, a reduction is observed of 5.93%, due to increased set aside and finally water demand decreased by 11.65%.

Additionally, the results of the model for Magnisia agricultural area, suggest the abandonment of sugar beets cultivation (Table 12). It is observed that there is a decrease of 73.6% in the cultivated area of maize, and 8.57 in the cultivated area of cotton. In addition, there is an increase of 349% in the cultivated area of soft wheat, 47% in the area of vines, 42% in the area of oat, 17% of tomatoes, 9.4% of vetch, 5% of hard wheat, 4.88% of barley, 6.23% of alfalfa, 1.97% of olives and an increase of 1.84% in the cultivated area of apples. The participation of set aside in the optimal production plan decreases, as compared with the existent production plan, by 3.94% of the total cultivated area of Magnisia. From the comparison of the existent and

Table 11 Comparison between observed values and MCDM model for Larisa (from Manos et al. 2010a)

		MCDM model	
	Existent Plan	Model values	% deviation
Gross margin (€)	165,607,253	167,430,154	+1.10
Fertilizer use (kg)	71,089,746	67,254,693	−5.39
Total labour (h)	29,273,622	27,538,865	−5.93
Water demand (m^3)	685,135,467	605,341,091	−11.65
Soft wheat	6.14%	3.68%	−40.07
Hard wheat	40.21%	42.22%	+5.00
Barley	7.34%	7.7%	+4.90
Oat	0.76%	0.83%	+9.21
Rye	0.32%	0%	−100
Maize	5.26%	0.73%	−86.12
Sugarbeets	0.79%	0%	−100
Cotton	20.98%	22.02%	+4.96
Alfalfa	5.50%	5.77%	+4.91
Vetch	1.15%	1.20%	+4.35
Apples	0.98%	1.00%	+2.04
Tomatoes	1.77%	2.08%	+17.51
Vines	1.62%	1.78%	+9.88
Olives	2.95%	3.09%	+4.75
Set aside	4.23%	7.90%	+86.76
Total	100%	100%	

optimal production plans, it is observed that gross margin is increased by 1.60% (Fig. 12). In addition, a reduction is observed of fertilizers' use by 0.02%. Regarding water demand, a reduction is observed of 14.33% and finally labor use decreased by 0.01%. These results are expected, due to the high participation of the trees in the crop plan, which are intensive in labor use.

Finally, the results of the MCDM model for Trikala agricultural area, suggest the abandonment of vetch cultivation (Table 13). There is a decrease of 45.55% in the cultivated area of soft wheat, and 33.27% in the cultivated area of maize. In addition, there is an increase of 36.49% in the cultivated area of oat, 20% in the area of barley, 19.75% in the area of vines, 8.43% of hard wheat, 9.97% of cotton, 15.64% of alfalfa, and an increase of 6.78% in the cultivated area of olives. The participation of set aside in the optimal production plan increases, as compared with the existent production plan, by 13.40% of the total cultivated area of Trikala. From the comparison of the existent and optimal production plans, it is observed that gross margin is increased by 1.66% (Fig. 13). In addition, a reduction is observed of fertilizers' use by 5.54%. Regarding labor use, a reduction is observed of 4.12%, due to increased set aside and finally water demand decreased by 6.74%.

Table 12 Comparison between observed values and MCDM model for Magnisia (from Manos et al. 2010a)

	Existent plan	MCDM model	
		Model values	% deviation
Gross margin (€)	231,188,315	234,877,176	+1.60
Fertilizer use (kg)	26,854,148	26,848,280	−0.02
Total labour (h)	16,025,611	16,023,371	−0.01
Water demand (m^3)	110,471,869	94,636,780	−14.33
Soft wheat	0.55%	2.47%	+349
Hard wheat	31.01%	32.56%	+5.00
Barley	6.56%	6.88%	+4.88
Oat	0.42%	0.60%	+42.86
Maize	2.20%	0.58%	−73.64
Sugarbeets	0.39%	0.00%	−100
Cotton	8.52%	7.79%	−8.57
Alfalfa	2.89%	3.07%	+6.23
Vetch	1.48%	1.62%	+9.46
Apples	3.26%	3.32%	+1.84
Tomatoes	0.85%	1.00%	+17.65
Vines	0.68%	1.00%	+47.06
Olives	34.47%	35.15%	+1.97
Set aside	6.72%	3.94%	−41.37
Total	100%	100%	

Table 13 Comparison between observed values and MCDM model for Trikala (from Manos et al. 2010a)

	Existent plan	MCDM model	
		Model values	% deviation
Gross margin (€)	36,019,785	36,619,490	+1.66
Fertilizer use (kg)	16,734,628	15,807,634	−5.54
Total labour (h)	7,755,291	7,435,470	−4.12
Water demand (m^3)	248,167,137	231,433,771	−6.74
Soft wheat	7.53%	4.10%	−45.55
Hard wheat	15.77%	17.10%	+8.43
Barley	4.75%	5.70%	+20.00
Oat	0.74%	1.01%	+36.49
Maize	22.51%	15.02%	−33.27
Cotton	22.46%	24.70%	+9.97
Alfalfa	11.57%	13.38%	+15.64
Vetch	1.31%	0.00%	−100
Vines	1.62%	1.94%	+19.75
Olives	2.95%	3.15%	+6.78
Set aside	8.79%	13.40%	+52.45
Total	100%	100%	

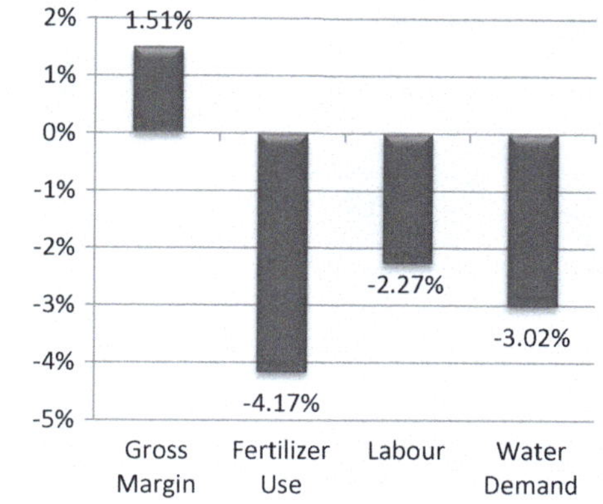

Fig. 10 Optimum production plan for Karditsa (from Manos et al. 2010a)

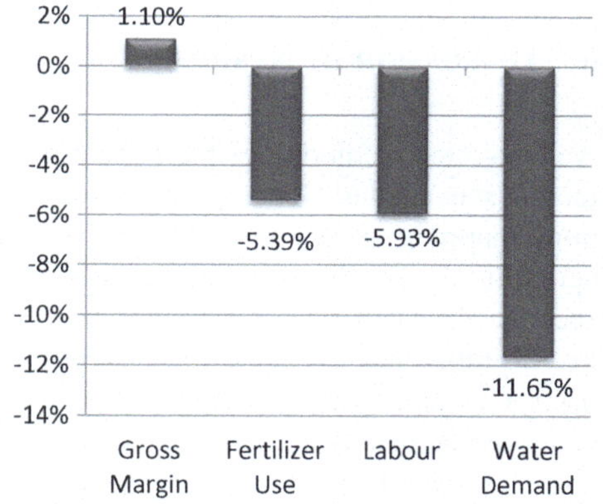

Fig. 11 Optimum production plan for Larisa (from Manos et al. 2010a)

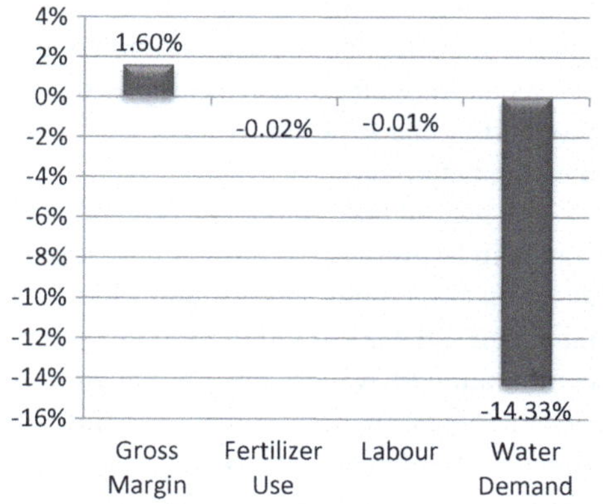

Fig. 12 Optimum production plan for Magnisia (from Manos et al. 2010a)

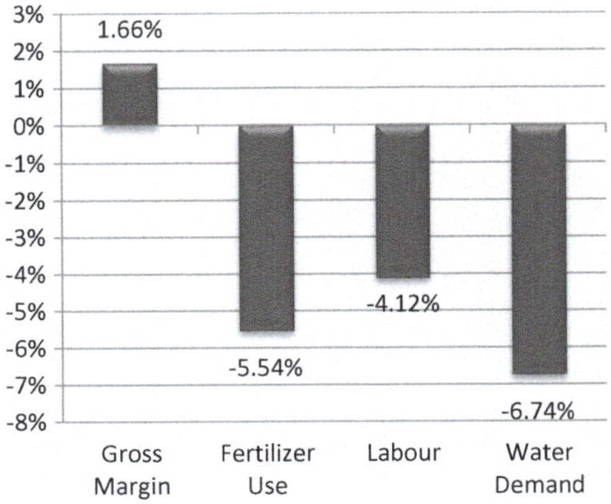

Fig. 13 Optimum production plan for Trikala (from Manos et al. 2010a)

6 Discussion of Results

An integrated methodology has been developed to identify agroclimatic zones using remote sensing and Geographic Information System (GIS). The agroclimatic conditions mapping shown that 59% of Thessaly water district is not appropriate for agricultural use. Agroclimatic classifications are divided into general and crop specific. The first task is to define a general methodology (not crop specific) for identifying zones adequate for sustainable farming according to water limitations using GIS and remote sensing. As crop growth is affected by water supply, these zones are named Water Limited Growth Environment (WLGE) zones. The WLGE zones are significant, since they delineate areas, where plant growth is limited by water availability. The results of the application justify the use of AI and VHI. Specifically, using VHI, areas frequently affected by agricultural drought are identified and excluded. The combination of the frequency of occurrence of such extreme events along with climatic aridity is useful for identifying areas unsuitable for crop production due to water availability. Such areas are excluded from any sustainable management plan.

The second task is to identify sustainable production zones (general agroclimatic zones) in terms of water sufficiency, fertility (appropriate or not for agricultural use), desertification vulnerability and altitude restrictions. Specifically, the use of soil maps and DEMs excludes areas inappropriate for agricultural activities. Thus, the combination of WLGE zones along with soil maps and DEMs can be used to identify sustainable productions zones. The third task is to identify crop specific agroclimatic zones with reference to existing crops in Thessaly, such as cotton and wheat. Such zones are essential in developing any sustainable development/farming plan, since they can be combined with crop specific agroclimatic indices in order to obtain agroclimatic zones.

The set of weights for the Prefecture of Larisa is compatible with a type of behaviour that combines profit maximisation (weighted by 97%) and labour minimisation (3% weight). It is important to note that although minimization is proposed of fertilizers as an objective taken into account by farmers, the results have shown that actually fertilizers are not considered as a relevant criterion in this particular agricultural system.

Similarly in the Prefecture of Magnisia, farmers' behaviour combines profit maximization and labour minimization but in different levels than in Larisa. In the Prefecture of Magnisia, farmers give more attention to labour minimization than in Prefecture of Larisa (it is weighted by 7.4%). The profit maximization is also considered an important criterion, since it is weighted by 92.6%. In this Prefecture the minimization of fertilizers is also not considered as a relevant criterion for the farmers' decisions. In the Prefecture of Karditsa the minimization of total labour is not considered a relevant criterion, since it is weighted by only 0.02%. This criterion is combined with the profit maximization criterion, which presents a very high weight (99.98%). On the contrary, in the Region of Trikala, the minimization of total labour is considered an important criterion, since it is weighted by 15%. This criterion is combined with the profit maximization criterion, which presents a high weight (85%) Also in the case of Trikala, minimization of fertilizers is not considered a relevant criterion in this agricultural system. As mentioned before, fertilizer minimization is a public objective. The utility function explains the farmers' behavior, so it can be concluded that, for this reason, fertilizer minimization is not considered in the decision process by the farmers.

Moreover, it is essential to compare the real (observed) situation with the situation predicted with the help of the estimated utility function for each Prefecture. It can be concluded that the adopted methodology provides a better approximation to observed values at the present (2009). Trying to combine the two objectives, profit maximization and total labour minimization, the MCDM model gives a farm plan that achieves an increase in gross margin in all Prefectures of Thessaly. On the issue of the minimization of total labour, the MCDM model achieves an important reduction in all Prefectures of Thessaly, except Magnisia, where it achieves a minor reduction. It can be concluded that labour use remains in the same level in Magnisia without significant differences due to the intensive labour demand of the trees cultivation (olives and apples). The results have also revealed that the MCDM model achieves a minor decrease in fertilizer use in Magnisia and an important reduction of fertilizer use in Trikala. Additionally, in Larisa and in Karditsa, regarding the minimization of fertilizers, the model achieves also an important reduction, respectively. Finally, from the comparison of the existent and optimal production plans, a significant reduction of water demand in all Prefectures of Thessaly can also be observed.

7 Summary and Conclusions

The first objective of this chapter has been to identify sustainable production zones by conducting contemporary agroclimatic classification. The innovation of the proposed methodology for this first objective consists of the joint use of the previously described steps, as well as the classification of areas escalating the suitability of agricultural activities. The main advantage of the methodology is that it uses satellite and raster data, providing continuous spatial and temporal information. In this way, there are no fuzzy borders regarding the derived zones. Methods that use conventional data are lacking the above advantages. The proposed methodology is not crop specific and has the advantage of providing total spatial coverage of the area under investigation. Lastly, the proposed methodology is transferable and integrates all the climatic variables that are important in order to identify agroclimatic zones.

The second objective is based on the outcome of the agroclimatic classification and develops a decision support system (DSS) by using multi-criteria analysis in each production class combining different criteria to a utility function under a set of constraints concerning different categories of agroclimatic, social, cultural and economic conditions. Specifically, a weighted goal programming model was applied, that combines the advantages of Linear Programming (LP), i.e. simplicity and flexibility, with the integrative ability of Multi-criteria Decision Making (MCDM) models. This methodology estimates the farmers' utility function taking in account various conflicting criteria that can explain the farmers' behavior (maximization of gross margin, minimization of fertilizers and minimization of total labor). The MCDM model was then applied in each Prefecture of Thessaly separately. From the results, for each Prefecture the set of weights is obtained that best reflects farmers' preferences. These weights show a type of farmers' behavior that combines profit maximization and labor minimization.

The described methodology presents new optimal solutions to existing international environmental issues. Furthermore, this methodology capitalizes the recent continuous technological and computational advances. Specifically, it can be stated that this innovative integrated methodology, which satisfies the above two objectives, can significantly contribute to the global food production adequacy problem due to the expected population increase globally. Moreover, this methodology can also contribute to the new research trend of water-energy-food (WEF) nexus in several regions around the world facing water availability problems, such as the whole Mediterranean basin. Nevertheless, the presented methodology constitutes a unique combination of the above two objectives, since it incorporates the latest technological and scientific achievements in remote sensing, as well as in optimization methods and crop yield modelling. However, as expected, there are certain limitations, which are essentially based on the availability of satellite and conventional environmental and agronomic data. There is also need for training of farmers and stakeholders, as well as increase of the public awareness on the above issues.

The future trend of the presented research seems promising. At first, during the forthcoming years the number of environmental satellites is expected to increase providing additional data with finer spatial and temporal resolution, as well as more bands. As a result, agroclimatic zoning could reach mesoscale and fine-scale classification of the agricultural land. Moreover, the continuous computational advancements are expected to lead to the development of web environmental and production platforms, as well as further improvement in optimization, namely multi-criteria analysis, methods and crop yield modelling and systems.

Acknowledgements The conventional meteorological data was provided by the National Meteorological Service. The satellite data was provided by NASA-NOAA from the USA. The rainfall maps were provided by EU-JRC at Ispra-Varese, Italy. The technical and economic coefficients of crops in each Prefecture resulted from the Regional Authority of Thessaly and from the Department of Agriculture and Veterinary of each Prefecture. Additional data have been provided by the Department of Agricultural Economics of Aristotle University of Thessaloniki, and from the National Agricultural Research Foundation (NAGREF—National Agricultural Research Foundation, n.d.).

References

Al-Bakri, J. T., & Taylor, J. C. (2003). Application of NOAA–AVHRR for monitoring vegetation conditions and biomass in Jordan. *Journal of Arid Environments, 54*, 579–593.

Amador, F., Sumpsi, M. J., & Romero, C. (1998). A non-interactive methodology to assess farmers' utility functions: An application to large farms in Andalusia, Spain. *European Review of Agricultural Economics, 25*, 92–109.

Arneth, A., Brown, C., & Rounsevell, M. D. A. (2014). Global models of human decision-making for land-based mitigation and adaptation assessment. *Nature Climate Change, 4*(7), 550–557.

Badini, O., Stocle, C. O., & Franz, E. H. (1997). Application of crop simulation modeling and GIS to agroclimatic assessment in Burkina Faso. *Agriculture, Ecosystems and Environment, 64*, 233–244.

Bartolini, F., Bazzani, G. M., Gallerani, V., Raggi, M., & Viaggi, D. (2007). The impact of water and agricultural policy scenarios on irrigated farming systems in Italy: An analysis based on farm level multi-attribute linear programming models. *Agricultural Systems, 93*(1–3), 90–114.

Bartolini, F., Gallerani, V., Raggi, M., & Viaggi, D. (2007). Implementing the water framework directive: Contract design and the cost of measures to reduce nitrogen pollution from agriculture. *Environmental Management, 40*(4), 567–577.

Becker, F., & Li, Z. (1990). Towards a local split window method over land surfaces. *International Journal of Remote Sensing, 11*, 369–393.

Berbel, J., & Rodriguez, A. O. (1998). An MCDM approach to production analysis: An application to irrigated farms in southern Spain. *European Journal of Operational Research, 107*, 108–118.

Bhuiyan, C., Singh, R. P., & Kogan, F. N. (2006). Monitoring drought dynamics in the Aravalli region (India) using different indices based on ground and remote sensing data. *International Journal of Applied Earth Observation and Geoinformation, 8*, 289–302.

Blaney, H. F., & Criddle, W. D. (1950). *Determining water requirements in irrigated areas from climatological and irrigation data*. USDA Soil Conservation Service, Technical Paper, No. 96, 48 pp.

Botts, M. (2012). *OGC® SensorML: Model and XML encoding standard*. Open Geospatial Consortium 12-000. Retrieved from https://portal.opengeospatial.org/files/?artifact_id=55939

Bröring, A., Echterhoff, J., Jirka, S., Simonis, I., Everding, T., Stasch, C., et al. (2011). New generation sensor web enablement. *Sensors, 11*(3), 2652–2699.

Bröring, A., Stasch, C., & Echterhoff, J. (2012). *OGC® sensor observation service interface standard*. Open Geospatial Consortium, 12-006. Retrieved from https://portal.opengeospatial.org/files/?artifact_id=47599.

Brown, J. F., Wardlow, B. D., Tadesse, T., Hayes, J. J., & Reed, B. C. (2008). The Vegetation Drought Response Index (VegDri): An integrated approach for monitoring drought stress in vegetation. *GIScience and Remote Sensing, 45*(1), 16–46.

Bruce, J. P. (1994). Natural disaster reduction and global change. *AMS Bulletin, 75*(10), 1831–1835.

Chen, P. P. (1976). The entity-relationship model–toward a unified view of data. *ACM Transactions on Database Systems, 1*(1), 9–36.

Dalezios, N. R. (2015, November). *AGROMETEOROLOGY: Analysis and simulation* (in Greek). KALLIPOS: Libraries of Hellenic Universities (also e-book), 481 pp. ISBN: 978-960-603-134-2.

Dalezios, N. R., Blanta, A., Spyropoulos, N. V., & Tarquis, A. M. (2014). Risk identification of agricultural drought for sustainable agroecosystems. *NHESS, 14*, 2435–2448.

Dalezios, N. R., Spyropoulos, N. V., & Eslamian, S. (2017). Remote sensing in drought quantification and assessment (Chapter 21). In S. Eslamian & F. Eslamian (Eds.), *Handbook of drought and water scarcity (HDWS)* (Vol. 1, pp. 377–396). Boca Raton, FL: Taylor and Francis.

Danalatos, N. (2007). *An introduction in crop production simulation*. Volos, Greece: University of Thessaly Press.

Dilley, M., & Heyman, B. N. (1995). ENSO and disaster: Droughts, floods and El Niño-Southern Oscillation warm events. *Disasters, 19*(2), 181–193.

EC COM 672. (2010, November 18). *The CAP towards 2020: Meeting the food, natural resources and territorial challenges of the future*. Brussels.

Echterhoff, J., & Everding, T. (2008). *OpenGIS® sensor event service interface specification* (proposed). Open Geospatial Consortium 08-133. Retrieved from http://portal.opengeospatial.org/files/?artifact_id=29576

Eitzinger, J., Utset, A., Trnka, M., Zalud, Z., Nikolaev, M., & Uskov, I. (2007). Weather and climate and optimization of farm technologies at different input levels. In M. V. K. Sivakumar & R. Motha (Eds.), *Managing weather and climate risks in agriculture* (Vol. 554). Berlin: Springer. ISBN 3540727442.

ESA. (2014). *Sentinel*. Earthonline – ESA. Retrieved April 6, 2014, from https://earth.esa.int/web/guest/missions/esa-future-missions/sentinel-1

European Commission. (2009). *Consultation on the future "EU 2020" strategy*. Brussels: Commission of the European Communities.

Gomez-Limon, J. A., & Berbel, J. (2000). Multi-criterion analysis of derived water demand functions: A Spanish case study. *Agricultural Systems, 63*, 49–72.

Gomez-Limon, J. A., & Riesgo, L. (2004). Irrigation water pricing: Differential impacts on irrigated farms. *Agricultural Economics, 31*(1), 47–66.

Gomez-Limon, J. A., & Sanchez-Fernandez, G. (2010). Empirical evaluation of agricultural sustainability using composite indicators. *Ecological Economics, 69*, 1062–1075.

Gomez-Limon, J. A., Arriaza, M., & Berbel, J. (2002). Conflicting implementation of agricultural and water policies in irrigated areas in the EU. *Journal of Agricultural Economics, 53*, 259–277.

Gong, J., Geng, J., & Chen, Z. (2015). Real-time GIS data model and sensor web service platform for environmental data management. *International Journal of Health Geographics, 14*, 2. Retrieved from http://www.ij-healthgeographics.com/content/14/1/2

Gong, J., Wu, H., Gao, W., Yue, P., & Zhu, X. (2009). Geospatial service Web (Chapter 13). In D. Li, J. Shan, & J. Gong (Eds.), *Geospatial technology for earth observation* (1st ed., pp. 355–379). New York: Springer. Retrieved from http://www.barnesandnoble.com/w/geospatial-technology-for-earth-observation-deren-li/1117115908?ean=9781441900500

Goward, S. N., Markham, B., Dye, D. G., Dulaney, W., & Yang, J. (1991). Normalized difference vegetation index measurements from the advanced very high resolution radiometer. *Remote Sensing of Environment, 35*, 257–277.

Hatcher, G., & Maher, N. (2000). Real-time GIS for marine applications. In D. J. Wright & D. J. Barlett (Eds.), *Marine and coastal geographical information systems* (pp. 137–147). New York: Taylor & Francis.

Hong, S.-Y., & Lim, J.-O. J. (2006). The WRF single-moment 6-class microphysics scheme (WSM6). *Journal of the Korean Meteorological Society, 42*, 129–151.

IPCC. (2012). *Managing the risks of extreme events and disasters to advance climate change adaptation*. Special Report of IPCC, 582 p.

IPCC. (2013). Climate change 2013. In T. F. Stocker, D. Qin, G.-K. Plattner, M. Tignor, S. K. Allen, J. Boschung, A. Nauels, Y. Xia, V. Bex, & P. M. Midgley (Eds.), *The physical science basis. Contribution of Working Group I to the Fifth Assessment Report of the Intergovernmental Panel on Climate Change*. Cambridge: Cambridge University Press. 1535 p.

Janjic, Z. I. (2001). A nonhydrostatic model based on a new approach. *Meteorology and Atmospheric Physics, 82*, 271–285. https://doi.org/10.1007/s00703-001-0587-6.

Kain, J. S. (2004). The Kain–Fritsch convective parameterization: An update. *Journal of Applied Meteorology, 43*, 170–181. https://doi.org/10.1175/1520-0450(2004)043<0170:TKCPAU>2.0.CO;2.

Kioutsioukis, I., Melas, D., & Zerefos, C. (2010). Statistical assessment of changes in climate extremes over Greece (1955–2002). *International Journal of Climatology, 30*(11), 1723–1737.

Kogan, F. N. (1995). Application of vegetation index and brightness temperature for drought detection. *Advances in Space Research, 15*, 91–100.

Kogan, F. N. (2001). Operational space technology for global vegetation assessment. *Bulletin of the American Meteorological Society, 82*, 1949–1964.

Malheiro, A. C., Santos, J. A., Fraga, H., & Pinto, J. G. (2010). Climate change scenarios applied to viticultural zoning in Europe. *Climate Research, 83*, 163–177.

Manos, B., Bournaris, T., Kamruzzaman, M., Begum, A. A., & Papathanasiou, J. (2006). The regional impact of irrigation water pricing in Greece under alternative scenarios of European policy: A multicriteria analysis. *Regional Studies, 40*(9), 1055–1068.

Manos, B., Begum, A. A., Kamruzzaman, M., Nakou, I., & Papathanasiou, J. (2007). Fertilizer Price Policy, the environment and farms behavior. *Journal of Policy Modeling, 29*(1), 87–97.

Manos, B., Bournaris, T., Papathanasiou, J., & Chatzinikolaou, P. (2008). Evaluation of tobacco cultivation alternatives under the EU common agricultural policy (CAP). *Journal of Policy Modeling, 31*(2), 225–238.

Manos, B., Papathanasiou, J., Bournaris, T., & Voudouris, K. (2010a). A multicriteria model for planning agricultural regions within a context of groundwater rational management. *Journal of Environmental Management, 91*, 1593–1600.

Manos, B., Papathanasiou, J., Bournaris, T., & Voudouris, K. (2010b). A DSS for sustainable development and environmental protection of agricultural region. *Environmental Monitoring and Assessment, 164*(1–4), 43–52.

Manos, B., Bournaris, T., & Chatzinikolaou, P. (2011). Impact assessment of CAP policies on social sustainability in rural areas: An application in Northern Greece. *Operational Research, 11* (1), 77–92.

Mavi, H. S., & Tupper, G. J. (2004). *Agrometeorology: Principles and applications of climate studies in agriculture* (364 p.). Australia: Food Products Press.

McVicar, T. R., & Jupp, D. L. B. (1998). The current and potential operational uses of remote sensing to aid decisions on drought exceptional circumstances in Australia. A review. *Agricultural Systems, 57*(3), 399–468.

Nastos, P., Dalezios, N. R., & Ulbrich, U. (Eds.). (2016). *Advances in meteorological hazards and extreme events*. Special Issue of NHESS.

Niemeyer, S. (2008). New drought indices. *Options Méditerranéennes. Série A: Séminaires Méditerranéens, 80*, 267–274.

Olesen, J. E., & Bindi, B. (2002). Consequences of climate change for European agricultural productivity, land use and policy. *European Journal of Agronomy, 16*, 239–262.

Pereira, L. S. (2017). Water, agriculture and food: Challenges and issues. *Water Resources Management, 31*, 2985–2999.

Politi, N., Nastos, P. T., Sfetsos, A., Vlachgiannis, D., Dalezios, N. R., Gounaris, N., Cardoso, M. R., & Soares, M. M. P. (2016). *Comparison and evaluation of WRF model physics parameterizations over the domain of Greece*. Proceedings, COMECAP2016. AUTH, Thessaloniki, Greece, September 19–21 (in CD).

Schut, P. *OpenGIS® web processing service*. Retrieved from http://portal.opengeospatial.org/files/?artifact_id=24151

Simonis, I., & Echterhoff, J. (2009). *Sensor planning service implementation standard*. Open Geospatial Consortium 09-000. Retrieved from http://portal.opengeospatial.org/files/?artifact_id=38478

Sivakumar, M. V. K., Motha, R. P., & Das, H. P. (Eds.). (2005). *Natural disaster and extreme events in agriculture* (367 p.). Berlin: Springer. ISBN-10 3-540-22490-4.

Skamarock, W. C., Klemp, J. B., Dudhia, J., Gill, D. O., Barker, D. M., Wang, W., & Powers, J. G., (2006). *A description of the advanced research WRF version 3*. NCAR tech. note TN-475_STR, 113 p.

Soares, P. M. M., Cardoso, R. M., Miranda, P. M. A., Medeiros, J., Belo-Pereira, M., & Espirito-Santo, F. (2012). *WRF high resolution dynamical downscaling of ERA-interim for Portugal. Climate Dynamics, 39*, 2497–2522. https://doi.org/10.1007/s00382-012-1315-2

Steinemann, A., Hayes, M. A., & Cavalcanti, L. (2005). Drought indicators and triggers. In D. A. Wilhite (Ed.), *Drought and water crises: Science technology and management issues* (pp. 71–90). New York: Marcer Dekker.

Sumpsi, M. J., Amador, F., & Romero, C. (1997). On farmers objectives: A multi-criteria approach. *European Journal of Operational Research, 96*, 64–71.

Tarquis, A. M., Gobin, A., Ulbrich, U., & Dalezios, N. R. (Eds.). (2013). *Weather related hazards and risks in agriculture*. Special Issue of NHESS journal.

Thenkabail, P. S., Gamage, M. S. D. N., & Smakhtin, V. U. (2004). *The use of remote sensing data for drought assessment and monitoring in Southwest Asia*. Research Report, International Water Management Institute, No. 85, 1–25.

Thornthwaite, C. W. (1948). An approach toward a rational classification of climate. *Geographical Review, 38*, 55–94.

Tsiros, E. C. (2009). *Satellite methodologies of agroclimatic-hydroclimatic zoning for agricultural production*. Ph.D. Thesis, School of Agricultural Sciences, University of Thessaly, Volos, Greece, 259 p.

Tsiros E., Domenikiotis, C., Kanellou, E., & Dalezios, N. R. (2008). *Identification of water limited growth environment zones using NOAA/AVHRR data* (pp. 150–155). In 4th International Conference on Information and Communication Technologies in Bio and Earth Sciences HAICTA 2008, Athens, Greece, September 18–20.

Tsiros, E., Domenikiotis, C., & Dalezios, N. R. (2009). Sustainable production zoning for agroclimatic classification using GIS and remote sensing. *IDŐJÁRÁS, 113*(1–2), 55–68.

UNESCO. (1979). *Map of the world distribution of Arid regions: Explanatory note*. Man and the Biosphere. Technical Notes 7, Paris.

Van De Griend, A. A., & Owe, M. (1993). On the relationship between thermal emissivity and the normalized difference vegetation index for natural surfaces. *International Journal of Remote Sensing, 14*, 1119–1131.

Van Dijk, A., Callis, S. L., & Sakamoto, C. M. (1987). Smoothing vegetation index profiles: An alternative method for reducing radiometric disturbance in NOAA/AVHRR data. *Journal of Photogrammetric Engineering and Remote Sensing, 63*, 1059–1067.

Van Westen, C. J. (2013). Remote sensing and GIS for natural hazards assessment and disaster risk management. In J. F. Shroder (Ed.), *Treatise on geomorphology* (Vol. 3, pp. 259–298). San Diego: Academic Press.

Waseem, M., Ajimal, M., & Kim, T.-W. (2015). Development of a new composite drought index for multivariate drought assessment. *Journal of Hydrology, 527*, 30–37.

Weiss, J. L., Gutzler, D. S., Coonrod, J. E., & Dahm, C. N. (2004). Seasonal and inter-annual relationships between vegetation and climate in central New Mexico, USA. *Journal of Arid Environment, 57*, 507–534.

Yassoglou, N. J. (2004). *Soil associations map of Greece*. Greek National Committee for Combating Desertification. Athens: Agricultural University of Athens Press.

Zargar, A., Sadiq, R., Naser, B., & Khan, F. I. (2011). A review of drought indices. *Environmental Reviews, 19*, 333–349.

Zittis, G., Hadjinicolaou, P., & Lelieveld, J. (2014). Comparison of WRF model physics parameterizations over the MENA-CORDEX domain. *American Journal of Climate Change, 3*, 490–511. https://doi.org/10.4236/ajcc.2014.35042

Review of Multicriteria Methodologies and Tools for the Evaluation of the Provision of Ecosystem Services

Parthena Chatzinikolaou, Davide Viaggi, and Meri Raggi

Abstract An ecosystem is defined as an area, place or environment where organisms interact with the physical and chemical environment. Ecosystems provide a variety of benefits to people that are divided into market and non-market ecosystem goods or ecosystem services (ES) and classified in multiple ways. A wide range of methodologies is available to value changes in ES. The type of valuation technique chosen depends on the type of ES to be valued, as well as the quantity and quality of data available. Since ES provide multiple benefits, they are valued for a range of reasons and researchers must employ valuation methods that better match this diversity. How to compare objects with multiple characteristics has been the focus of Multi-Criteria Decision Analysis (MCDA). In this paper, a literature review has been performed that covers an overview of various methodologies that seek to improve the knowledge base of existing tools and methodologies in ES evaluation. The focus will be on studies that used MCDA for an ES assessment; attempting to cover a representative sample of case studies of ES assessments through MCDA. We also discuss advantages and disadvantages of different methodological choices in ecosystem service evaluation. We conclude that the ES concept is currently used in a range of studies with widely differing aims creating difficulties for policy makers as well as researchers since it makes it difficult to assess the credibility of assessment results. MCDA techniques can indeed help structuring the problem and supporting a more transparent decision-making.

Keywords Ecosystem services · Literature review · Multi-criteria analysis · Assessment · Evaluation

P. Chatzinikolaou (✉) · D. Viaggi
Department of Agricultural Sciences, University of Bologna, Bologna, Italy
e-mail: parth.chatzinikolao2@unibo.it

M. Raggi
Department of Statistical Sciences, University of Bologna, Bologna, Italy

J. Berbel et al. (eds.), *Multicriteria Analysis in Agriculture*, Multiple Criteria Decision Making, https://doi.org/10.1007/978-3-319-76929-5_2

1 Introduction and Objective

The term Ecosystem Services (ES) was originally conceived to highlight humanity's dependence to nature, and the fact that the multitude of plant and animal species in the forests, oceans, lakes, wetlands and other ecosystems provides humanity with a wide selection of goods. The concept of ES has been used in research since the 1980s and involved the framing of beneficial ecosystem functions as services in order to increase public interest in biodiversity conservation (de Groot 1987). The next step during the 1990s, was the mainstreaming of ES in the literature (Costanza et al. 1997; Perrings et al. 1992). In 2001, the Millennium Ecosystem Assessment (MEA) launched the ES in the global policy agenda (Gómez-Baggethun et al. 2010). The objective was to assess the consequences of ecosystem change for human well-being. The series of MEA publications described the condition and trends of the world's ecosystems and the services they provide, and the options available to restore, conserve or enhance their sustainable use. At present, the ES concept has become a central issue in conservation planning and environmental impact assessment (Burkhard et al. 2010).

The ability of ecosystems to yield ES is largely connected to biodiversity; many researchers agree about the positive effect of biodiversity on ecosystem functions (Hooper et al. 2005; Schneiders et al. 2012). Biological diversity, or biodiversity, refers to the variety of life forms at all levels of organization. Biodiversity is generated and maintained in natural ecosystems, where organisms encounter a wide variety of living conditions that shape their evolution in unique ways. Biodiversity is usually quantified in terms of numbers of species, and this perspective has greatly influenced conservation goals. It is important to remember, however, that the benefits that biodiversity supplies to humanity are delivered through populations of species residing in living communities within specific physical settings in other words, through complex ecological systems, or ecosystems (Luck et al. 2003). Natural ecosystems provide fundamental services which humanity needs. These include the production and maintenance of biodiversity; purification of air and water; decomposition of wastes and regulation of climate; services until recently have been less appreciated. The consequences of population loss for species conservation are well recognized, but have been little addressed from the point of the functioning of ecosystems and the provision of ES (Hughes et al. 1997). Because threats are increasing, there is a critical need for identification and monitoring of ES both locally and globally, and for the incorporation of their value into decision-making processes. There is a need for policies that achieve a balance between sustaining ES and pursuing the short-term goals of economic development.

Following the publication of the Millennium Ecosystem Assessment in 2005 (MEA 2005), ES have been included for the first time into the international environmental policy agenda. This agenda included efforts to develop integrated systems of ecosystems (Weber 2007). In current policies the ES concept is being integrated at global and European level (EC 2009; Perrings et al. 2011). The global strategic plan for biodiversity for the period 2011–2020 (EC 2011) of the Convention of Biological

Diversity complements previous conservation biodiversity targets with the addition of ES. The EU has adopted an ambitious strategy to halt the loss of biodiversity and ES in the EU by 2020 (EC 2010). There are 6 main targets, and 20 actions to help Europe reach its goal. Target 2 focuses on maintaining and enhancing ES and restoring degraded ecosystems by incorporating green infrastructure in spatial planning. This will contribute to the EU's sustainable growth objectives and to mitigating and adapting to climate change, while promoting economic, territorial and social cohesion and safeguarding the EU's cultural heritage. It will also ensure better functional connectivity between ecosystems within and between Natura 2000 areas and in the wider countryside. According to the EU Biodiversity Strategy, by 2050, biodiversity and the ES, its natural capital, are protected, valued and appropriately restored for their essential contribution to human wellbeing and economic prosperity, and changes caused by the loss of biodiversity are avoided.

Biodiversity and agricultural production are connected and their capacity to be mutually supportive is increasingly recognized. On the one hand, maintaining biodiversity makes agricultural production and related practices more sustainable. On the other hand, it is recognized that changing agricultural land use is a major cause of the decline of biodiversity. As a result, the Common Agricultural Policy (CAP), the largest agricultural support system worldwide has been reformed in order to meet the Europe 2020 Strategy goals. The European Commission highlighted the importance of using the CAP to halt the decline of biodiversity, and various efforts have been made to merge biodiversity conservation into agricultural policy. The reformed CAP has shifted from an agricultural production support system towards a broader focus including the inventory of public goods and ecosystems services provided by agriculture (EC 2009). Both CAP pillars contribute to biodiversity conservation, pillar 1 through direct payments and pillar 2, through agri-environmental measures. Importantly, also the EU Water, Regional and Cohesion Policy recognize the importance of investing in natural ecosystems, in particular urban green areas, floodplains and nature for recreation, as a source of economic development. However, it is further necessary to develop a policy framework that considers the most recent research on multidimensional ES and enhance the provision of ES in order to preserve social and cultural landscape values and maintain the multifunctionality of agricultural ecosystems. Both agriculture and regional development contribute to over 80% the annual EU budget, so the inclusion of ES in these policies is considered an important step towards a more sustainable economy.

In the last decades, in the field of ES, there is a rise of concern for the valuation of ecosystem functions, goods and services. Early references refer to the concept of ecosystem functions, services and their economic value (Helliwell 1969; Odum and Odum 1972). More recently, there is a growth in publications on the benefits of natural ecosystems to human society (Daily 1997; de Groot 1992; Pearce 1993; Wilson and Carpenter 1999). Boyd and Banzhaf (2007) and Fisher and Kerry Turner (2008) focused on the quantification of ES and their value to stakeholders and suggested various classification schemes. Turner and Daily (2008) proposed that ecosystem service research should address the various stages in decision-making, from problem identification to policy evaluation and capacity building. The measurement and monitoring of ES requires also relating ecosystem functioning to ecosystem service indicators. Different perspectives are taken to describe the relationships underlying the supply of ES, which representation has been approached using derived land use or land cover data (Lautenbach et al. 2011),experiments (Sandhu et al. 2008), expert opinion or modelling (Carpenter et al. 2009). Other approaches to map ecosystem services were based on land cover (Nelson and Kennedy 2009). Willemen et al. (2010) estimated landscape functions which were related to a number of ecosystem functions and ecosystem services based on proxy variables.

Since ecosystems are multifunctional, complex systems, described by a multitude of characteristics from the point of view of multiple criteria. The multidimensional logic of ES seems highly consistent with multicriteria approaches. According to Chan et al. (2012), ES provide multiple benefits, valued for a range of reasons, and researchers must employ valuation methods that better match the diversity of values in question. How to compare objects with multiple characteristics is focus of Multi-Criteria Decision Analysis (MCDA). MCDA is a general framework for supporting complex decision-making situations with multiple and often conflicting objectives. The considered goals are usually too complex to be properly assessed by a single criterion or indicator. Therefore, multiple relevant criteria or indicators are considered at the same time (Kiker et al. 2005).

The objective of this paper is to review existing methodologies and tools for the evaluation of the provision of the ES with a focus on studies that used MCDA for an ES evaluation. We attempt to cover a representative, though not comprehensive, sample of case studies of ES assessments using MCDA. The literature review performed highlights the scope of MCDA as a decision support tool for ES evaluation. The structure of the paper is as follows. Section 2 describes the Ecosystem Services concept and Sect. 3 presents a review of the relevant literature and a review of different methods for evaluation, mapping and assessment of the Ecosystem Services. In Sect. 4, the paper is presenting a review of multicriteria approaches for Ecosystem Services evaluation and in the final sections (Sects. 5 and 6) the discussion and conclusions are presented.

2 Ecosystem Services Concept

An ecosystem is usually defined as an area, place or environment where organisms interact with the physical and chemical environment. The ecosystem concept describes the interrelationships between living organisms and the non-living environment. *"An ecosystem is a dynamic complex of plant, animal, and microorganism communities and the non-living environment interacting as a functional unit"* (MEA 2005). There is a full range of ecosystems, from natural forests, to ecosystems managed and modified by humans, such as agricultural land. Ecosystems provide a variety of benefits to people that is divided into market and non-market ecosystem goods or ecosystem services (ES) and classified in multiple ways. It is common practice to refer to goods and services separately and to include the two concepts under the term services. Definitions of the ES concept through various publications, give attention to the ecological basis or the economic use, capturing environmental concerns in ecological and socio-economic terms (Diehl et al. 2016):

- Ecosystem Services are the benefits human populations derive, directly or indirectly, from ecosystem functions (Costanza et al. 1997).
- Ecosystem Services are the benefits people obtain from ecosystems (WRI 2005).
- Ecosystem Services are components of nature, directly enjoyed, consumed, or used to yield human well-being (Boyd and Banzhaf 2007).
- Ecosystem Services are the aspects of ecosystems utilized (actively or passively) to produce human well-being (Fisher et al. 2009).
- Ecosystem Services are the direct and indirect contributions of ecosystems to human well-being (TEEB 2010a).

ES have been categorized in a number of different ways, including:

- functional groupings, such as regulation, carrier, habitat, production, and information services (de Groot et al. 2002);
- organizational groupings, such as services that are associated with certain species, that regulate some exogenous input, or that are related to the organization of biotic entities (Norberg 1999); and,
- descriptive groupings, such as renewable resource goods, nonrenewable resource goods, physical structure services, biotic services, biogeochemical services, information services, and social and cultural services (Moberg and Folke 1999).

The Millennium Ecosystem Assessment framework (MEA 2003) categorizes the ES within four categories: provisioning, regulating, supporting and cultural services:

Provisioning services are the services that describe the material or energy outputs from ecosystems:

- food and fiber including the range of food products derived from plants, animals, and microbes; food comes principally from managed agro-ecosystems but marine and freshwater systems or forests also provide food for human consumption;
- raw materials, fuel, wood, and other biological materials that serve as sources of energy, also a great diversity of materials for construction and fuel including wood, biofuels and plant oils that are directly derived from wild and cultivated plant species;
- fresh water, because ecosystems regulate the flow and purification of water; also vegetation and forests influence the quantity of water available; fresh water is an example of linkages between categories, in this case, between provisioning and regulating services;
- genetic resources, that includes the genes and genetic information used for animal and plant breeding and biotechnology;
- biochemical, natural medicines, and pharmaceuticals, since many medicines, biocides, food additives such as biological materials derived from ecosystems and are potential source of medicinal resources;

- ornamental resources including animal products, such as skins and shells, and flowers used as ornaments, although the value of these resources is often also determined culturally.

Regulating services are the benefits obtained from the regulation of ecosystem processes, the services that ecosystems provide by acting as regulators:

- air quality maintenance, because ecosystems contribute to extract chemicals to the atmosphere, influencing many aspects of air quality;
- climate regulation, both locally and globally; for example, at a local scale, changes in land cover can affect both temperature and precipitation; at global scale, ecosystems play an important role in climate by either sequestering or emitting greenhouse gases;
- water regulation, the timing and magnitude of flooding, and aquifer recharge can be strongly influenced by changes in land cover, including, in particular, alterations that change the water storage potential of the system, such as the conversion of wetlands or the replacement of forests with croplands or croplands with urban areas;
- erosion prevention and maintenance of soil fertility: soil erosion is a key factor in the process of land degradation and desertification; vegetation cover provides a vital regulating service by preventing soil erosion. Soil fertility is essential for plant growth and agriculture and well-functioning ecosystems supply the soil with nutrients required to support plant growth;
- water purification and waste treatment: ecosystems can be a source of impurities in fresh water but also can help to filter out and decompose organic wastes introduced into inland waters and coastal and marine ecosystems;
- regulation of human diseases; changes in ecosystems can directly change the abundance of human pathogens, such as cholera, and can alter the abundance of disease vectors, such as mosquitoes;
- biological control: ecosystem changes affect the prevalence of crop and livestock pests and diseases; they regulate pests and diseases through the activities of predators and parasites that all act as natural controls;
- pollination: insects and wind pollinate plants and trees, which are essential for the development of fruits, vegetables and seeds; animal pollination is an ecosystem service mainly provided by insects but also by some birds and bats.

Cultural services are tightly bound to human values and behavior, as well as to human institutions and patterns of social, economic, and political organization. They are the non-material benefits people obtain from ecosystems through spiritual enrichment, cognitive development, reflection, recreation, and aesthetic experiences, including:

- recreation, mental and physical health: walking and playing sports in green space is not only a good form of physical exercise but also lets people relax; the role that green space plays in maintaining mental and physical health is increasingly being recognized, despite difficulties of measurement;

- spiritual and religious value: many religions attach spiritual and religious values to ecosystems or their components like natural features or specific forests, caves or mountains are considered sacred or have a religious meaning; nature is a common element of all major religions and traditional knowledge;
- educational values because ecosystems and their components and processes provide the basis for both formal and informal education in many societies;
- inspiration for culture, art and design: language, knowledge and the natural environment have been intimately related throughout human history; biodiversity, ecosystems and natural landscapes have been the source of inspiration for much of our art, culture and increasingly for science;
- aesthetic values: many people find beauty or aesthetic value in various aspects of ecosystems, as reflected in the support for parks or even the selection of housing locations;
- social relations, since ecosystems influence the types of social relations that are established in particular cultures; fishing societies, for example, differ in many respects in their social relations from nomadic herding or agricultural societies;
- cultural heritage values: many societies place high value on the maintenance of either historically important landscapes (cultural landscapes) or culturally significant species;
- recreation and ecotourism: ecosystems and biodiversity play an important role for many kinds of tourism which in turn provides considerable economic benefits and is a vital source of income; cultural and eco-tourism can also educate people about the importance of biological diversity.

Supporting services are those that are necessary for the production of all other ES. They differ from provisioning, regulating, and cultural services in that their impacts on people are either indirect or occur over a very long time, whereas changes in the other categories have relatively direct and short-term impacts on people:

- soil formation: humans do not directly use this as a service, but changes in soil formation would indirectly affect people through the impact on other services such as the provisioning service of food production;
- nutrient cycling: this indirect supporting service is required e.g. as the basis for crop production and plant growth;
- biomass production: primary production provides the basis of the food for all consumers;
- production of atmospheric oxygen through photosynthesis is often categorized as a supporting service since oxygen forms the basis for any animal life on Earth; any impacts on the concentration of oxygen in the atmosphere would only occur over an extremely long time;
- habitats for species: Habitats provide everything that an individual plant or animal needs to survive: food; water; and shelter; each ecosystem provides different habitats that can be essential for a species' lifecycle; migratory species including birds, fish, mammals and insects all depend upon different ecosystems during their movements.

Some services, like erosion control, can be categorized as both a supporting and a regulating service, depending on the time scale and immediacy of their impact on people. For example, humans do not directly use soil formation services, although changes in this would indirectly affect people through the impact on the provisioning service of food production. Fresh water is another example of linkages between categories, in this case, between provisioning and regulating services. Similarly, climate regulation is categorized as a regulating service since ecosystem changes can have an impact on local or global climate over time scales relevant to human decision-making. The production of oxygen gas (through photosynthesis) is categorized as a supporting service since any impacts on the concentration of oxygen in the atmosphere would only occur over an extremely long time.

A new classification of ES is under development at international level, by the Common International Classification of Ecosystem Services (CICES 2013). According to CICES, there are three types of services: (1) provisioning (products obtained from ecosystems, e.g. food, wood, water), (2) regulation and maintenance (moderation or control of environmental conditions, e.g. flood control, water purification), (3) cultural (non-material benefits obtained from ecosystems, e.g. recreation, education, aesthetics). CICES highlights the importance of making a clear distinction between final ES, ecosystem goods or products and ecosystem benefit, and recommend the following definitions:

- Final ecosystem services are the contributions that ecosystems make to human well-being. These services are final in that they are the outputs of ecosystems (whether natural, semi-natural or highly modified) that most directly affect the well-being of people. A fundamental characteristic is that they retain a connection to the underlying ecosystem functions, processes and structures that generate them.
- Ecosystem goods and benefits are things that people create or derive from final ecosystem services. These final outputs from ecosystems have been turned into products or experiences that are not functionally connected to the systems from which they were derived. Goods and benefits can be referred to collectively as "products".
- Human well-being is that which arises from adequate access to the basic materials for a good life needed to sustain freedom of choice and action, health, good social relations and security. The state of well-being is dependent on the aggregated output of ecosystem goods and benefits, the provision of which can change the status of well-being.

Ecosystem service capacity and service output are closely related to the notion of stocks and flows. Layke et al. (2012) define stocks of ES as the capacity of an ecosystem to deliver a service while the flow corresponds to the benefits people receive. Stocks may be expressed in total size area or the total biomass whereas the associated ecosystem service flow or output must have units per time period. The capacity of an ecosystem to provide a flow is not necessarily measured in hectares or tons since the capacity does not only contain a quantity aspect but also a quality aspect. For a given quantity, an ecosystem may provide more output if it is in a healthy state, or at least be able to provide a sustained flow of services. As a result, the capacity of such a system to produce services will be higher. Ecosystems in a healthy state are considered resilient systems, which are able to recover after disturbance, and high species diversity and a balanced trophic community characterize them. Benefits derived from ES are food, drinking water, clear air, fuel, fiber, construction materials, protection against disasters and stable climate. In the case of regulating services (e.g., climate regulation) and supporting services (e.g., nutrient cycling), aggregate stocks are most important. The "efficiency" with which stocks deliver specific ES will vary with ecology. Thus, a forest's capacity to provide climate regulation services will reflect broad patterns of species composition and architecture, but flows of services may not be closely linked to conservation value in terms of diversity (Carnus et al. 2006). Differences in harvesting technologies can lead to different flows of outputs from similar stocks. In timber concessions, chainsaws and powered haulage also change the scale and value of economic flows from a forest. In both cases, such technologies also, of course, have implications for sustainability of revenue flows. The value of provisioning service flows from a given stock of natural capital can therefore vary with the technology used. Technology is also important in demand for ES. The relationship between cultural services and stocks of biodiversity is a little more complex and can vary over time. Thus, the income from tourism based on the reintroduction of a charismatic locally extinct species might be expected to decline as the species becomes widespread. Moreover, different human actors may perceive the relationship between the size of stocks of biodiversity and the value of the flows of services that they provide differently. These relationships between biodiversity and valued ecosystem functions are an important area for further theoretical and experimental research (Vira and Adams 2009).

3 Literature Review on ES Concept

Improved ways and methods for ES quantification, mapping and assessment are needed to investigate the number and quality of ES produced by individual ecosystems and to increase the ability to feed such knowledge into policy design (TEEB 2010b). While provisioning ES can often be directly quantified thanks to the availability of primary data, for other ES the collection of such information is often impossible (Maes et al. 2015). Thus, for most regulating, supporting, and cultural services, researchers must rely on proxies for their quantification. As a result, altogether, data on quantifiable ES remain limited and only a small number of indicators are being used for those that cannot be measured directly (Feld et al. 2010). Reviews of indicators used for ES are available from the literature and contribute to developing reliable indicators for modelling and for bridging current data gaps (Cowling et al. 2008; Egoh et al. 2012).

Several studies have assessed changes in land use and their connection with the provision of ES (Carreño et al. 2012; Silvert 2000). In many cases, their output includes environmental and land use information that are connected to landscape features, although few yield a direct assessment of changes in ES provision (Burkhard et al. 2012; Swetnam et al. 2011). According to de Groot et al. (2010), ES approaches and ES valuation efforts have changed the terms of discussion on nature conservation, natural resource management, and other areas of public policy. These efforts have strengthened both public and private sector development strategies and improved environmental outcomes (de Groot 2006; de Groot et al. 2002).

In this paper, a literature review has been performed that covers an overview of various methodologies that seek to improve the knowledge base of the contribution of landscape management to the rural economy. This review of the literature is concerning methods relevant to the landscape management taking into account the policy strategies and linkages with environmental impacts and climate change. Based on the analysis of the literature performed, the methodological tools were classified into three main categories (Table 1):

- Identification and valuation of the ES and natural resource management;
- Sustainable land use, in terms of assessment of agricultural systems and linking socio-economic requirements with landscape potentials;
- Structure of the landscape, and linkages with environmental impacts and climate change.

In the first category, De Groot (2006) presented a comprehensive framework for integrated assessment of ecological services and socio-economic benefits of natural and semi-natural ecosystems and landscapes. The framework can be applied at different scale levels to different ecosystems or landscape-units and consists of three steps: (1) Function-analysis, which translates ecological complexity into a limited number of ES; (2) Function valuation, which includes ecological, socio-cultural and economic valuation methods; (3) Conflict analysis, to facilitate the application of function-analysis and valuation at different scale levels. Hein et al. (2006) established an enhanced framework for the valuation of ES, with specific attention for stakeholders. The framework included a procedure to assess the value of regulation services that avoids double counting of these services. Moreover, the study analyzed the spatial scales of ES. The analysis has shown that stakeholders at different spatial scales can have very different interests in ES, and it is highly important to consider the scales of ES when valuation of services is applied to support the formulation or implementation of ecosystem management plans.

Moreover, Fisher et al. (2009) offer a definition of ES that is likely to be operational for ecosystem service research and several classification schemes: "*ecosystem services are the aspects of ecosystems utilized (actively or passively) to produce human well-being*". Defined this way, ES include ecosystem organization or structure as well as process and/or functions if they are consumed or utilized by humanity either directly or indirectly. There is not one classification scheme that will be adequate for the many contexts in which ecosystem service research may be utilized. The paper discusses several examples of how classification schemes will be a function of both ecosystem and ecosystem service characteristics and the decision-making context. In addition, Wallace (2007) developed a classification of ES that

Table 1 Studies that estimate ecosystem services

Method	Used for/example	Key references
Ecosystem services approach	Identification of ES and valuation of them separately	de Groot (2006), de Groot et al. (2010), Hein et al. (2006)
	Classification schemes as functions of both ecosystem and ecosystem service characteristics and decision-making	Fisher and Kerry Turner (2008), Fisher et al. (2009)
	Classification of ES—framework for decision-making in natural resource management	Wallace (2007)
	Classifying and valuing ecosystem functions, goods and services—link ecosystem functions to the main ecological, socio-cultural and economic valuation methods	de Groot et al. (2002)
Sustainable land use	Potential effects, economic viability and social acceptability of Agri-environmental Regulation—AEI indicators	Onate et al. (2000)
	Conceptual framework for the economic valuation and prioritization of sustainability indicators—based on Bayesian decision theory	Pannell and Glenn (2000)
	Self-assessment tool based on the IDEA method to support sustainable agriculture—41 indicators covering the three dimensions of sustainability	Zahm et al. (2008)
	Indicators that cover the three components of the sustainability concept—evaluated the three dimensions sustainability with composite indicators	Gomez-Limon and Sanchez-Fernandez (2010)
	Indicators for multifunctional land use—Linking socio-economic requirements with landscape potentials	Wiggering et al. (2006)
	Framework for sustainability assessment of agricultural systems, encompassed the three dimensions of sustainability	Glaser and Diele (2004), Rasul and Thapa (2004)
Landscape structure	Selection of different input–output IOA systems as effective tools for Agri-environmental improvement—environmental indicators based on good agricultural practices	Halberg et al. (2005)
	Approach which measures environmental sustainability of urban water system, based on LCA methodology	Lundin and Morrison (2002)
	LCA method, for assessing the environmental impact of production processes	Haas et al. (2000, 2001)
	Quantitative measure of human disturbance based on land-use and solar energy (Energy Accounting) consumption per use	Brown and Vivas (2005)
	Evaluation of ecosystem health and its measurement at a variety of landscape scales—linkages between socioeconomic drivers, biogeochemical indicators	Patil et al. (2001)

(continued)

Table 1 (continued)

Method	Used for/example	Key references
	Stepwise downscaling procedure based on expert-judgement and pairwise comparison to obtain socio-economic parameters between the evolution of socio-economics and climate change	Abildtrup et al. (2006)

Source: Own elaboration

provides a framework for decisions in natural resource management. However, further work is still required to resolve particular issues, such as the classification of socio-cultural services. De Groot et al. (2002) presented a framework and typology for describing, classifying and valuing ecosystem functions, goods and services in a clear and consistent manner. In the second part of the paper, a checklist and matrix is provided, linking these ecosystem functions to the main ecological, socio-cultural and economic valuation methods.

As regards the second category, Onate et al. (2000) tried to evaluate the potential effects of Agri-environmental Regulation EC 2078/92 on European agricultural landscapes through the use of agri-environmental indicators (AEIs) on policy effects. The main effects may be catalogued as improvement effects or protection effects since they represent a change in participant over non-participant farmers' decisions. Finally, the importance of this type of policy evaluation approach is discussed in the light of the likely future development of AEP in the European Union. Pannell and Glenn (2000) presented a conceptual framework for the economic valuation and prioritization of sustainability indicators. The framework was based on Bayesian decision theory, particularly its use to calculate the value of information under conditions of uncertainty. They tried to fill the gap of a conceptual framework as basis for evaluation and sustainable development. Moreover, Zahm et al. (2008) based on 41 sustainability indicators covering the three dimensions of sustainability, tried to design a self-assessment tool not only for farmers but also for policy makers to support sustainable agriculture. Gomez-Limon and Sanchez-Fernandez (2010) developed a methodology for evaluating the sustainability of farms by means of composite indicators. This methodology was based on calculating 16 sustainability indicators that cover the three components of the sustainability concept (economic, social and environmental). The evaluation of farm sustainability using the methods suggested is a potentially useful tool for public decision-makers who are tasked with designing and implementing agricultural policy. Wiggering et al. (2006) presented an approach to merge different types of output by defining an indicator of social utility. Social utility in this sense includes environmental and economic services as long as society expresses a demand for them. Within this approach, the integrated indicator concept incorporates the approaches of both sustainability and multifunctionality in land use and management. In addition, Glaser and Diele (2004) presented some central aspects of a sustainability assessment for a North Brazilian mangrove crab fishery, based on a number of criteria from biology, economics and sociology. They intended to contribute to future resource management plans to improve the living conditions of current and future generations while ensuring the health and productivity of the crab population and the mangrove ecosystem they depend on. Rasul and Thapa (2004) examined the sustainability of two production systems in terms of their environmental soundness, economic viability and social acceptability, based on 12 indicators of sustainability.

In the last category, Halberg et al. (2005) selected 10 input–output accounting systems (IOA) covering the topics of the farm's use of nutrients, pesticides and energy, from a survey of 55 systems and compared them. The approaches and indicators used vary from systems based on good agricultural practices to accounts based systems that use physical input–output units. Haas et al. (2000, 2001) used the framework of a LCA in 18 grassland dairy farms covering three farming intensity levels. In this study, the selection of appropriate impact categories and functional units are emphasized, to fit specific agricultural and regional requirements in order to compare the impact of farms. The objective of this study was to adapt the LCA method, developed for assessing the environmental impact of production processes, to agriculture on the whole farm level, efficiently and feasibly assessing all relevant environmental impacts. In addition, Lundin and Morisson (2002) presented a procedure for the selection of indicators, which reflects the environmental sustainability of urban water system. The chosen indicators were evaluated in case studies in a developed and a developing region. This procedure combined empirical results with a theoretical framework based on LCA methodology. Brown and Vivas (2005) developed a method of quantitatively evaluating the human disturbance gradient that is applicable to landscapes of varying scales from watersheds to forest patches or isolated wetlands. Moreover, Patil et al. (2001) described the challenges of reporting on changes in an ecosystem health at the different landscape scales. The focus was to show how the integration of recent advances in quantitative techniques and tools will facilitate the evaluation of ecosystem health and its measurement at a variety of landscape scales. The challenge was to characterize, evaluate, and validate linkages between socioeconomic drivers, biogeochemical indicators, multiscale landscape metrics, and human life indicators. Finally, Abildrup et al. (2006) presented an integrated approach to the construction of socio-economic scenarios required for the analysis of climate change impacts on European agricultural land use. The chosen scenarios ensured internal consistency between the evolution of socioeconomics and climate change.

A range of methodologies are available to value changes in ES. The type of valuation technique chosen depends on the type of ES to be valued, as well as the quantity and quality of data available. Some valuation methods may be more suited to capturing the values of particular ES than others (Galimberti et al. 2014). The valuation methodologies reviewed are not new in themselves. The challenge is in their appropriate application to ES. The Ecosystem Services Framework emphasizes the need to consider the ecosystem as a whole and stresses that changes or impacts on one part of an ecosystem have consequences for the whole system (Jopke et al. 2015). Key challenges in the valuation of ES relate to how ecosystems interrelate to provide services and to dealing with issues of irreversibility and high levels of uncertainty in how ecosystems function. All of this suggests that, while valuation is an important and valuable tool for good policy-making, it should be seen as only one of the inputs in decision making (Mace et al. 2012).

4 Multicriteria Approaches in ES Evaluation

Multi-Criteria Decision Analysis (MCDA) is a set of methods that can be used to support the process of decision making by taking into consideration multiple criteria in a flexible manner, by means of a structured and intelligible framework. MCDA are integrative evaluation methods in the sense that they combine information about the performance of the alternatives with respect to the criteria (scoring) with subjective judgements about the relative importance of the evaluation criteria in the particular decision-making context (weighting). Due to their capacity to assess trade-offs and accommodate value pluralism, MCDA has been proposed as a tool for an integrated valuation of ES (Daily et al. 2009).

In order to valuing benefits provided by ecosystems and biodiversity social and economic aspects should be considered along with environmental issues. Multi-Criteria Decision tools allow simultaneous consideration of a wide range of economic, social and environmental decision criteria, representing different dimensions of sustainability. In the ecological domain, recent research by Gorshkov et al. (2000), show how the complexity of ecosystems and the ecological web and the biosphere in general can determine the climatic stability and resilience of the surrounding region or the global system. Recent research by Costanza et al. (2014) contributes to the debate on the evaluation of a multitude of ES. Regan et al. (2006) present a coherent set of environmental criteria for evaluating biodiversity. Moffett et al. (2005) offers an extensive overview of existing applications of multi-criteria methods to the problem of biodiversity evaluation.

A systematic literature review has been performed, in order to assess the knowledge on the use of MCDA methods in ES evaluation. The review was focused on studies that used MCDA for an ES assessment; attempting to cover a representative sample of study cases of ES assessments through MCDA. This section highlights and discusses the main results from the review. A summary of the main results is given in Table 2. As presented in the table below, there is a group of studies has attempted to build a framework in order to assess the ES. Another group has used different multicriteria evaluation tools and techniques, and the last group has used different multicriteria approaches for the evaluation of landscape functions.

MCDA methods have been applied by a wide range of studies, as decision support systems that integrate economic and noneconomic values (Newton et al. 2012), as approaches for cost-benefit analysis (Wegner and Pascual 2011) or as methodological framework for addressing value dimensions related to ES (Mendoza and Prabhu 2003). Based on the analysis of the literature performed, the methodological tools were classified into three main groups. The first group includes different studies that have applied frameworks for the assessment of ES. The objective of Comino et al. (2014) was to explore the use of a Multicriteria Spatial Decision

Table 2 Multicriteria methods for ES assessment and evaluation

Method	Problem definition/scope	Reference
Assessment frameworks	Multicriteria Spatial Decision Support System tool for the analysis of the environmental quality with reference to nature conservation	Comino et al. (2014)
	Ecosystem function-based planning and management approach	Oikonomou et al. (2011)
	Development of sustainability indicators for analyzing their interactions and effects on ecosystems	Mendoza and Prabhu (2003)
	Multi-scale framework, for the assessment of European regions in terms of sustainability and ex ante impact assessment of policy scenarios	Paracchini et al. (2011)
	MCA for a comparative assessment of ES for alternative land and water management scenarios	Posthumus et al. (2010)
	Framework in which MCDA tools are used for evaluating strategic planning options based on policy scenarios	Stewart and Scott (1995)
	Framework for sustainability assessment of agricultural systems related to the multiple functions of the agro-ecosystem	Van Cauwenberg et al. (2007)
Evaluation techniques	PROMETHEE-based methods for the evaluation of potential water resources/ to obtain new indicators for ES/ to compare land use alternatives considering ES as criteria/ to compare municipalities based on ES indicators	Abu-Taleb and Mareshal (1995), Segura et al. (2015), Fontana et al. (2013), Chatzinikolaou et al. (2015)
	Three different evaluation techniques to compare alternative plans for the socioeconomic development	Bodini and Giavelli (1992)
	ELECTRE III for the evaluation and selection of a solid waste management system	Hokkanen and Salminen (1997)
	Interaction matrix to evaluate the effects of farmer production practices on the agro-ecosystem	Girardin et al. (2000)
Landscape scales	Multicriteria assessment framework for the evaluation of land cover change scenarios and the estimation of regional potentials to provide ES	Koschke et al. (2012)
	Qualitative approach to assess landscape functions using four land use scenarios	Kienast et al. (2009)
	Spatial multi-criteria assessment methodology to construct maps of regional conservation priorities	Zerger et al. (2011)
	Qualitative tool for the quantification and spatial modelling of multiple ES in different landscapes	Burkhard et al. (2009)
	Multicriteria evaluation approach to assess how changes in cultural landscapes might affect people	Martínez-Sastre et al. (2017)

Source: Own elaboration

Support System tool for the analysis of the environmental quality of the river basin with particular reference to the issue of nature conservation and to support both the planning and management processes. Oikonomou et al. (2011) proposed a conceptual framework that combines ecosystem function analysis, multi-criteria evaluation and social research methodologies for introducing an ecosystem function-based planning and management approach. Moreover, Mendoza and Prabhu (2003) applied two general MCA methodologies for developing sustainability indicators and for analyzing their interactions. This analysis allowed a more holistic assessment of the indicators by examining how they affect each other directly and indirectly, like individual elements of an ecosystem. Paracchini et al. (2011) proposed a further advancement in integrated assessment procedures by setting up an operational multi-scale and transparent framework, which comprised the assessment of European regions in terms of sustainability, and the identification of the impact that policy options might have on the sustainability of these regions. The framework was designed for use in ex ante sustainability impact assessment of policy scenarios on multifunctionality of land use and integrates economic, environmental and social issues (Paracchini and Capitani 2011). Additionally, Posthumus et al. (2010) applied a multi-criteria analysis that enables a comparative assessment of ES for alternative land and water management scenarios, especially regarding the assessment of non-monetary values. They focused on a methodology in order to measure and value ES under different land management scenarios that reflect different priorities for food plain areas. Stewart and Scott (1995) introduced a framework in which MCDA tools are used for evaluating strategic planning options. This framework was based on direct evaluation of sequences of policy scenarios. Finally, Van Cauwenberg et al. (2007) proposed a framework for sustainability assessment of agricultural systems, (Sustainability Assessment of Farming and the Environment). The framework was composed of principles, criteria, indicators and reference values in a structured way. Principles were related to the multiple functions of the agro-ecosystem and consistent approaches were used for the identification and selection of the indicators.

The second group of the reviewed methods has applied evaluation or ranking techniques as tools to evaluate ES. Abu-Taleb and Mareshal (1995) have applied the PROMETHEE V multicriteria method to evaluate potential water resources and select from a variety of potentially feasible water resources development options, so that the allocation of limited funds to alternative development projects and programs can proceed in the most efficient manner. Segura et al. (2015) applied a PROMETHEE-based method to obtain new composite indicators for provisioning, maintenance and "direct to citizen services". Fontana et al. (2013) have used PROMETHEE to compare land use alternatives considering ES as criteria. Chatzinikolaou et al. (2015) applied the PROMETHEE III method, for the comparison of the 26 municipalities of the province of Ferrara, based on a set of ES indicators. Moreover, Bodini and Giavelli (1992) have applied three different evaluation techniques, multicriteria weighted concordance, discordance analysis and a qualitative procedure. These evaluation techniques were used to compare four alternative plans for the socioeconomic development of Salina Island, based on 14 criteria that reflect the socio environmental perception of the inhabitants. Hokkanen and Salminen (1997) applied the multicriteria ELECTRE III decision-aid in the context of choosing a solid waste management system in Finland. Girardin et al. (2000) have adopted an interaction matrix to evaluate the effects of farmer production practices on the agro-ecosystem. The evaluation matrix provided the raw material both for the development of Agro-Ecological Indicators (AEI) and Indicators of Environmental Impact (IEI) and for the use of multicriteria methods for sorting, selecting, or classifying cropping or farming systems.

The last group of studies is focused on a regional or landscape scale, have attempted a spatial distribution of specific ES. Availability of data for an assessment of ES provision on a regional scale is often very limited. Up-scaling of detailed data from lower scales does not always contribute to an improvement in the data base on a regional scale (Egoh et al. 2008). Koschke et al. (2012) have applied a multicriteria assessment framework for the evaluation of land cover change scenarios and for the qualitative estimation of regional potentials to provide ES as a prerequisite to support regional development planning. The objective was to present a conceptual framework for how to assess the actual and potential future capacity of a region to provide ES. Moreover, Kienast et al. (2009) have applied a framework to analyze the relationships between the ecosystem properties of large portions of land, and their capacities to provide goods and services using four land use scenarios. Burkhard et al. (2009) have used quantitative and qualitative assessment data in combination with land cover and land use information for the assessment of multiple ES. The proposed framework was applied in different case studies and is proposed as a useful tool for the quantification and spatial modelling of multiple ES in different landscapes. Zerger et al. (2011) have focused on integrated sustainability and impact assessments in conservation planning. More specifically, they apply a spatial multi-criteria assessment methodology to construct maps of regional conservation priorities and assesses how these maps map influence farm-scale actions. Finally, Martínez-Sastre et al. (2017) tried to assess, through an empirical case study, how changes in cultural landscapes might affect people living in and making use of the ES associated with those landscapes. They also discuss the usefulness of combining a multicriteria evaluation approach with the ES framework, as a methodological tool to support complex decision-making in situations associated with land use planning where multiple and conflicting interests are involved.

5 Discussion

Different studies have tried to classify, quantify, map and value ES in order to integrate the concept into decision-making processes (e.g. Costanza et al. 1997; de Groot et al. 2010). In general, frameworks include three main parts: (i) measuring the provision of ES; (ii) determining the monetary value of ES; (iii) designing policy tools for managing ES (Polasky 2008). This variety of methodological approaches is, on which ES to measure, which indicators to use and in which scale. Seppelt et al. (2011) provided a quantitative review of 153 ecosystem service studies that are using a set of indicators. More than 75% excluded scenario analysis and more than 60% of the studies did not involve stakeholders. Between 45% and 80% of the studies also did not give sufficient information, concerning the results' uncertainty and validation. However, this does not mean that these studies were not reproducible. This variety of methodological approaches is, on the one hand, a creative scientific process and typical of the development of new concepts, however on the other hand, it can deliver confusing messages to the policy makers and researchers.

As assessing and mapping of ES is mainly dependent on data availability and finding the appropriate indicators, most publications focused either on selected ES and/or only on one landscape scale. Ecosystems have the ability to produce multiple ES simultaneously, which is referred to as multifunctionality. Efforts that focus on the provision of single services (e.g. production of a crop) can have a negative impact on other services; for example, water and often pollination are needed for agricultural crops. Enhancing important provisioning services, such as food and timber, often leads to trade-offs between regulating and cultural services, such as nutrient cycling, flood protection, and tourism. As an example, Van Zanten et al. (2016) tried to measure the relative importance of landscape features across agricultural landscapes in order to better understanding the cross-regional variation of aesthetic and recreational values and how these values relate to characteristics of the agricultural landscape. Moreover, Guo et al. (2016) proposed a modelling framework which considers land-competition across different land types and sectors and accounts for ecosystem service changes due to changes in land use transitions over multiple time periods without exploring options based on policy scenarios. They considered a number of quantitative and semi-quantitative indicators of ES, focusing on provisioning (e.g. bioenergy, livestock) and biodiversity.

Scientific progress is also being made in developing socio-economic scenarios and models of global change impacts on biodiversity and ES (Cheung et al. 2009; Rodriguez et al. 2006; Sitch et al. 2008; Thomas et al. 2004; Xenopoulos et al. 2005). Currently a major challenge in this field of research, is improving the relevance and value of these advances for decision makers at multiple scales (Donner et al. 2005). Scenarios are widely used in land use planning, climate change analysis and conservation planning (Verburg et al. 2006; Xiang and Clarke 2003), and, increasingly, in ecosystem service assessment (Castella et al. 2005; Duinker and Greig 2007; Kirchner et al. 2015). The CAP has recently entered a new programming period and the new objectives are oriented towards the sustainable management of natural resources and climate action (Viaggi 2015). The CAP 2014–2020 has presented policies centered on efficient provision of ES from agricultural land. However, only a few studies have tried to assess the value of changes in ES with the implementation of the reformed CAP measures that are addressed on ecosystems. As an example, Lupp et al. (2015) applied an ES approach as a framework to assess the impact of increasing energy crop production (with a focus on biogas production) on the quality and services of those ecosystems concerned, taking into account the European and national energy production targets and action plans about biomass and total energy consumption by 2020. Kirchner et al. (2016) tried to assess the impacts of alternative policy pathways on the supply of ES, considering the aggregate and spatial impacts of the latest CAP reform. Applying different policy scenarios is a key component of forward-looking decision making in some instances at local and national levels (e.g., climate change impacts on forests and protected areas, management of fisheries) but covers only a small range of sectors and cases (Jetz et al. 2007).

6 Conclusions

The objective of this paper is to review existing methodologies and tools for the evaluation of the provision of ES with a focus on studies that used MCDA for the assessment and evaluation of ES. We attempt to cover a representative, though not comprehensive, sample of case studies of ES assessments using MCDA. The rising demand for ES measurement, modelling and evaluation is the main driver for development of ES research. A key challenge of ecosystem management is determining how to manage multiple ecosystem services across landscapes. The review performed highlights the scope of MCDA as a decision support tool for ES evaluation, since it can structure an assessment of a complex problem like evaluating ES and allows comparison of ecological objectives with socio-cultural and economic ones in a structured framework.

Different studies have tried to classify, quantify, map and value ES in order to integrate the concept into decision-making processes. These efforts have suggested measures to better evaluate ES and to improve the knowledge base of the value of ES, as well as proposed measures that will mainstream the importance of biodiversity and the value of ES. However, there are still open questions to fully integrate the ecosystem service concept in landscape research and decision making (Seppelt et al. 2012). Despite the increase in publications on ecosystem goods and services, a comprehensive framework for integrated assessment and valuation of ES is still missing (Armsworth et al. 2007; van Zanten et al. 2014). Improved ways and methods for ES quantification and assessment are needed to investigate the number and quality of ES produced by the individual ecosystems and to increase the ability to feed such knowledge into policy design.

According to the review performed, one of the main findings was that there does not exist one standard method or approach to map and assess the ES and their value. MCDA methods can structure an assessment of a complex problem along both cognitive and normative dimensions, both of which are essential in evaluating and assessing ES. There exists a wide variety of MCDA approaches put in practice at different geographical scales or ongoing efforts at to harmonize the classification of ES and their valuation. There are a number of choices to be made between classifications, methods and approaches. These choices involve firstly defining what the purpose of the ES valuation is; determine which ES are of highest relevance; defining the types of value information that are required; and finally select the relevant and appropriate valuation methods. Although MCDA techniques can help structuring the problem and supporting a more transparent decision-making, the ES concept is currently used in a range of studies with widely differing aims creating difficulties for policy makers as well as researchers since it makes it difficult to assess the credibility of assessment results.

References

Abildtrup, J., Audsley, E., Fekete-Farkas, M., Giupponi, C., Gylling, M., Rosato, P., & Rounsevell, M. (2006). Socio-economic scenario development for the assessment of climate change impacts on agricultural land use: A pairwise comparison approach. *Environmental Science and Policy, 9*, 101–115.

Abu-Taleb, M. F., & Mareschal, B. (1995). Water resources planning in the Middle East: Application of the PROMETHEE V multicriteria method. *European Journal of Operational Research, 81*, 500–511.

Armsworth, P. R., Chan, K. M. A., Daily, G. C., Ehrlich, P. R., Kremen, C., Ricketts, T. H., & Sanjayan, M. A. (2007). Ecosystem-service science and the way forward for conservation. *Conservation Biology, 21*, 1383–1384.

Bodini, A., & Giavelli, G. (1992). Multicriteria analysis as a tool to investigate compatibility between conservation and development on Salina Island, Aeolian Archipelago, Italy. *Environmental Management, 16*, 633–652.

Boyd, J., & Banzhaf, S. (2007). What are ecosystem services? The need for standardized environmental accounting units. *Ecological Economics, 63*, 616–626.

Brown, M., & Vivas, M. B. (2005). Landscape Development Intensity Index. *Environmental Monitoring and Assessment, 101*, 289–309.

Burkhard, F., Kroll, F., Müller, F., & Windhorst, W. (2009). Landscapes' capacities to provide ecosystem services – A concept for land-cover based assessments. *Landscape Online, 15*, 1–22.

Burkhard, B., Kroll, F., Nedkov, S., & Müller, F. (2012). Mapping ecosystem service supply, demand and budgets. *Ecological Indicators, 21*, 17–29.

Burkhard, B., Petrosillo, I., & Costanza, R. (2010). Ecosystem services – Bridging ecology, economy and social sciences. *Ecological Complexity, 7*, 257–259.

Carnus, J. M., Parrotta, J., Brockerhoff, E., Arbez, M., Jactel, H., Kremer, A., Lamb, D., OHara, K., & Walters, B. (2006). Planted forests and biodiversity. *Journal of Forestry, 104*, 65–77.

Carpenter, S. R., Mooney, H. A., Agard, J., Capistrano, D., Defries, R. S., Diaz, S., Dietz, T., Duraiappah, A. K., Oteng-Yeboah, A., Pereira, H. M., et al. (2009). Science for managing ecosystem services: Beyond the millennium ecosystem assessment. *Proceedings of the National Academy of Sciences of the United States of America, 106*, 1305–1312.

Carreño, L., Frank, F. C., & Viglizzo, E. F. (2012). Tradeoffs between economic and ecosystem services in Argentina during 50 years of land-use change. *Agriculture, Ecosystems and Environment, 154*, 68–77.

Castella, J.-C., Trung, T. N., & Boissau, S. (2005). Participatory simulation of land-use changes in the Northern Mountains of Vietnam: The combined use of an agent-based model, a role-playing game, and a geographic information system. *Ecology and Society, 10*, 27.

Chan, K. M. A., Satterfield, T., & Goldstein, J. (2012). Rethinking ecosystem services to better address and navigate cultural values. *Ecological Economics, 74*, 8–18.

Chatzinikolaou, P., Viaggi, D., & Raggi, M. (2015). The evaluation of ecosystem services production: An application in the Province of Ferrara. *Bio-based and Applied Economics, 4*(3), 235–259.

Cheung, W. W. L., Lam, V. W. Y., Sarmiento, J. L., Kearney, K., Watson, R., & Pauly, D. (2009). Projecting global marine biodiversity impacts under climate change scenarios. *Fish and Fisheries, 10*, 235–251.

CICES. (2013). *The common international classification of ecosystem services*. Report to the European Environment Agency.

Comino, E., Bottero, M., Pomarico, S., & Rosso, M. (2014). Exploring the environmental value of ecosystem services for a river basin through a spatial multicriteria analysis. *Land Use Policy, 36*, 381–395.

Costanza, R., d'Arge, R., & de Groot, R. (1997). The value of the world's ecosystem services and natural capital. *Nature, 387*, 253–260.

Costanza, R., de Groot, R., Sutton, P., van der Ploeg, S., Anderson, S. J., Kubiszewski, I., Farber, S., & Turner, R. K. (2014). Changes in the global value of ecosystem services. *Global Environmental Change, 26*, 152–158.

Cowling, R. M., Egoh, B., Knight, A. T., O'Farrell, P. J., Reyers, B., Rouget, M., Roux, D. J., Welz, A., & Wilhelm-Rechman, A. (2008). An operational model for mainstreaming ecosystem services for implementation. *Proceedings of the National Academy of Sciences, 105*, 9483–9488.

Daily, G. (1997). *Nature's services: Societal dependence on natural ecosystems*. Washington, DC: Island Press.

Daily, G. C., Polasky, S., Goldstein, J., Kareiva, P. M., Mooney, H. A., Pejchar, L., Ricketts, T. H., Salzman, J., & Shallenberger, R. (2009). Ecosystem services in decision making: Time to deliver. *Frontiers in Ecology and the Environment, 7*, 21–28.

de Groot, R. (1987). Environmental functions as a unifying concept for ecology and economics. *Environmentalist, 7*, 105–109.

de Groot, R. (1992). *Functions of nature: Evaluation of nature in environmental planning, management and decision making*. Groningen: Wolters Noordhoff. 315 pp.

de Groot, R. (2006). Function-analysis and valuation as a tool to assess land use conflicts in planning for sustainable, multi-functional landscapes. *Landscape and Urban Planning, 75*, 175–186.

de Groot, R. S., Alkemade, R., Braat, L., Hein, L., & Willemen, L. (2010). Challenges in integrating the concept of ecosystem services and values in landscape planning, management and decision making. *Ecological Complexity, 7*, 260–272.

de Groot, R. S., Wilson, M. A., & Boumans, R. M. J. (2002). A typology for the classification, description and valuation of ecosystem functions, goods and services. *Ecological Economics, 41*, 393–408.

Diehl, K., Burkhard, B., & Jacob, K. (2016). Should the ecosystem services concept be used in European Commission impact assessment? *Ecological Indicators, 61*(Part 1), 6–17.

Donner, S. D., Skirving, W. J., Little, C. M., Oppenheimer, M., & Hoegh-Guldberg, O. (2005). Global assessment of coral bleaching and required rates of adaptation under climate change. *Global Change Biology, 11*, 2251–2265.

Duinker, P. N., & Greig, L. A. (2007). Scenario analysis in environmental impact assessment: Improving explorations of the future. *Environmental Impact Assessment Review, 27*, 206–219.

EC. (2009). *Consultation on the future "EU 2020" strategy*. Brussels: Commission of the European Communities.

EC. (2010). *European Commission: Europe 2020. A strategy for smart, sustainable and inclusive growth*. COM(2010) 2020, Brussels.

EC. (2011). *European Commission: Our life insurance, our natural capital: An EU biodiversity strategy to 2020*. COM(2011) 244, Brussels.

Egoh, B., Drakou, E.G., Dunbar, M., Maes, J., & Willemen, L. (2012). *Indicators for mapping ecosystem services: A review*. Joint Research Centre Institute for Environment and Sustainability.

Egoh, B., Reyers, B., Rouget, M., Richardson, D. M., Le Maitre, D. C., & van Jaarsveld, A. S. (2008). Mapping ecosystem services for planning and management. *Agriculture, Ecosystems and Environment, 127*, 135–140.

Feld, C., Sousa, J., da Silva, P., & Dawson, T. (2010). Indicators for biodiversity and ecosystem services: Towards an improved framework for ecosystems assessment. *Biodiversity and Conservation, 19*, 2895–2919.

Fisher, B., & Kerry Turner, R. (2008). Ecosystem services: Classification for valuation. *Biological Conservation, 141*, 1167–1169.

Fisher, B., Turner, R. K., & Morling, P. (2009). Defining and classifying ecosystem services for decision making. *Ecological Economics, 68*, 643–653.

Fontana, V., Radtke, A., Bossi Fedrigotti, V., Tappeiner, U., Tasser, E., Zerbe, S., & Buchholz, T. (2013). Comparing land-use alternatives: Using the ecosystem services concept to define a multi-criteria decision analysis. *Ecological Economics, 93*, 128–136.

Galimberti, G., Raggi, M., & Viaggi, D. (2014). *Landscape perception and ecosystem service uses: Some results from surveys and latent variable models*. Deliverable 4.20 "Summary Report on Case Study Activity".

Girardin, P., Bockstaller, C., & Van der Werf, H. (2000). Assessment of potential impacts of agricultural practices on the environment: The AGRO*ECO method. *Environmental Impact Assessment Review, 20*, 227–239.

Glaser, M., & Diele, K. (2004). Asymmetric outcomes: Assessing central aspects of the biological, economic and social sustainability of a mangrove crab fishery, *Ucides cordatus* (Ocypodidae), in North Brazil. *Ecological Economics, 49*, 361–373.

Gómez-Baggethun, E., de Groot, R., Lomas, P. L., & Montes, C. (2010). The history of ecosystem services in economic theory and practice: From early notions to markets and payment schemes. *Ecological Economics, 69*, 1209–1218.

Gomez-Limon, J. A., & Sanchez-Fernandez, G. (2010). Empirical evaluation of agricultural sustainability using composite indicators. *Ecological Economics, 69*, 1062–1075.

Gorshkov, V. G., Gorshkov, V. V., & Makarieva, A. M. (2000). *Biotic regulation of the environment*. London: Springer, Published in association with Praxis Publishing.

Guo, M., Richter, G. M., Holland, R. A., Eigenbrod, F., Taylor, G., & Shah, N. (2016). Implementing land-use and ecosystem service effects into an integrated bioenergy value chain optimisation framework. *Computers and Chemical Engineering, 91*, 392–406.

Haas, G., Wetterich, F., & Geier, U. (2000). Life cycle assessment framework in agriculture on the farm level. *International Journal of Life Cycle Assessment, 5*, 345–348.

Haas, G., Wetterich, F., & Köpke, U. (2001). Comparing intensive, extensified and organic grassland farming in southern Germany by process life cycle assessment. *Agriculture, Ecosystems and Environment, 83*, 43–53.

Halberg, N., Verschuur, G., & Goodlass, G. (2005). Farm level environmental indicators; are they useful? An overview of green accounting systems for European farms. *Agriculture, Ecosystems and Environment, 105*, 195–212.

Hein, L., van Koppen, K., de Groot, R. S., & van Ierland, E. C. (2006). Spatial scales, stakeholders and the valuation of ecosystem services. *Ecological Economics, 57*(2), 209–228.

Helliwell, D. R. (1969). Valuation of wildlife resources. *Regional Studies, 3*, 41–49.

Hokkanen, J., & Salminen, P. (1997). Choosing a solid waste management system using multicriteria decision analysis. *European Journal of Operational Research, 98*, 19–36.

Hooper, D. U., Chapin, F. S., Ewel, J. J., Hector, A., Inchausti, P., Lavorel, S., Lawton, J. H., Lodge, D. M., Loreau, M., Naeem, S., et al. (2005). Effects of biodiversity on ecosystem functioning: A consensus of current knowledge. *Ecological Monographs, 75*, 3–35.

Hughes, J. B., Daily, G. C., & Ehrlich, P. R. (1997). Population diversity: Its extent and extinction. *Science, 278*, 689–692.

Jetz, W., Wilcove, D. S., & Dobson, A. P. (2007). Projected impacts of climate and land-use change on the global diversity of birds. *PLoS Biology, 5*, e157.

Jopke, C., Kreyling, J., Maes, J., & Koellner, T. (2015). Interactions among ecosystem services across Europe: Bagplots and cumulative correlation coefficients reveal synergies, trade-offs, and regional patterns. *Ecological Indicators, 49*, 46–52.

Kienast, F., Bolliger, J., Potschin, M., de Groot, R., Verburg, P., Heller, I., Wascher, D., & Haines-Young, R. (2009). Assessing landscape functions with broad-scale environmental data: Insights gained from a prototype development for Europe. *Environmental Management, 44*, 1099–1120.

Kiker, G. A., Bridges, T. S., Varghese, A., Seager, T. P., & Linkov, I. (2005). Application of multi-criteria decision analysis in environmental decision making. *Integrated Environmental Assessment and Management, 1*, 95–108.

Kirchner, M., Schmidt, J., Kindermann, G., Kulmer, V., Mitter, H., Prettenthaler, F., Rüdisser, J., Schauppenlehner, T., Schönhart, M., Strauss, F., et al. (2015). Ecosystem services and economic development in Austrian agricultural landscapes — The impact of policy and climate change scenarios on trade-offs and synergies. *Ecological Economics, 109*, 161–174.

Kirchner, M., Schönhart, M., & Schmid, E. (2016). Spatial impacts of the CAP post-2013 and climate change scenarios on agricultural intensification and environment in Austria. *Ecological Economics, 123*, 35–56.

Koschke, L., Fürst, C., Frank, S., & Makeschin, F. (2012). A multi-criteria approach for an integrated land-cover-based assessment of ecosystem services provision to support landscape planning. *Ecological Indicators, 21*, 54–66.
Lautenbach, S., Kugel, C., Lausch, A., & Seppelt, R. (2011). Analysis of historic changes in regional ecosystem service provisioning using land use data. *Ecological Indicators, 11*, 676–687.
Layke, C., Mapendembe, A., Brown, C., Walpole, M., & Winn, J. (2012). Indicators from the global and sub-global millennium ecosystem assessments: An analysis and next steps. *Ecological Indicators, 17*, 77–87.
Luck, G. W., Daily, G. C., & Ehrlich, P. R. (2003). Population diversity and ecosystem services. *Trends in Ecology and Evolution, 18*, 331–336.
Lundin, M., & Morrison, G. M. (2002). A life cycle assessment based procedure for development of environmental sustainability indicators for urban water systems. *Urban Water, 4*, 145–152.
Lupp, G., Steinhäußer, R., Bastian, O., & Syrbe, R.-U. (2015). Impacts of increasing bioenergy use on ecosystem services on nature and society exemplified in the German district of Görlitz. *Biomass and Bioenergy, 83*, 131–140.
Mace, G. M., Norris, K., & Fitter, A. H. (2012). Biodiversity and ecosystem services: A multilayered relationship. *Trends in Ecology and Evolution, 27*, 19–26.
Maes, J., Egoh, B., Willemen, L., Liquete, C., Vihervaara, P., Schägner, J. P., Grizzetti, B., Drakou, E. G., Notte, A. L., Zulian, G., et al. (2015). Mapping ecosystem services for policy support and decision making in the European Union. *Ecosystem Services, 1*, 31–39.
Martínez-Sastre, R., Ravera, F., González, J. A., López Santiago, C., Bidegain, I., & Munda, G. (2017). Mediterranean landscapes under change: Combining social multicriteria evaluation and the ecosystem services framework for land use planning. *Land Use Policy, 67*, 472–486.
Mendoza, G. A., & Prabhu, R. (2003). Qualitative multi-criteria approaches to assessing indicators of sustainable forest resource management. *Forest Ecology and Management, 174*, 329–343.
Millennium Ecosystem Assessment (MEA). (2003). *Ecosystems and human well-being. In A framework for assessment*. Washington, DC: Island Press.
Millennium Ecosystem Assessment (MEA). (2005). *Ecosystems and human well-being: Wetlands and water synthesis report*. Washington, DC: World Resources Institute.
Moberg, F., & Folke, C. (1999). Ecological goods and services of coral reef ecosystems. *Ecological Economics, 29*, 215–233.
Moffett, A., Garson, J., & Sarkar, S. (2005). MultCSync: A software package for incorporating multiple criteria in conservation planning. *Environmental Modelling and Software, 20*, 1315–1322.
Nelson, J. P., & Kennedy, P. E. (2009). The use (and abuse) of meta-analysis in environmental and natural resource economics: An assessment. *Environmental and Resource Economics, 42*, 345–377.
Newton, A. C., Hodder, K., Cantarello, E., Perrella, L., Birch, J. C., Robins, J., Douglas, S., Moody, C., & Cordingley, J. (2012). Cost–benefit analysis of ecological networks assessed through spatial analysis of ecosystem services. *Journal of Applied Ecology, 49*, 571–580.
Norberg, J. (1999). Linking Nature's services to ecosystems: Some general ecological concepts. *Ecological Economics, 29*, 183–202.
Odum, E. P., & Odum, H. T. (1972). *Natural areas as necessary components of man's total environment* (Vol. 37, pp. 178–189). In Transactions of the 37th North American Wildlife and Natural Resources Conference, Wildlife Management Institute, Washington, DC.
Oikonomou, V., Dimitrakopoulos, P., & Troumbis, A. (2011). Incorporating ecosystem function concept in environmental planning and decision making by means of multi-criteria evaluation: The case-study of Kalloni, Lesbos, Greece. *Environmental Management, 47*, 77–92.
Onate, J., Andersen, E., Peco, B., & Primdahl, J. (2000). Agri-environmental schemes and the European agricultural landscapes: The role of indicators as valuing tools for evaluation. *Landscape Ecology, 15*, 271–280.

Pannell, D. J., & Glenn, N. A. (2000). A framework for the economic evaluation and selection of sustainability indicators in agriculture. *Ecological Economics, 33*, 135–149.

Paracchini, M. L., & Capitani, C. (2011). *Implementation of a EU wide indicator for the rural-agrarian landscape*. In support of COM(2006)508 "Development of agri-environmental indicators for monitoring the integration of environmental concerns into the Common Agricultural Policy".

Paracchini, M. L., Pacini, C., Jones, M. L. M., & Pérez-Soba, M. (2011). An aggregation framework to link indicators associated with multifunctional land use to the stakeholder evaluation of policy options. *Ecological Indicators, 11*, 71–80.

Patil, G. P., Brooks, R. P., Myers, W. L., Rapport, D. J., & Taillie, C. (2001). Ecosystem health and its measurement at landscape scale: Toward the next generation of quantitative assessments. *Ecosystem Health, 7*, 307–316.

Pearce, D. W. (1993). *Economic values and the natural world*. London: Earthscan.

Perrings, C., Folke, C., & Maler, K.-G. (1992). The ecology and economics of biodiversity loss: The research agenda. *Ambio, 21*, 201–211.

Perrings, C., Naeem, S., Ahrestani, F. S., Bunker, D. E., Burkill, P., Canziani, G., Elmqvist, T., Fuhrman, J. A., Jaksic, F. M., Kawabata, Z., et al. (2011). Ecosystem services, targets, and indicators for the conservation and sustainable use of biodiversity. *Frontiers in Ecology and the Environment, 9*, 512–520.

Polasky, S. (2008). What's nature done for you lately: Measuring the value of ecosystem services. *Choices, 23*, 42–46.

Posthumus, H., Rouquette, J. R., Morris, J., Gowing, D. J. G., & Hess, T. M. (2010). A framework for the assessment of ecosystem goods and services; a case study on lowland floodplains in England. *Ecological Economics, 69*, 1510–1523.

Rasul, G., & Thapa, G. B. (2004). Sustainability of ecological and conventional agricultural systems in Bangladesh: An assessment based on environmental, economic and social perspectives. *Agricultural Systems, 79*, 327–351.

Regan, H. M., Davis, F. W., Andelman, S. J., Widyanata, A., & Freese, M. (2006). Comprehensive criteria for biodiversity evaluation in conservation planning. *Biodiversity and Conservation, 16*, 2715–2728.

Rodriguez, J. P., Beard, J., Douglas, T., Bennett, E. M., Cumming, G. S., Cork, S. J., Agard, J., Dobson, A. P., & Peterson, G. D. (2006). Trade-offs across space, time, and ecosystem services. *Ecology and Society, 11*, 28.

Sandhu, H. S., Wratten, S. D., Cullen, R., & Case, B. (2008). The future of farming: The value of ecosystem services in conventional and organic arable land. An experimental approach. *Ecological Economics, 64*, 835–848.

Schneiders, A., Van Daele, T., Van Landuyt, W., & Van Reeth, W. (2012). Biodiversity and ecosystem services: Complementary approaches for ecosystem management? *Ecological Indicators, 21*, 123–133.

Segura, M., Maroto, C., Belton, V., & Ginestar, C. (2015). A new collaborative methodology for assessment and management of ecosystem services. *Forests, 6*, 1696.

Seppelt, R., Dormann, C. F., Eppink, F. V., Lautenbach, S., & Schmidt, S. (2011). A quantitative review of ecosystem service studies: Approaches, shortcomings and the road ahead. *Journal of Applied Ecology, 48*, 630–636.

Seppelt, R., Fath, B., Burkhard, B., Fisher, J. L., Grêt-Regamey, A., Lautenbach, S., Pert, P., Hotes, S., Spangenberg, J., Verburg, P. H., et al. (2012). Form follows function? Proposing a blueprint for ecosystem service assessments based on reviews and case studies. *Ecological Indicators, 21*, 145–154.

Silvert, W. (2000). Fuzzy indices of environmental conditions. *Ecological Modelling, 130*, 111–119.

Sitch, S., Huntingford, C., Gedney, N., Levy, P. E., Lomas, M., Piao, S. L., Betts, R., Ciais, P., Cox, P., Friedlingstein, P., et al. (2008). Evaluation of the terrestrial carbon cycle, future plant geography and climate-carbon cycle feedbacks using five Dynamic Global Vegetation Models (DGVMs). *Global Change Biology, 14*, 2015–2039.

Stewart, T. J., & Scott, L. (1995). A scenario-based framework for multicriteria decision analysis in water resources planning. *Water Resources Research, 31*, 2835–2843.

Swetnam, R. D., Fisher, B., Mbilinyi, B. P., Munishi, P. K. T., Willcock, S., Ricketts, T., Mwakalila, S., Balmford, A., Burgess, N. D., Marshall, A. R., et al. (2011). Mapping socio-economic scenarios of land cover change: A GIS method to enable ecosystem service modelling. *Journal of Environmental Management, 92*, 563–574.

TEEB. (2010a). *The economics of ecosystems and biodiversity: mainstreaming the economics of nature: A synthesis of the approach conclusions and recommendations of TEEB*. London: Earthscan.

TEEB. (2010b). A quick guide to the economics of ecosystems and biodiversity for local and regional policy makers. www.teebweb.org.

Thomas, C. D., Cameron, A., Green, R. E., Bakkenes, M., Beaumont, L. J., Collingham, Y. C., Erasmus, B. F. N., de Siqueira, M. F., Grainger, A., Hannah, L., et al. (2004). Extinction risk from climate change. *Nature, 427*, 145–148.

Turner, R. K., & Daily, G. C. (2008). The ecosystem services framework and natural capital conservation. *Environmental and Resource Economics, 39*, 25–35.

Van Cauwenbergh, N., Biala, K., Bielders, C., Brouckaert, V., Franchois, L., Garcia Cidad, V., Hermy, M., Mathijs, E., Muys, B., Reijnders, J., et al. (2007). SAFE—A hierarchical framework for assessing the sustainability of agricultural systems. *Agriculture, Ecosystems and Environment, 120*, 229–242.

van Zanten, B., Verburg, P., Espinosa, M., Gomez-y-Paloma, S., Galimberti, G., Kantelhardt, J., Kapfer, M., Lefebvre, M., Manrique, R., Piorr, A., et al. (2014). European agricultural landscapes, common agricultural policy and ecosystem services: A review. *Agronomy for Sustainable Development, 34*, 309–325.

van Zanten, B. T., Zasada, I., Koetse, M. J., Ungaro, F., Häfner, K., & Verburg, P. H. (2016). A comparative approach to assess the contribution of landscape features to aesthetic and recreational values in agricultural landscapes. *Ecosystem Services, 17*, 87–98.

Verburg, P. H., Schulp, C. J. E., Witte, N., & Veldkamp, A. (2006). Downscaling of land use change scenarios to assess the dynamics of European landscapes. *Agriculture, Ecosystems and Environment, 114*, 39–56.

Viaggi, D. (2015). Special section: Exploring the contribution of landscape management to the rural economy. *Journal of Environmental Planning and Management, 58*, 2082–2087.

Vira, B., & Adams, W. M. (2009). Ecosystem services and conservation strategy: Beware the silver bullet. *Conservation Letters, 2*, 158–162.

Wallace, K. J. (2007). Classification of ecosystem services: Problems and solutions. *Biological Conservation, 139*, 235–246.

Weber, J.-L. (2007). Implementation of land and ecosystem accounts at the European Environment Agency. *Ecological Economics, 61*, 695–707.

Wegner, G., & Pascual, U. (2011). Cost-benefit analysis in the context of ecosystem services for human well-being: A multidisciplinary critique. *Global Environmental Change, 21*, 492–504.

Wiggering, H., Dalchow, C., Glemnitz, M., Helming, K., Müller, K., Schultz, A., Stachow, U., & Zander, P. (2006). Indicators for multifunctional land use—Linking socio-economic requirements with landscape potentials. *Ecological Indicators, 6*, 238–249.

Willemen, L., Hein, L., van Mensvoort, M. E. F., & Verburg, P. H. (2010). Space for people, plants, and livestock? Quantifying interactions among multiple landscape functions in a Dutch rural region. *Ecological Indicators, 10*, 62–73.

Wilson, M. A., & Carpenter, S. R. (1999). Economic valuation of freshwater ecosystem services in the United States: 1971–1997. *Ecological Applications, 9*, 772–783.

WRI. (2005). *Managing ecosystems to fight poverty*. Washington, DC: World Resources Institute.

Xenopoulos, M. A., Lodge, D. M., Alcamo, J., Märker, M., Schulze, K., & Van Vuuren, D. P. (2005). Scenarios of freshwater fish extinctions from climate change and water withdrawal. *Global Change Biology, 11*, 1557–1564.

Xiang, W.-N., & Clarke, K. C. (2003). The use of scenarios in land-use planning. *Environment and Planning B: Planning and Design, 30*, 885–909.

Zahm, F., Viaux, P., Vilain, L., Girardin, P., & Mouchet, C. (2008). Assessing farm sustainability with the IDEA method – from the concept of agriculture sustainability to case studies on farms. *Sustainable Development, 16*, 271–281.

Zerger, A., Warren, G., Hill, P., Robertson, D., Weidemann, A., & Lawton, K. (2011). Multi-criteria assessment for linking regional conservation planning and farm-scale actions. *Environmental Modelling and Software, 26*, 103–110.

Integrating AHP and GIS Techniques for Rural Landscape and Agricultural Activities Planning

Massimo Rovai and Maria Andreoli

Abstract This chapter aims at providing some insights on the usefulness of the analytic hierarchy process (AHP) in the context of geographic multi-criteria analysis applied to GIS techniques for empirical applications. The increasing complexity in planning and programming applied to rural landscape and territories asks for multidisciplinary and transdisciplinary approaches based on a holistic knowledge system. The AHP allows organizing in a hierarchic way both quantitative and qualitative information related to different disciplines, usually expressed in incommensurable measure units. Participatory approaches can be included either through information based on the perception of the value of indicators (criteria) or by providing weights on the relative importance of the elements included in each hierarchical level. When applied to GIS techniques, the AHP allows taking into account both spatial distribution of elements/information and their physical relations, which are paramount for the analysis of interventions about landscape, biodiversity, etc. This chapter illustrates four case studies from Tuscany Region (Italy) where this approach has been applied. Results highlight the flexibility of this approach in planning, programming and designing specific interventions where several biophysical characteristics of a territory or landscape have to be integrated with socioeconomic information both at territorial and farm levels. Results show that it is possible to increase the effectiveness and efficiency of tools for the territorial governance by applying a scientifically sound approach that does not ask for complex mathematical models and provides a methodology and results that can be understood also by "non-experts", improving participation processes.

Keywords Rural planning and programming · Governance model · DSS · Spatial multicriteria analysis · Tuscany

M. Rovai (✉) · M. Andreoli
Department of Agricultural, Food and Agro-Environmental Sciences (DAFE)
University of Pisa, Pisa, Italy
e-mail: massimo.rovai@unipi.it

J. Berbel et al. (eds.), *Multicriteria Analysis in Agriculture*, Multiple Criteria Decision Making, https://doi.org/10.1007/978-3-319-76929-5_3

1 Introduction

The case studies presented in this chapter have been developed through a 10-year time span and have as a common denominator the use of the Analytic Hierarchy Process (Saaty 1980, 2004, 2008) as a decision support tool.

Many policy instruments and specific interventions are based on concepts such as the ones of sustainability (see, e.g. Graymore et al. 2009; Gómez-Limón and Riesgo 2009; Sands and Podmore 2000), landscape quality (see, e.g. Daniel and Vining 1983; Fry et al. 2009); vulnerability (see, e.g. Villa and McLeod 2002; Barnett et al. 2008), risk of negative effects, e.g. due to forest fires (see, e.g. Van Wagner 1987; Chuvieco et al. 2010), level of disadvantage based on natural constraints (see, e.g. Eliasson et al. 2010), which can be associated to composite indexes (OECD 2008) or criteria depending on several elementary indicators, usually organized hierarchically. These phenomena can be studied by analyzing; (1) current situation, (2) past trends, (3) drivers of change, (4) most likely future evolutions and (5) policy responses in the case that either the current situation or the expected trends ask for them. AHP can help to manage data and to organize them in several information layers which can be related to each other. As regards rural areas, in order to design proper interventions, it is important that analyses take into account not only bio-physical characteristics of a territory (related to land suitability to agricultural uses) and land cover or land use, but also the role of farmers' decisions in determining a bundle of potential effects (e.g. in term of Ecosystem Service provision) in specific locations, since farmers' strategies are the main drivers of agricultural and rural landscape changes (Primdahl and Kristensen 2011; Van Zanten et al. 2013). In their turn, farmers' decisions are influenced by market trends and policy tools, but depend also on farm structure and bio-physical features of the land farmers manage (see, e.g. Fastelli et al. 2017).

Higher profits deriving from land development compared with the income from cultivation have brought about inefficient and persistent soil consumption (ISPRA 2017) and the creation of relics of farmland enclosed by built-up areas. This approach driven by economic interests has determined an underestimation of the importance of open spaces especially in areas of urban-rural fringe that have not been properly managed, and consequently it should be abandoned. In this case it is important to have a clear vision not only of the productive function of open spaces, but of the whole bundle of Ecosystem Services (ESs) they are able to provide. On the other hand, as stated by Cooper et al. (2006) and Pelorosso et al. (2011), the abandonment of agricultural activities may result both in a loss of ESs provision, such as hydrologic regulation, and of landscape value, especially in the case of cultural landscapes. Indeed, woodland abandonment can bring about several problems related to: (a) landslides, due to the excessive weight of trees that are no longer cut; (b) the development of vegetation more suitable to ignition; (c) the abandonment of the minor road systems, causing problems of accessibility; and, (d) the lack of custodianship and maintenance of the ancient hydraulic works protecting from floods. Thus policy tools influencing agricultural activities should take into account the whole spectrum of ESs that these activities can provide. Due to the negative effects of abandonment, in some of the case studies presented, the economic aspects have been considered as prerequisites for the permanence of agricultural activities in a territory, rather than as ESs belonging to the productive function.

Where the spatial distribution is relevant, it is possible to integrate AHP with a GIS approach (see, e.g., Malczewski 1999, 2006a; Greene et al. 2011). In this case, the analysis often aims at individuating spatial ambits that are homogeneous as regards relevant criteria. Indeed, ex post analyses of the effects of policies impacting on agriculture and environment have highlighted that spatial models are more suitable to achieve an effective policy, if compared to non-spatial regression models (Yang et al. 2014). After the adoption of the European Landscape Convention (Council of Europe 2000) innovative policy directions focusing on designing measures appropriate for different contexts and scales have increased (Conrad et al. 2011).

According to Magnaghi (2005) a territorial project should be considered as the reference scenario that should guide both specific actions and projects, and the strategic assessment of operational projects and policies, by referring to evaluation parameters which can be continuously reformulated in relation to the information acquired in the process. The need of information at local/detailed level and of updating them frequently makes it difficult to base policy decisions on information that are exclusively depending on external sources and that often are also quite difficult to be interpreted, as in the case of complex mathematical models. The lack of awareness about the impact of policies can encourage decision-makers to take decisions on the basis of lobbying or economic pressures, rather than of objective analyses (Marson 2010).

Recently, there has been a growing diffusion of web-GIS tools based on Multicriteria Analyses (Labiosa et al. 2010, 2013; Jackson et al. 2013; Tallis and Polasky 2011) allowing to integrate a great amount of information and models typical to different disciplines by homogenizing the information. They provide a useful decision support system to problems that are not completely structured and aim to determine a final score (a-dimensional and expressed in relative terms) allowing to individuate a hierarchical order of the alternatives, e.g. from the best to the worst one. These tools can be useful especially when analyses and policies are related to regional/local ambits and to operational tools.

In this framework, after describing the main features of the methodological approach integrating AHP and GIS, this chapter presents some empirical studies at increasing level of complexity, in which at least one of the Authors has been directly involved, in order to highlight how the general methodology has been adapted to each specific problem and the results obtained.

The chapter is organized as follows. After this introduction (Sect. 1), a section on the methodology (Sect. 2) is given, distinguishing the general aspects by the specific ones which are characterizing the four case studies presented. In this section also a short description of Tuscany, the Italian Region where case studies are located, is given. Section 3 presents the results of the case studies, while Section 4 provides a general analysis of strengths and critical points of the proposed approach.

2 Methodology

All the case studies presented in this contribution have employed Geographic Multicriteria Aiding Techniques able to rank the spatial alternatives under study according to their specific and often conflicting evaluation criteria, which are represented through standardized map layers (Malczewski 1999, 2006a, b). Among the several multicriteria analysis techniques described in literature (Beinat and Nijkamp 1998; Mendoza and Martins 2006), the multiattribute Saaty's Analytic Hierarchy Process (AHP) has been chosen. AHP constructs the evaluation process through distinct phases assuming as a principle the possibility to segment a complex decision-making problem into smaller and simpler sub-problems composing a hierarchical structure, within which it is always possible to measure the influence each part has on the whole system. The hierarchical structure is organized in three levels: goals, criteria, and alternatives, where criteria can be detailed in attributes and sub-attributes, in order to reach elementary indices, represented by cardinal or ordinal values that can be reliably measured or assessed. In the case of AHP integrated with GIS, alternatives are spatial and they can be represented by points, lines, polygons or pixels that contain the attribute values. The presented case studies deal with situations where the relevant spatial unit varies from polygons referring to a regular grid in which the territory has been subdivided, to plots of land with homogeneous characteristics, up to the land managed[1] by an agricultural enterprise, considered as a single decisional business unit.

The advantages of AHP are related to the following features:

- It is possible to use not only quantitative but also qualitative information, provided that they can be ordered. The transformation of information in Saaty's semantic scale allows comparing attributes expressed in different measure units.
- The AHP hierarchic structure allows analyzing phenomena described by criteria that in their turn are related to attributes and sub-attributes, thus allowing the representation of complex problems, as the ones that have to be faced in dealing with sustainability or integrate territorial planning.
- AHP allows confronting the performances of alternatives, represented by decisions, e.g. in terms of alternative policy designs, or by territorial units. In the case of integration of AHP with GIS techniques the result of the analysis is usually a characterization of the space through its subdivision in ambits which are homogeneous as regards the goals/criteria included in the analysis.

All the case studies presented are located in Tuscany, an administrative Region in the Central-Western part of Italy. Tuscany has a territory of about 23,000 km^2, which

[1]Through databases related to Common Agricultural Policy aid it is possible to individuate the cadastral parcels that are managed by each farm. Cadastral data refers to the Italian inventory of agricultural land (Catasto Terreni), where the elementary unit is a parcel of land belonging to the same Municipality, holder, category of agricultural utilization and class of productivity that is not divided by roads, rivers, railways, etc.

is mainly hilly (66.5%) and mountainous, while plains account only for 8.4% of the whole territory, and it has a population of ca. 3.7 million inhabitants. Tuscany is famous worldwide, due to the beauty of its rural landscape, often characterized by the presence of urban structures dating back to the medieval and Renaissance age, together with fortified villages, scattered rural and religious buildings, and to its cities of art (Rovai et al. 2016). According to Ciampi et al. (2015) in Tuscany on average in 2013 artificial areas accounted for about 8.6%, agricultural areas for 38.1%, natural and semi natural areas for 52.3%, and water bodies and wetlands for 0.9%; these average figures hide a high variability among areas; artificial areas, e.g., account only for 5.3% in the mountains and for 7.5% in hilly areas, but raise up to 33.9% in the lowlands. Agricultural land use is the most important one in hills and lowlands, where it accounts for 68.0% and 57.4%, respectively, while natural and semi natural areas are prevalent in the mountains where they account for 75.3%. In the period 2007–2013 agricultural areas decreased by 5800 ha, mainly due to artificialization processes and, in some areas, due to an increase of woodlands (Ciampi et al. 2015). Furthermore, the urbanization model is very often characterized by low density, with a high consumption of soil for each inhabitant.

The case studies are presented in order of increasing complexity. The first one deals with a specific and technical problem, i.e. the spatial analysis of the level of risk of forest fires, in order to plan the best ways to prevent and face them. The second one deals with the problem of land abandonment in a rural area and the individuation of policy instruments that could lead either to a recover in terms of productive use or, when this is not economically sustainable, to reduce the risks coming from a "sudden" renaturalization after a long period of anthropic pressure. The third one deals with problems of open spaces in areas of urban sprawls and in the characterization of the main ESs provided by different portions of a territory. The last one proposes a model for the sustainable governance of the rural cultural landscape of Val d'Orcia, which is one of the UNESCO heritage sites.

2.1 Territorial Assessment of the Risk of Forest Fires in Livorno (Leghorn) Province

This first case study[2] concerns the design of a plan for forest fires in Livorno province (Italy). Italy, especially in its central and southern areas, is very prone to forest fires due to its bio-physical characteristics and hazards of anthropogenic causes, often related to arson with the aim of allowing the development of burnt land. Italian legislation requires that every Region adopts a Plan for forecasting, preventing and actively counteracting forest fires. In this application the goal of the

[2]The methodology applied for this case study is described in details in Candura (2005), although some improvements to the initial methodology have been introduced.

GIS-AHP analysis was to assess and map the risk of forest fires in order to provide a scientifically sound basis for actions aiming to reduce and mitigate this risk. The methods that are more frequently employed for simulating the frequency or the probability of fire events use GIS-based MCA tools to produce risk maps. In other cases, it is proposed to employ geoprocessing routines based on algorithms able to interpret input maps (Gai et al. 2011). While many studies regarding forest fire risk (Arpaci et al. 2014; You et al. 2017) deal with factors affecting the probability of fire, in our model, as in Hardy (2005) the level of risk, for each area, has been assumed to depend on:

- The probability (*P*) to have a forest fire in a predefined time period,
- The entity of consequences of forest fires, i.e. damage (*D*) that they will cause in terms both of economic losses of productive activities and of environmental negative effects.

In our model the Risk (*R*), in analogy to the definition of risk in the case of safety problems, is given by the product of the probability of an event (*P*) and the damage (*D*) it would cause, i.e.

$$R = P \times D.$$

Thus, while Probability measures the chance of a forest fire in each area in a predefined time period, Damage estimates the consequences in terms of loss of economic and environmental values in case of forest fire.

The provincial territory of Livorno has been subdivided in 119.922 units by a grid of 100×100 m and each elementary unit has been described by means of the data of the Regional GIS through criteria and attributes related to fire-related risks. Probability and Damage criteria have been described by separate hierarchical trees, by considering the attributes derived from literature and by interviews to key informants, and initially mapped as separated criteria. Then a composite index of forest fires risk has been computed, under different scenarios as regards the weights to be assigned to each attribute.

In accordance with other contributions (Jaiswal et al. 2002; Arpaci et al. 2014; You et al. 2017) the criterion assessing the probability of forest fires has been related to the following attributes:

1. Topographical factors, described by: Altitude, Slope and Orientation (sub-attributes)
2. Climate factors, described by: Temperature, Precipitation, Wind (sub-attributes)
3. Land cover factors, described by Land cover ignitability (sub-attribute)
4. Human activity factors, described by: Proximity to roads and Number of tourists (sub-attributes)

The criterion assessing the entity of damage has been related to the following attributes:

5. Economic Value, described by: Average value of agricultural land (VAM)[3] (Ciancio et al. 2007; Blahut et al. 2014); and Presence of infrastructures, such as power lines and methane pipelines (sub-attributes);
6. Environmental Value, described by: Presence of areas with Environmental constraints; Presence of Areas with Archaeological constraints, Presence of areas with Hydrological constraints; Presence of Protected areas (e.g. Natura 2000 areas).

For more details on data sources and how sub-attributes have been calculated, see Candura Master's Degree thesis on Building Engineering (2005).

In order to compare attributes expressed in different measure units, they have been normalized to a 0–1 range. According to the type of attribute, the following three normalization methods have been used:

1. Linear scale transformation, using a direct or indirect relation between raw data and normalized ones depending from the fact that a high value of the raw datum was increasing or decreasing the probability or the damage level of forest fires;
2. Expert evaluations, when functions relating raw data and probability or damage level of forest fires were not available;
3. Probability method, when the relation has been estimated on the base of a statistical analysis relating forest fires and the features of the areas where they have developed. This analysis was performed on a random sample of 50% of the elementary units.

After normalization, the computation of criteria has been performed under the hypothesis that all sub-attributes and attributes had the same weight, thus obtaining two separate maps: the first one measuring the probability of forest fires and the second one measuring the extent of economic and environmental damages in case of forest fires.

Lastly, in order to assess the risk in its components of probability and damage level, four different scenarios were built under the following hypotheses about weights:

- Scenario 1: all criteria and attributes have the same weights;
- Scenario 2: human activity and land cover factors have a higher importance than topographical and climate factors in influencing forest fire probability and environmental values are more important than economic values when assessing the entity of damages;
- Scenario 3: human activity and land cover factors have a higher importance than topographical and climate factors in influencing forest fire probability and economic values are more important than environmental values when assessing the entity of damages;

[3]VAM (Valori Agricoli Medi) are the average real estate prices for land with agricultural destination and have been mostly used in case of compulsory purchase, i.e. when a state or a national government takes private property for public use. They are individuated at provincial level.

- Scenario 4: human activity and land cover factors have a much higher importance than topographical and climate factors in influencing forest fire probability and environmental values are slightly more important than economic values when assessing the entity of damages.

The ability of the different scenarios to fit real data (model testing) has been checked by using the sample of 50% of the elementary units that were not selected for building the model (see normalization phase), thus searching an independent validation of the model.

2.2 *Productive and Landscape Restoration of a Hilly Rural Area (Pieve di Compito, Lucca)*

The second case study[4] deals with a territory with strong historical, identity and landscape values, and it is located in a hilly area of Tuscany traditionally characterized by olive groves and currently suffering for severe abandonment problems, partly caused by land fragmentation. Rural areas have not been deemed as very important in the cultural debate on landscape until 2000, when the European Landscape Convention (Council of Europe 2000) caused a new and increasing interest for the maintenance and restoration of rural landscapes. Currently, rural landscapes are not only appreciated for their value from a historic and identity point of view, but also for the economic role they can play for tourism and the high quality produce reputation of an area. The new role of agro-forestry landscapes as a common good to be protected asks for new models aiming to individuate the best intervention strategies for territories characterized by both high landscape value and widespread phenomena of abandonment, and for appropriate governance approaches to implement them. In this case study, AHP has been used to build an "expeditious" model to individuate intervention priority for the productive and landscape restoration of olive groves or, as an alternative, for the transition to woodland, in the framework of the Integrated Territorial Projects (ITPs). ITPs are one of the "collective measures" included in the Rural Development Plan (RDP) 2014–2020 by Tuscany. In this case, as it has been partly done for the analysis of Val d'Orcia landscape described in the last case study, the elementary unit is represented by cadastral parcels. Each elementary unit (i.e. cadastral parcel) has been characterized by means of attributes about the bio-physical features of the plot, the way it is managed by the farmer and its importance in terms of landscape and environmental values. The model has been tested on a small area (81 ha ca) in the "Colline Lucchesi" (hills of Lucca province), characterized by small farms where production for self-consumption is prevalent and the main agricultural use is the cultivation of olive trees which, due to an average slope of 45%, are grown on terraces. The model integrates all these aspects with the

[4]This case study has as a main sources (Bonelli 2017) and Rovai et al. 2017.

aim to provide useful information to policy-makers on the specific interventions that would be most favorable for each type of cadastral parcel. The case-study area, as most of the hilly areas in Lucca province with olive groves, has shown increasing phenomena of abandonment since the 1960s, due both to fires and to the reduction of the profitability of olive cultivation, together with a cultural change in the local community, where the new generation shows a lower attachment to rural values. A diachronic analysis of the evolution of land use highlights how the area suffered for a noticeable landscape change, due to the abandonment of olive groves, to which renaturalization processes followed (Fig. 1).

The percentage of abandoned agricultural land during each period is described in Table 1.

Since the focus of the analysis was on the problem of abandonment, the spatial MCA model has been built:

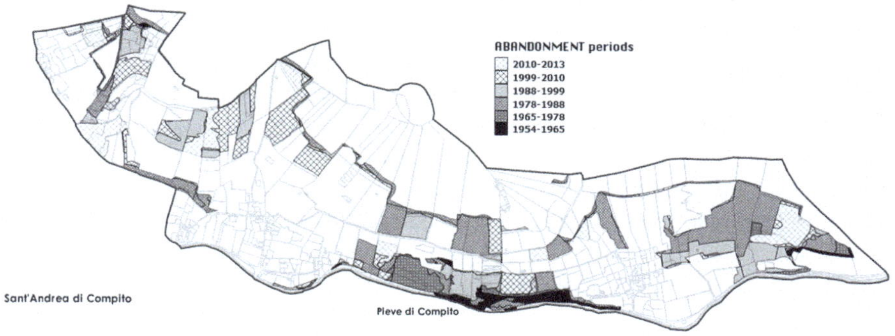

Fig. 1 Abandoned agricultural land by period of abandonment—spatial distribution. Source: Bonelli Master's Degree thesis in Architecture and Building Engineering (2017), modified

Table 1 Abandoned agricultural land by period of abandonment

Period	Abandoned agricultural land (ha)	Abandoned agricultural land on total agricultural land (%)
1954–1965	0.96	1.2
1965–1978	1.31	1.6
1978–1988	7.32	9.0
1988–1999	4.49	5.5
1999–2010	4.65	5.7
2010–2013	2.99	3.7
Total (1954–2013)	21.72	26.8

Source: Bonelli (2017, modified)

1. From a spatial point of view, on the layer of cadastral parcels where the agricultural use has been abandoned,
2. Using as a goal the assessment of the possibility of restoration of olive groves and their cultivation,
3. Defining attributes and criteria related to the goal, which could be calculated from available data sources. The creation of the MCA matrix has been done directly in GIS environment in order to use the data collected during the territorial analysis in the more effective way and to profit of the computing power of QGIS software (QGIS Development Team, 2012).

Table 2 describes the 12 criteria considered in the spatial MCA: among them, the first eight have been classified as "Cost criteria" since they deal with the costs for olive groves restoration and management, while the last four criteria have been classified as "Benefit criteria", since they are related to the degree of quality improvement from an environmental and landscape point of view and, consequently, with an increase of community well-being.

A score in the 1–5 range—where the lowest score is related to the lowest criterion value, both for cost and benefit criteria—has been given to attributes belonging to each criterion. Table 3 gives an example of the scores for the attribute "Average slope", whose spatial representation is given in Fig. 2.

Finally, in order to compute the goal value for each spatial elementary unit, weights were assigned to the 12 criteria. This has been done separately for the two subsets of Cost and Benefit Criteria. The ranking of the alternatives, aiming to assess

Table 2 Criteria related to the suitability to productive rehabilitation of abandoned parcels

Cost criteria	Benefit criteria
1. Period of land abandonment	9. Architectural value of open ditches drainage
2. Size of the abandoned areas	10. Proximity to cultivated areas
3. Soil average slope	11. Proximity to inhabited zones
4. Vegetation cover on abandoned land	12. Aesthetic-perceptive significance
5. State of maintenance of terraces	
6. Accessibility (from outside farm)	
7. Access to farm and forest land (internal)	
8. Management type	

Source: Bonelli (2017)

Table 3 Score attribution to the criterion "Soil Average Slope"

Description	Score
Soil average slope $\leq$ 15%	1
Soil average slope > 15% and $\leq$ 30%	2
Soil average slope > 30% and $\leq$ 45%	3
Soil average slope > 45% and $\leq$ 60%	4
Soil average slope > 60%	5

Source: Bonelli (2017)

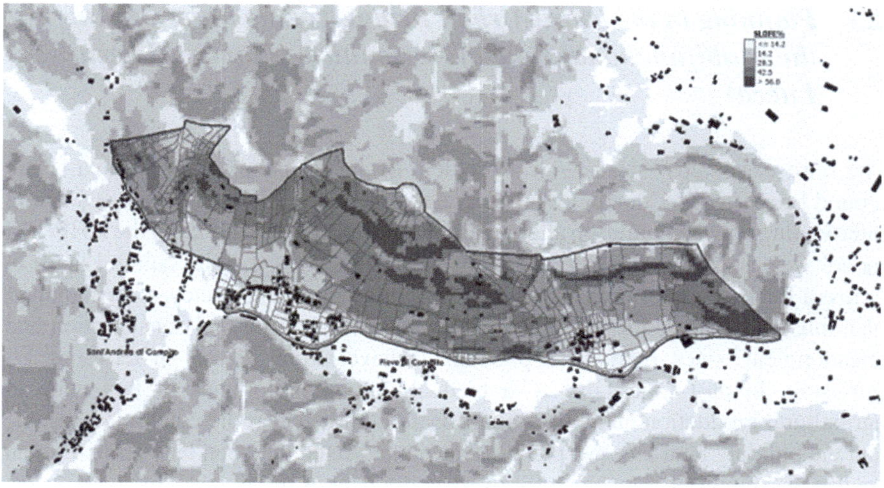

Fig. 2 Spatial representation of average slope in the case-study area. Source: Bonelli (2017), modified

the degree of parcels suitability to rehabilitation for land cultivation, has been carried out with two methods in order to compare strengths and weaknesses of both of them. The methods were the following:

1. Weighted Sum: for each cadastral parcel the total score has been computed as sum of the weighted values of each criterion. In this case, due to the fact that costs are inversely related with utility, the highest utility level has been assigned to the score 1 of Cost criteria. In this way, all criteria are summarized in a unique cardinal value (goal value), that is available for each spatial alternative (cadastral parcel). While cadastral parcels with a very high value for the Goal show a high suitability to rehabilitation since farm costs are low and community benefits are high, this method has the drawback of not allowing to know if the suitability is mostly due to low costs, high benefits or a combination of good scores for criteria belonging to both classes.
2. Miles's Value Index: this index has been defined in the framework of the Value Analysis (Miles 1961). When defining the concept of value, Miles introduces a composite index called "Value Index", or *IV*, computed as the ratio between the utility of an entity, i.e. "Worth" or *W*, and its global cost, i.e. "Cost" or *C*.

$$IV = W/C.$$

In this case study, *IV* has been computed as a ratio between the weighted sums of criteria classified as Benefits and Cost criteria; consequently, the higher is the ratio, the higher is the priority that parcels should have in rehabilitation strategies.

2.3 *Planning in the Rural-Urban Fringe Taking into Account the Multifunctional Role of Agriculture (Piana di Lucca, Lucca)*

The third case study[5] deals with a problem of spatial planning in an area (Plain around Lucca) where, due to low density and scattered settlements, there is a strong intertwining between build-up and open agricultural areas. In this case, the AHP-GIS model was used to evaluate the value of the open spaces (i.e. agricultural land) in terms of Ecosystem Services (ESs), in order to give support to policy-makers in their planning decisions and to promote strategies aiming to limit soil consumption and abandonment, since both these phenomena have strong effects in terms of ES provision. In past times, due to a gradual loss of importance of the value of agricultural production and a disproportionate ratio between the cultivation income and the building land income, planners have interpreted the open land around urban centers as an unlimited resource for the development of settlements. In recent years, as a result of the growth of a new cultural sensitivity, the use of open spaces, and in particular of agricultural soils, for building purposes is increasingly perceived as a waste, if not an abuse, when not justified by the actual need for improving the quality of life and welfare of citizens. The strengthening of public awareness about the importance of open and rural land areas to ensure the reproducibility of resources and vital functions, should be a key objective for planning actions for the coming years. This objective should be reflected in the implementation of proper strategies to enhance the value of the open territory, and in particular in the rural areas, as a place of co-production between man and environment where to make the best use of reproductive capabilities of nature (Rovai et al. 2013). In this framework, it is paramount to assess ESs provided by open spaces and the negative impacts on community well-being deriving from the waste of them.

The methodology allowed not only mapping the spatial distribution of each specific function, but also finding the relative intensity with which each type of ES or ES category was provided by an area.

Following the Common International Classification of Ecosystem Services (Haines-Young and Potschin 2013), in this contribution the categories of ESs considered have been reduced by the initial four, i.e. provisioning, regulating, supporting and cultural (Costanza et al. 1997; de Groot et al. 2002; MEA 2005) to only three relating respectively to: productive function, protective function, and recreational and cultural function.

The case-study area is "Piana di Lucca" (lowland around Lucca town), where remains of "Lucca Court", a peculiar rural settlement organization in the open territory that has a great value both from an aesthetic and from a historic and cultural point of view, are still present.

[5]This case study has as a main source (Rovai et al. 2013).

While in the marginal hilly areas vineyards and olive groves have been replaced by natural and semi natural areas, the expansion of lowland settlement typical of urban sprawl is leading to a strong fragmentation of the ecological network and exacerbating hydraulic problems. From 1995 to 2007 the area around Lucca suffered for a gradual decrease of agricultural land, mainly due to the process of urban expansion; this phenomenon caused the loss of 37,000 ha at province level. Besides, agricultural activities suffer for the increased fragmentation of farms in the areas of urban sprawl, which in its turn causes increased costs and organizational problems. For the above reasons it was deemed as important to include in local planning an assessment of the values (in broad sense) related to rural areas and to highlight costs and benefits of their conservation.

This case-study area was analyzed through a GIS-AHP analysis aiming at spatially assessing the three following categories of ESs:

1. Productive services, related to the ability to produce food and agricultural goods and, consequently, economic value;
2. Protective services, related to the ability of open spaces to ensure the preservation and reproduction of natural resources;
3. Cultural and social services, related to the rural environment ability to contribute to the physical and psychological well-being of the community.

Productive services have been estimated through the identification of areas:

1. Which are managed by farmers who have applied for EU contribution, considered as "professional";
2. Where there are Protected Designation of Origin (PDO) products; and
3. By taking into account the degree of estimated profitability of crops.

Protective services have been estimated through the identification of areas:

1. Of ecological connectivity (giving the maximum score to "nodes" and the minimum to "secondary connections");
2. CO_2 sequestration capacity, and;
3. Ability to recharge the aquifers.

Cultural Services have been estimated through:

1. The persistence of the historical settlement system;
2. The topological relationship to the settlement system, i.e. the Intervisibility between settlements and agricultural land portions, and;
3. The proximity to high population areas, considered as a proxy of the number of people who can benefit from these services.

The elementary spatial unit was a hexagon having a size of about 1000 sq. m.

While it would have been possible, through AMC, to summarize the total importance in terms of provision of ESs belonging to all the three categories described above, the Authors deemed more important and useful, as a cognitive framework for planning, to individuate the main suitability in terms of ES provision of each homogeneous ambit of the territory. This result was obtained by using a three-dimensional color gradient built through the integration of the three RGB channels of a 24-bit image, where red was related to Cultural services, green to Protective Services and blue to Productive Services.

2.4 Sustainable Governance of a Cultural-Historic Landscape (Val d'Orcia, Siena)

The fourth and last case study[6] deals with the demand for new landscape planning tools deriving from the adoption of the European Landscape Convention in 2000 (Council of Europe 2000). This case study analyses the situation of Val d'Orcia, one of the Italian UNESCO heritage sites, and it identifies future landscape scenarios by integrating past evolution (historical analysis), landscape sensitivity (territorial analysis) and farmers' adaptation to market and policy changes (farm analysis). In this framework, policies are considered both as drivers, since Common Agricultural Policy (CAP) has often had a negative influence on landscape, and as responses to landscape change. This case study presents a model for the governance of rural landscapes (Fig. 3) that is based on the integration of a geographical multi-criteria analysis and advanced GIS-based geo-processing tools (Rovai et al. 2016). The spatial elementary unit considered varies depending on the features that are analyzed; thus, there are attributes described on the base of a regular grid; attributes related to plots as an elementary unit with homogeneous agricultural use, and attributes related to the farm as an economic unit.

AHP integrated with GIS techniques has been used in the analyses at territorial and farm level. The territorial analysis is necessary to classify a territory into homogeneous ambits as regards the level of ecosystem services that they are able to provide and the risk that anthropic activities might negatively impact on them. As in many other studies concerning the risk of loss of ecologic, environmental or cultural values, the problem was faced with a twofold approach. First of all, the areas characterized by higher values were identified, in order to be able to rank territorial units from the ones most deserving interventions to the ones whose value was deemed as negligible; in this case the aim was to classify areas in homogeneous ambit in terms of "Intrinsic Value" and the spatial elementary unit was a regular square grid. Then, the same areas were classified in order to assess their level of "Vulnerability". Crossing the two above classifications it is possible to obtain a matrix giving priorities for the interventions: the highest priority relates to territories with high value and high vulnerability, while at the opposite end of the hierarchy there are territories with negligible value and low vulnerability.

In both cases, criteria were organized in attributes and sub-attributes chosen among those commonly used in landscape ecology approaches (Farina 2006). The criterion "Intrinsic Value" included two attributes, i.e. "Historical and

[6]This case study has as a main source (Rovai et al. 2016).

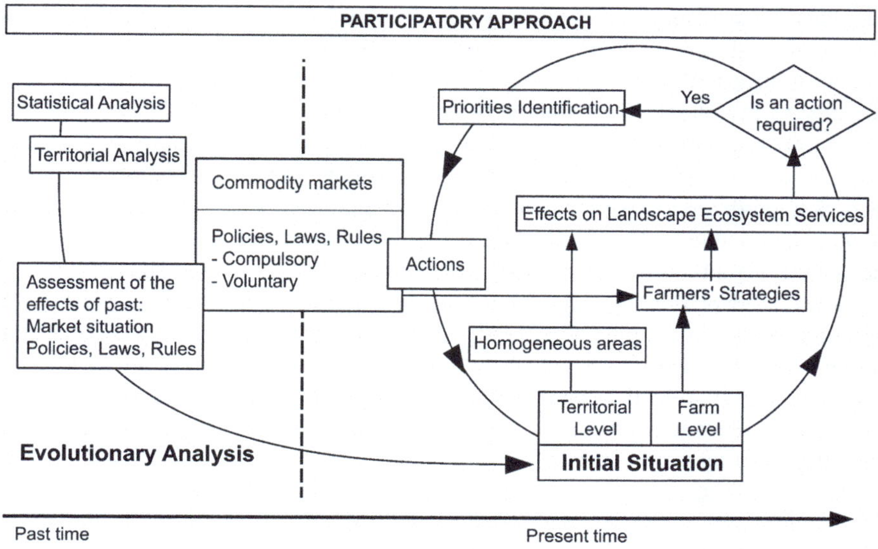

Fig. 3 Flow chart of the model for the governance of sustainable rural landscape. Source: Rovai et al. (2016)

Environmental Value" and "Frequentation"; this latter was introduced under the hypothesis that Historical and Environmental values were not only important for themselves but also insofar as they could be enjoyed by somebody. In its turn, the "Historical and Environmental value" of each ambit depended on the values of several sub-attributes, i.e. Fragmentation, Persistence, Uniqueness, Integrity and Visibility, for which rules for attributing a score were given, based on literature and participatory approach. Vice versa, in the case of "Frequentation" it was possible to directly attribute a score. For a detailed description of criteria and attributes used in the analysis, see Rovai et al. (2016). The territorial analysis allowed individuating and locating zones with similar level of importance for community wellbeing and vulnerability to the modifications induced by anthropic activities.

While the territorial analysis aims to classify a territory in homogeneous ambits, farm analysis aims to highlight the most likely effects of farmers' strategies, seen as the main drivers of change in agricultural landscape. Intensification (Stoate et al. 2001) and abandonment (Cooper et al. 2006; Pelorosso et al. 2011) of agricultural activities may result in a loss of ES provision and of landscape value, but in this case—due to the specific bio-physical characteristics of the area—we deemed abandonment to be more likely and more dangerous than intensification. Farmers' strategies are heavily depending on farm structural, socio-economic and management characteristics but also on the quality level of available resources in terms of land suitability to cultivation. While the first characteristics have to be analyzed at farm level, intended as the business unit were decisions are taken, the second characteristics mainly depend on bio-physical features and consequently had to be analyzed at plot level. Plot and farm levels were related by attributing to each farm the plots it was managing. The analysis at farm level had the aim to understand the degree of resilience of business units in relation to policy and adverse market trends, due to its structural and organizational features, while the analysis at plot level had the aim to investigate their suitability to cultivation. This latter differs from the commonly used concept of land capability insofar as it depends from characteristics, such as scale/morphology, that are more influencing farm organization and costs than land productivity, while variables like climate, that was considered to be quite homogeneous due to the small scale of the case-study area, were not taken into account. By combining the analyses at farm and plot levels it was possible to forecast the level of adaptation and capacity to survive either as business unit or productive resource of local agriculture when confronted with external negative impacts; indeed at territorial and landscape levels the exit of a business unit from the market has a different impact from the abandonment of land as a productive resource. Thus a decrease in farmland could have a very different and worse impact than a decrease in the number of farms.

The analysis at farm level aims to provide information on the areas that are at risk of abandonment or strong extensification, with negative impacts due to the abandonment of the agricultural activity and its custodianship role. Moreover, farms with different resilience to adverse trends react in a different way to the possible policy tools that could be implemented in order to promote landscape sustainability; consequently a good knowledge of their features and how these could affect farm's strategies is very important when designing policy tools.

The final values of plots' suitability to cultivation and farms' adaptation capacity to negative impacts have been parted into three classes (low, medium, high) with the aim to compute, in accordance with fuzzy logic (Boroushaki and Malczewski 2010; Karsak 2004) the probability of each alternative to belong to each class (see Rovai et al. 2016). Then, a cross-reading phase has been carried out via a fuzzy logic in order to compare and integrate the two composite indices and identify a set of farm evolutionary paths/strategies deriving from crossing the classes (L, M, and H) of plots and farms. These strategies are described and discussed in the section of results.

It is important to stress that the setup of the model and the validation of intermediate results have been carried out with the involvement of local stakeholders through individual interviews to key informants and focus groups. Interviews have been based on open-answer questionnaires and mainly used to gather information for the analyses, such as attribute and sub-attribute to be considered and their scores. Focus groups, which have involved representatives of associations of the territory and farmers, have been mainly used for checking the results of the analyses and for discussing actions to be taken. In particular, focus groups have explored the willingness/motivation to develop actions/products within a participative approach with the aim to promote an "environmental requalification" of Orcia Valley landscape; this with the aim to promote a participatory and transdisciplinary approach.

3 Results

3.1 Territorial Assessment of the Risk of Forest Fires in Livorno (Leghorn) Province

The analysis of forest fires in Livorno province allowed to produce a series of thematic maps for sub-attributes, attributes and criteria, as the one reported in Fig. 4, which represents the maps of the four attributes describing the criterion "Environmental value".

The methodology allowed producing also the mapping of the composite index estimating the risk of forest fires, according to the four different scenarios described in the methodology. These were tested by using the random sample of 50% of elementary cells that had not been used in the normalization process. The scenario which best fitted statistical data, i.e. with the highest number of coincidences between the model and the effective situation, was Scenario 4.

Figure 5 presents the maps of the estimated level of risk under the four Scenarios.

The spatial AHP model applied to GIS techniques has allowed building an integrated model of Analysis of Forest Fire Risk, as required by the current Italian legislation. The model takes into account the complexity of the relations existing among the component of the geographic, environmental, social and economic contexts and their influence on forest fires risk. While at research level there are analyses that have used more complex mathematical models to approach the problem of forest fires (Arpaci et al. 2014; You et al. 2017), in operational planning the studies on which Italian institutions in charge of territorial government have based their plans had a less transparent formalization of the concept of risk and used only a few attributes, with the consequence of a worse representation of the risk. The map of Scenario 4 can be considered as a good compromise between model complexity

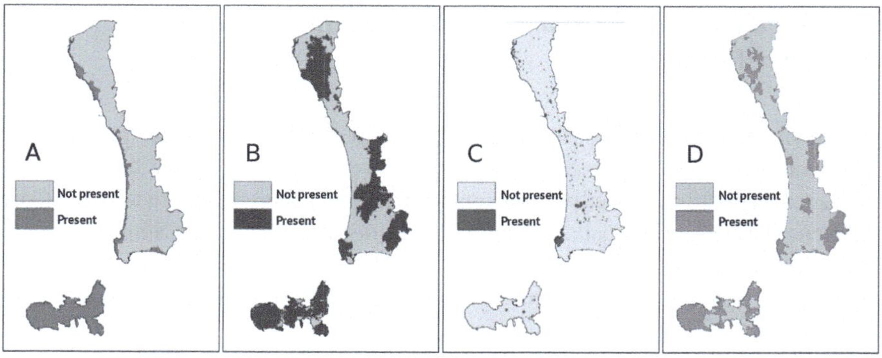

Fig. 4 Livorno province—maps of the distribution of the scores related to the four attributes of the Environmental Value criterion. Legend: Presence or absence of (**a**) landscape constraints, (**b**) hydrogeological constraints, (**c**) archaeological constraints, (**d**) Presence/absence of Protected areas. Source: Candura (2005)

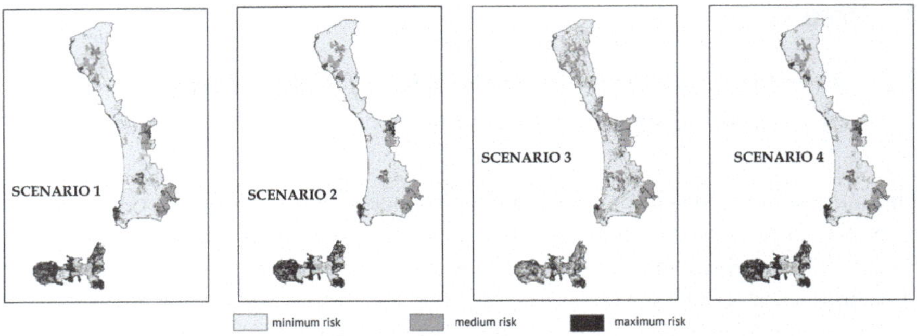

Fig. 5 Livorno province—maps of the level of forest fire risk, under the hypotheses of four scenarios. Source: Candura (2005, modified)

and utility for planning, being a sound tool from a technical and scientific viewpoint for organizing prevention activity and structure and drawing plans for facing problems in case of fires. The model shows a high ease and flexibility of use, being implemented through the same GIS software employed by Tuscany Region for its thematic mapping and having a plug-in for MCDA analyses. Therefore, it could be improved in time by introducing new available data sources or better integrating anthropic aspects as factors determining a higher risk of forest fires. Another area of improvement could be that of the assessment of damage, where a more complex framework for estimating economic and environmental impacts might be built. Finally, it is important to highlight the role of the archive of areas where there have been forest fires that must be created and updated by law, since historical data can be very useful for testing the effectiveness of models estimating forest fire risk, as it has been done for choosing the best scenario.

3.2 Productive and Landscape Restoration of a Hilly Rural Area

The main results of this case study are represented by the maps providing a prioritization of the areas in need of policy interventions, which have been built on the base of two different methods for summarizing the performances in terms of Cost and Benefit criteria. The spatialization of the results of the Weighted Sum and Miles' Value Index analyses are shown in Figs. 6 and 7; in both cases, according to the value of the goal, parcels have been classified as suitable for the following intervention strategies:

- Class 4—rehabilitation to cultivation uses through extraordinary maintenance interventions;
- Class 3—rehabilitation to cultivation uses through land improvement interventions;

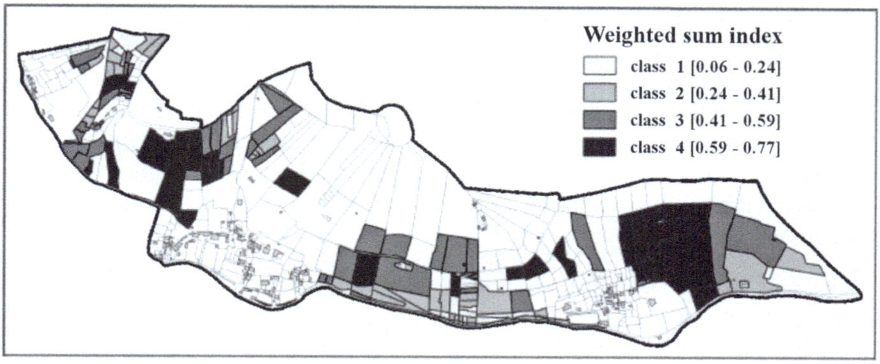

Fig. 6 Ranking and classification of abandoned cadastral parcels in terms of rehabilitation suitability according the Weighted Sum method. Source: Bonelli (2017), modified

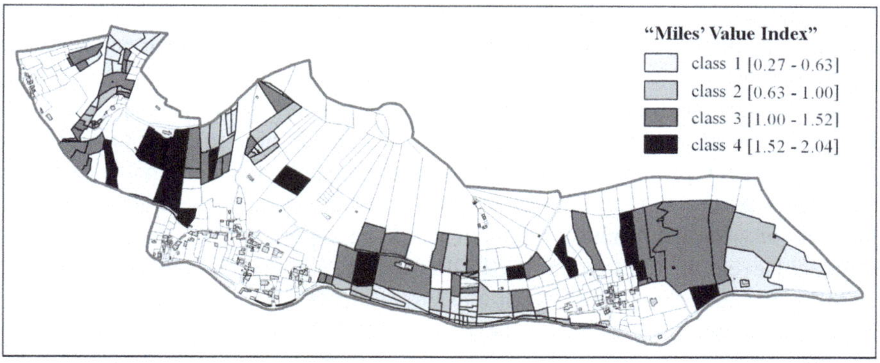

Fig. 7 Ranking and classification of abandoned cadastral parcels in terms of rehabilitation suitability according the Miles' Value Index method. Source: Bonelli (2017), modified

- Class 2—managed conversion into forests and woodland;
- Class 1—natural conversion into forests and woodland (null hypothesis).

According to transacts and experts' opinion, the Weighted Sum method (Fig. 6) allows a better spatial distribution of classes 3 and 4 in comparison with the Value Index method (Fig. 7). Nevertheless, both methods allow reaching good results as regards the goal of assessing the suitability to a productive rehabilitation, which has to be interpreted in relative terms, i.e. providing a rank in terms of priority, rather than in absolute ones.

The main results of this case study are:

1. The identification of abandonment processes as the main driver of the transformations landscape has undergone since 1954 and a detailed territorial analysis of the areas that have been interested by these negative changes;

2. The definition of a geographic multicriteria analysis model based on the awareness that the rehabilitation to productive uses of previously cultivated areas is not always possible or cost-effective. The model, on the basis of the characteristics of abandoned land, of the cost needed for its rehabilitation and of the benefits that this latter will bring to the community well-being, especially in terms of landscape functions, identifies a set of intervention strategies ranging from the productive rehabilitation to the so called "null hypothesis", i.e. the one of "non-intervention".

These results are consistent with the two main phases in which the research has been organized, i.e. an initial phase aiming to gather knowledge, mainly through information taken from Tuscany Region Geo-portal, and then a phase aiming to build and run the model. Both the phases have been implemented through QGIS that has allowed to coordinate and integrate in the same database all the relevant data and to archive the indices derived from data processing. This database and the processing power of QGIS have made it easy the subsequent phase of assessment of the criteria deemed relevant for the model. Indeed, QGIS plug-in "Vector MCDA" has allowed to easily and quickly obtaining both the results of the analyses and their cartographic representation.

The strategies that have been individuated by the model should be applied to all the territory or at least to the majority of the case-study area in order to achieve a rehabilitation and promotion of its landscape, since landscape is not made of single elements or portions of a territory, but have to be considered as a whole unit. From this point of view, the results of the model could be important in setting priorities for Regional Development Programmes (RDPs) actions, e.g. in highlighting the role that could be played by Integrated Territorial Projects (ITPs), i.e. voluntary and collective private-public projects, under the Tuscany RDP 2014–2020.

Although the case-study area is quite small, the above described approach can be applied to larger rural territories having the same features, i.e. a high historic and identity value of landscape and widely spread abandonment phenomena, tailoring the criteria—when necessary—to the specific context to be analyzed. This methodology can be very useful for the public decision maker both in the phase of definition of planning rules and standards for rural territories and to channel in a more effective way the use of public resources aiming to promote and protect cultural landscapes located in hilly and mountainous areas, which are widely spread not only in Tuscany, but in the whole Italian territory.

3.3 Planning in the Rural-Urban Fringe Taking into Account the Multifunctional Role of Agriculture

The results of this case study mainly consist on the production of maps with specific themes (or thematisms). In particular, for each category of ESs, i.e. those concerning productive, protective, and cultural/recreational functions, a specific map was produced representing the areas providing the ESs of this category and their intensity (related to the shade intensity of the relevant color) of provision.

As regards the production of ESs on the whole, i.e. without distinguishing among categories, the problem has been faced with two different approaches. In the first case, through a new AHP, the three thematic maps have been summarized into a new map, where the intensity of color (different shades of the same color) was related to the whole production of ESs by each portion of the territory. Summarizing ESs in a composite index, by expressing them in a common monetary unit or by means of a weighted sum, would allow for compensations, meaning that a low production of ESs belonging to a category can be compensated by a high production of ESs belonging to another category. Although the classes with very high or very low scores have a significant role in highlighting the areas where the situation of ES provision on the whole is highly positive or negative, the Authors deemed that this kind of information could be misleading in terms of planning actions, for the following reasons. First of all, summarizing ESs by the use of a "monetary evaluation" and additive methods might generate difficult and questionable results both from a scientific point of view, due to the complexity of the factors involved, and from an ethical point of view, due to the difficulties of interpretation of a total value obtained by the sum of values related to the provision of very different functions, affecting different groups. Then, in the case that model results are employed for spatial and urban planning, by summing up ESs and functions that are quite different, the information about the suitability of some territorial portion to produce specific ESs or ESs related to a specific function would be lost. Last, but not least, there is the risk that by summarizing ESs in a single composite index, a low score on it could be read as an index of suitability to artificialization, i.e. that the areas which have a relatively low production of ESs can be transformed in built-up or in other artificial areas without problems.

For these reasons, it was decided to avoid aggregating the values of the three categories of ESs and to produce a map with a 'three-dimensional' evaluation that keeps them separated. The map (Fig. 8) provides a spatial representation of the performances for the provision of ESs belonging to the three classes, which are represented by different colors. Consequently, the final shade of color depends on the mix of functions provided by each territorial ambit, so that it is possible to highlight if an area is specialized, or not, in the provision of a specific ESs category. In other words, this second map highlights the suitability of a territory to provide a specific class of ESs or a mix of them, rather than the intensity of ESs provision.

Figure 8 should be interpreted more as a spatial distribution of the different categories of ESs than in terms of intensity of ESs provision, since the assessment of ESs for each category is significant in relative terms more than in absolute ones. It provides an immediate and intuitive overview of areas with greater or lower suitability to the provision of specific services or categories. This should consent to support planning by better individuating where to localize strategies aiming to promote ES provisions, i.e. the areas with high priority in terms of protection, rehabilitation, or other interventions. In other words, Fig. 8 can be very useful when deciding where to concentrate aid to agricultural production, where to

Fig. 8 Lucca lowland—Map of the ESs provided by agricultural land by suitability to specific category of ESs (color shadow represents the prevalence of each ESs category). Legend: green: ESs related to ecological functions; blue: ESs related to productive functions; red: ESs related to cultural/recreational functions. Source: Rovai et al. (2013)

subsidize or promote the protection and improvement of ecological connectivity and where to promote actions related to the social and cultural role of open spaces. Last but not least, by changing the set of weights assigned to criteria, attributes and sub-attributes it is possible to build maps according to scenarios based on different sets of instances coming from stakeholders.

3.4 *Sustainable Governance of a Cultural-Historic Landscape*

The governance model described in Sect. 2.4 has been tested on the Municipality of Castiglione d'Orcia, one of the five municipalities belonging to Val d'Orcia. In this section, for the sake of brevity, we will focus on the farm analysis and in its utility when integrated with the results of the territorial analysis. In particular, we describe two matrices that can be very useful in designing policy tailored on the specific context of an area, the first one summarizing the results of the analyses at farm level in terms of evolutionary paths; the second one setting the more suitable policy actions according to farm evolutionary paths and territorial characteristics.

The first matrix has been built in the framework of the "farm analysis" and crosses the results in terms of Farm structural, socio-economic and management characteristics with those of Plots' suitability to cultivation. The result is the identification of seven classes of evolutionary paths of farms ranging from maintenance strategies of the current situation, to multifunctionality strategies or, in the worst cases, to the abandonment of agriculture activities (Fig. 9).

For example, Class VH includes farms characterized by a high adaptation capacity and by plots characterized on average by a high suitability to cultivation. These farms will very likely maintain their present cultivation systems without relevant changes in their organizational structure. Class MH represents farms with a high capability of adaptation to the external changes, but managing plots that have a low suitability to agricultural use. The evolution of these farms is almost inevitably directed to multifunctional agriculture (i.e. farm holidays, educational farms, eno-gastronomic tourism, etc.) or, alternatively, to the production of value-added products; otherwise the low suitability to agricultural use could cause land abandonment with negative repercussions on both environmental and landscape resources. Class VL is characterized by farms with low adaptation capacity and plots with a low suitability to cultivation. These are the weakest farms, which will probably abandon

		Plot "resistance" or suitability to cultivation		
		High (H)	***Medium (M)***	***Low (L)***
Farm "resistance" or reaction capacity to external solicitations	***High (H)***	**VH** *maintenance*	**H** *maintenance / multifunctionality*	**MH** *multifunctionality*
	Medium (M)		**M** *uncertain strategies*	
	Low (L)	**LM** *transfer to stronger farms*	**L** *transfer to stronger farms/abandonment*	**VL** *abandonment*

Fig. 9 Fuzzy classes defining farm evolutionary paths for the case study according to their adaptation capacity. Source: Rovai et al. (2016)

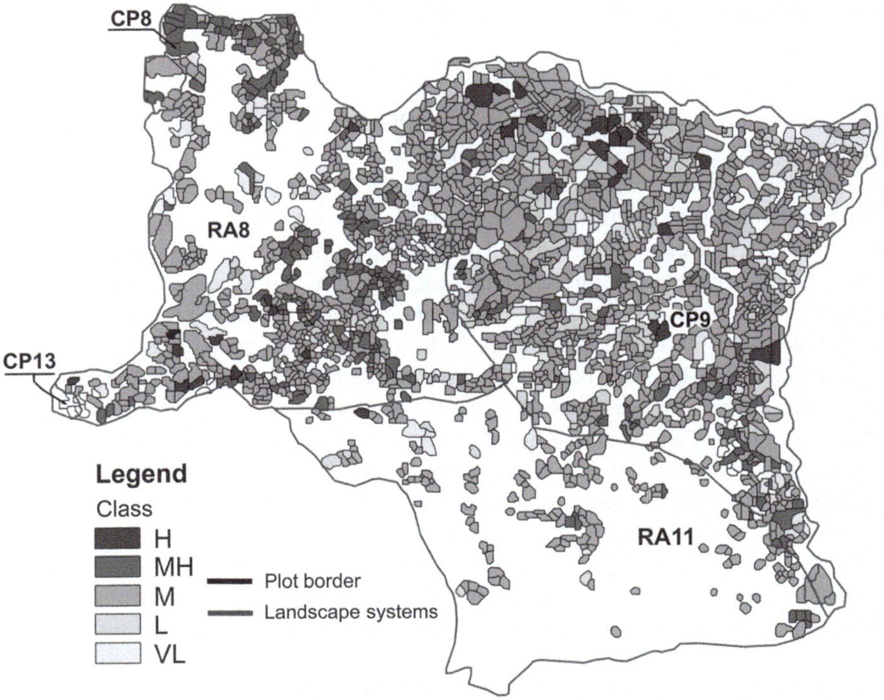

Fig. 10 Spatial distribution of classes related to farms evolutionary paths (*H* Maintenance/Multifunctionality, *MH* Multifunctionality, *M* Uncertain evolution, *L* Sale or renting of land to stronger farms/abandonment, *VL* Abandonment). VH and LM classes have been omitted, since they are not present in the case-study area. Source: Rovai et al. (2016)

all agricultural activities in the medium term, unless some supporting actions cause a rupture and a deep change of the farm development path.

By using GIS, it was possible to represent the spatial distribution of these farms in the municipality of Castiglione d'Orcia (Fig. 10) and to obtain important information for territorial policies: for example, farms directed to multifunctionality are mainly located on the western part of the study area, whereas the ones with a high risk of agricultural abandonment are concentrated in the central and in the eastern part of Castiglione d'Orcia municipality. Finally, those farms whose prevalent evolution will likely be the 'status quo' maintenance are mainly located in the northern part of the study area.

Results of the analyses, which have been validated by focus groups with local stakeholders, show that:

1. The local entrepreneurial class appears quite stable as a considerable percentage of farms presents a high adaptation capacity to the external transformations;
2. The potential impacts on environmental and landscape resources are anyway very critical as 4356 ha, on a total surface of 5600, belong to farms run by farmers that

Farm adaptation capacity (resulting from farm analysis)	Landscape sensitivity (resulting from territorial analysis)		
	High	**Medium**	**Low**
H *Maintenance*	Public promotion actions	Incentives aiming to improve quality of landscape	Incentives aiming to improve quality of landscape
MH *Multi-functionality*	Incentives aiming to promote multifunctionality	Incentives aiming to promote multifunctionality and to improve quality of landscape	Incentives aiming to improve quality of landscape
M *Uncertain strategies*	Incentives aiming to maintain agriculture activities and / or promote multifunctionality	Incentives aiming to maintain agriculture activities and to improve landscape quality	Incentives aiming to maintain agriculture activities and to improve landscape quality
L *Transfer to stronger farms/abandonment*	Incentives aiming to promote entrepreneurial "cultural change"	Incentives aiming to promote entrepreneurial "cultural change" and to improve landscape quality	Low priority interventions
VL *Abandonment*	Incentives to promote entrepreneurial "cultural change"	Incentives aiming to promote entrepreneurial "cultural change" and to improve landscape quality	Low priority interventions

Fig. 11 Response actions for the sustainable governance of the case-study rural landscape. Source: Rovai et al. (2016)

may opt for agricultural abandonment if the external conditions become too adverse, i.e. due to changes on CAP, market trends and climate.

The second matrix (Fig. 11) combines the results of the territorial analysis, where areas were classified on the base of their level of landscape sensitivity with the ones of farms strategies. Landscape sensitivity has been obtained under the hypotheses that in the specific case-study area Intrinsic Value and Vulnerability were directly related and consequently could be summarized in a unique composite index. Priorities and actions should take into account not only the results of the analysis but also the institutional and cultural context of the area.

As Fig. 11 shows, e.g. in a subsystem with high landscape sensitivity, depending on farms probable evolution, the response actions to counteract negative effects may vary from public institution actions for the promotion of the territory, to specific regulations and incentives for the preservation of landscape's current state, up to incentives to improve farm viability in the specific territorial context. Public promotion could, e.g. include such interventions as the creation or improvement of existing countryside walking paths in order to improve territory and landscape fruition. The need for a GIS able to collect and update all farm changes is a consequence of the above defined response actions, since it is essential, especially for the monitoring of the most critical areas. Due to the adopted spatial scale, all those actions can be calibrated not only at farm level, but also at plot level, which makes such interventions potentially very effective.

For example, from the spatial analysis emerges that the majority of plots present high landscape sensitivity and belong to the class M (Uncertain strategies) in relation to the farm analysis. In such circumstances the definition of rules and incentives for the preservation of the current landscape configuration, together with public promotion actions for landscape valorization, becomes absolutely necessary.

Although the definition of proper actions for promoting a sustainable rural landscape is very important, it is not possible to remain in the "scientific domain" but these actions have to be seen in the normative and administrative context where they have to be implemented. Nevertheless, in the Authors' opinion, the proposed governance model approach constitute a sound scientific tool for a sustainable governance of the rural landscape, since it is able to give suggestions about the most appropriate tools for intervention in the framework of an integrated planning approach and a prioritization of areas in need of interventions.

4 Concluding Remarks

Planning, programming and assessing the functions of rural territories are complex problems that ask for a multidisciplinary and transdiciplinary approach based on a holistic knowledge system. In this framework, the integration of MCDA techniques and GIS represents a very effective and promising solution for the following reasons:

1. Multicriteria Analysis supplies many technical solutions for building a model, assessing criteria and attributes, assigning weights, testing the results; allowing a high flexibility and adaptability of the approach to a vast range of specific problems;
2. GIS software has reached a high power of processing data and information rendering, allows having a high interactivity with the user, and it can be used also with computer of normal capacity. Furthermore, due to the ability to create maps and spatial representations, it is a powerful tool for communicating with stakeholders and citizens. This characteristic is very important due to the increasing role of participative processes in the setting and implementation of territorial policies and in the assessment of their results.

In the specific case, the choice to use the same GIS software used by the Tuscany Region for producing and making available geographical information, from the one hand made it easier for the Authors to access and integrate data sources, from the other hand it made the interaction with regional employees easier, since they have a good knowledge of this software.

This contribution aims to highlight these features through the analysis of the main features of four case studies characterized by an integrated approach of AHP and GIS techniques, but which are different in terms of size of the case-study area, goals, complexity of the problems, spatial elementary units considered, etc. Indeed, each approach was tailored to the specific context of application and had to be adapted to take into account data availability.

In each case of study the ability of the method to integrate information and knowledge belonging to different fields has clearly emerged as much as the capacity to represent a fundamental, scientifically robust and flexible tool providing results able to effectively support public decision-makers for territorial planning interventions.

The use of spatial multi-criteria analysis models would make it possible to organize planning and programming as continuous and flexible processes. This is very important in all the institutional contexts, as the one of Italy, where plans and programmes have a long life time-span and for which the "ad hoc" concertation process requires too much time, so that when a plan or a program is implemented, it is already "out of date".

Finally, this tool is useful to overcome the lack of coordination among policy tools. Indeed, what is still missing in the Italian context, as in the ones of other countries, is an operative policy approach able to integrate and coordinate instances coming from stakeholders, agricultural (or other sectoral) policies and territorial and landscape governance. Consequently, efforts should be directed towards innovative policies and governance instruments, such as collective voluntary actions, integrated projects, etc., which are needed for an adequate implementation of sustainable landscape and rural development policies.

References

Arpaci, A., et al. (2014). Using multi variate data mining techniques for estimating fire susceptibility of Tyrolean forests. *Applied Geography, 53*, 258–270. https://doi.org/10.1016/j.apgeog.2014.05.015

Barnett, J., Lambert, S., & Fry, I. (2008). The hazards of indicators: Insights from the environmental vulnerability index. *Annals of the Association of American Geographers, 98*(1), 102–119. https://doi.org/10.1080/00045600701734315

Beinat, E., & Nijkamp, P. (Eds.). (1998). *Multicriteria analysis for land-use management*. Dordrecht: Springer Science & Business Media.

Blahut, J., Glade, T., & Sterlacchini, S. (2014). Debris flows risk analysis and direct loss estimation: The case study of Valtellina di Tirano, Italy. *Journal of Mountain Science, 11*(2), 288–307.

Bonelli, S. (2017). *Riqualificazione e valorizzazione del paesaggio del Compitese. Proposta di un metodo di analisi multi criteri in ambiente GIS per la definizione di strategie di intervento*. Thesis. Pisa: University of Pisa. Accessed January 18, 2017, from https://etd.adm.unipi.it/t/etd-06202017-174259/

Boroushaki, S., & Malczewski, J. (2010). Using the fuzzy majority approach for GIS-based multicriteria group decision-making. *Computers and Geosciences, 36*(3), 302–312. https://doi.org/10.1016/j.cageo.2009.05.011

Candura, D. (2005). *Valutazione territoriale del rischio incendi nella Provincia di Livorno e piano delle infrastrutture di prevenzione e di soccorso di un'area periurbana*. Thesis. Pisa: Università di Pisa.

Chuvieco, E., et al. (2010). Development of a framework for fire risk assessment using remote sensing and geographic information system technologies. *Ecological Modelling, 221*(1), 46–58. https://doi.org/10.1016/j.ecolmodel.2008.11.017
Ciampi, C., et al. (2015). Tendenze recenti del cambiamento. La banca dati Uso e Copertura del Suolo di Regione Toscana. *Atti Conferenza Nazionale ASITA* (pp. 263–270). Retrieved September 14, 2017, from http://atti.asita.it/ASITA2015/Pdf/081.pdf
Ciancio, O., Corona, P., Marinelli, M., & Pettenella, D. (Eds.). (2007). *Metodologia per la valutazione economica dei danni da incendi boschivi*. Firenze: Accademia Italiana di Scienze Forestali-Corpo Forestale dello Stato. Accessed April 9, 2018, from https://aisfdotit.files.wordpress.com/2013/06/incendiitaliano.pdf
Conrad, E., Christie, M., & Fazey, I. (2011). Is research keeping up with changes in landscape policy? A review of the literature. *Journal of Environmental Management, 92*(9), 2097–2108. https://doi.org/10.1016/j.jenvman.2011.04.003
Cooper, T., Baldock, D., Rayment, M., Kuhmonen, T., Terluin, I., Swales, V., & Farmer, M. (2006). *An evaluation of the less favoured area measure in the 25 member states of the European Union*. Institute for European Environmental Policy, London, 249 pp. Retrieved September 14, 2017, from https://ieep.eu/uploads/articles/attachments/cd7c77fb-a873-4267-a650-7c3bed409998/LFA_evaluation_report_full_text_en.pdf?v=63664509706
Costanza, R., et al. (1997). The value of the world's ecosystem services and natural capital. *Nature, 387*(6630), 253–260. https://doi.org/10.1038/387253a0
Council of Europe. (2000). *European landscape convention*. Report and Convention Florence (ETS No. 17(176), p. 8). Available at: http://conventions.coe.int/Treaty/en/Treaties/Html/176.htm
Daniel, T. C., & Vining, J. (1983). Methodological issues in the assessment of landscape quality. In I. Altman & J. F. Wohlwill (Eds.), *Behavior and the natural environment. Human behavior and environment (Advances in theory and research)* (Vol. 6). Boston, MA: Springer. https://doi.org/10.1007/978-1-4613-3539-9_3
De Groot, R. S., Wilson, M. A., & Boumans, R. M. J. (2002). A typology for the classification, description and valuation of ecosystem functions, goods and services. *Ecological Economics, 41*(3), 393–408. https://doi.org/10.1016/S0921-8009(02)00089-7
Eliasson, Å., et al. (2010). Common criteria for the redefinition of intermediate less favoured areas in the European Union. *Environmental Science and Policy, 13*(8), 766–777. https://doi.org/10.1016/j.envsci.2010.08.003
Farina, A. (2006). *Principles and methods in landscape ecology – Towards a science of the landscape*. Available at: http://www.springer.com/life+sciences/ecology/book/978-1-4020-3327-8
Fastelli, L., et al. (2017). A spatial analysis of terrain features and farming styles in a disadvantaged area of Tuscany (Mugello): Implications for the evaluation and the design of CAP payments. *Bio-based and Applied Economics, 6*(1). https://doi.org/10.13128/BAE-14625
Fry, G., et al. (2009). The ecology of visual landscapes: Exploring the conceptual common ground of visual and ecological landscape indicators. *Ecological Indicators, 9*(5), 933–947. https://doi.org/10.1016/j.ecolind.2008.11.008
Gai, C., Weng, W., & Yuan, H. (2011). GIS-based forest fire risk assessment and mapping. *Proceedings – 4th International Joint Conference on Computational Sciences and Optimization, CSO 2011* (pp. 1240–1244). https://doi.org/10.1109/CSO.2011.140
Gómez-Limón, J. A., & Riesgo, L. (2009). Alternative approaches to the construction of a composite indicator of agricultural sustainability: An application to irrigated agriculture in the Duero basin in Spain. *Journal of Environmental Management, 90*(11), 3345–3362.
Graymore, M. L. M., Wallis, A. M., & Richards, A. J. (2009). An index of regional sustainability: A GIS-based multiple criteria analysis decision support system for progressing sustainability. *Ecological Complexity, 6*(4), 453–462. https://doi.org/10.1016/j.ecocom.2009.08.006
Greene, R., et al. (2011). GIS-based multiple-criteria decision analysis. *Geography Compass, 5*(6), 412–432. https://doi.org/10.1111/j.1749-8198.2011.00431.x

Haines-Young, R. and Potschin, M. (2013). Common international classification of ecosystem services (CICES): Consultation on Version 4, August–December 2012. Retrieved September 14, 2017, from https://cices.eu/content/uploads/sites/8/2012/07/CICES-V43_Revised-Final_Report_29012013.pdf

Hardy, C. C. (2005). Wildland fire hazard and risk: Problems, definitions, and context. *Forest Ecology and Management, 211*(1–2), 73–82.

ISPRA. (2017). *Consumo di suolo, dinamiche territoriali e servizi ecosistemici – Italiano, Roma*. Retrieved September 14, 2017, from http://www.isprambiente.gov.it/it/pubblicazioni/rapporti/consumo-di-suolo-dinamiche-territoriali-e-servizi-ecosistemici

Jackson, B., et al. (2013). Polyscape: A GIS mapping framework providing efficient and spatially explicit landscape-scale valuation of multiple ecosystem services. *Landscape and Urban Planning, 112*(1), 74–88. https://doi.org/10.1016/j.landurbplan.2012.12.014

Jaiswal, R. K., et al. (2002). Forest fire risk zone mapping from satellite imagery and GIS. *International Journal of Applied Earth Observation and Geoinformation, 4*(1), 1–10. https://doi.org/10.1016/S0303-2434(02)00006-5

Karsak, E. E. (2004). Fuzzy multiple objective programming framework to prioritize design requirements in quality function deployment. *Computers & Industrial Engineering, 47*(2–3), 149–163. https://doi.org/10.1016/j.cie.2004.06.001

Labiosa, W., et al. (2010). The South Florida ecosystem portfolio model: A web-enabled multicriteria land use planning decision support system. *Proceedings of the Annual Hawaii International Conference on System Sciences.* https://doi.org/10.1109/HICSS.2010.482

Labiosa, W. B., et al. (2013). An integrated multi-criteria scenario evaluation web tool for participatory land-use planning in urbanized areas: The ecosystem portfolio model. *Environmental Modelling and Software, 41*, 210–222. https://doi.org/10.1016/j.envsoft.2012.10.012

Magnaghi, A. (2005). *The Urban village: A charter for democracy and local self-sustainable development*. London: Zed Books. 220 pp.

Malczewski, J. (1999). *GIS and multicriteria decision analysis* (392 pages). New York: Wiley.

Malczewski, J. (2006a). GIS-based multicriteria decision analysis: a survey of the literature. *International Journal of Geographical Information Science, 20*(7), 703–726. https://doi.org/10.1080/13658810600661508

Malczewski, J. (2006b). Ordered weighted averaging with fuzzy quantifiers: GIS-based multicriteria evaluation for land-use suitability analysis. *International Journal of Applied Earth Observation and Geoinformation, 8*(4), 270–277. https://doi.org/10.1016/j.jag.2006.01.003

Marson, A. (2010). *Land-use planning scandals in Tuscany. Mismanagement or underestimation of general public interests?* Gießen. Retrieved September 14, 2017, from http://www.italienforschung.de

Mendoza, G. A., & Martins, H. (2006). Multi-criteria decision analysis in natural resource management: A critical review of methods and new modelling paradigms. *Forest Ecology and Management, 230*(1–3), 1–22. https://doi.org/10.1016/j.foreco.2006.03.023

Miles, L. D. (1961). *Techniques of value analysis and engineering*. New York: McGraw-Hill.

Millenium Ecosystem Assessment. (2005). *Ecosystem and human well-being: Synthesis* (pp. 1–59). Washington, DC: Island Press. Retrieved September 14, 2017, from https://www.millenniumassessment.org/documents/document.356.aspx.pdf

OECD. (2008). *Handbook on constructing composite indicators methodology and user guide*. Paris: OECD. Retrieved September 14, 2017, from http://www.oecd.org/std/42495745.pdf

Pelorosso, R., et al. (2011). Stability analysis for defining management strategies in abandoned mountain landscapes of the Mediterranean basin. *Landscape and Urban Planning, 103*, 335–346. https://doi.org/10.1016/j.landurbplan.2011.08.007

Primdahl, J., & Kristensen, L. S. (2011). The farmer as a landscape manager: Management roles and change patterns in a Danish region. *Geografisk Tidsskrift-Danish Journal of Geography, 111*(2), 107–116. https://doi.org/10.1080/00167223.2011.10669527

Rovai, M., et al. (2013). Un indicatore multidimensionale per la valutazione della distribuzione spaziale dei servizi agro-ecosistemici nei territori periurbani. *Atti 17ma Conferenza Nazionale*

ASITA – Riva del Garda, 5–7 novembre (pp.1157–1164). Retrieved September 14, 2017, from http://atti.asita.it/ASITA2013/Pdf/017.pdf

Rovai, M., et al. (2016). A DSS model for the governance of sustainable rural landscape: A first application to the cultural landscape of Orcia Valley (Tuscany, Italy). *Land Use Policy, 56*, 217–237. https://doi.org/10.1016/j.landusepol.2016.04.038

Rovai, M., et al. (2017). *Un modello di AMC spaziale per la definizione di strategie di intervento per gli oliveti abbandonati*. Atti 20ma Conferenza Nazionale ASITA - Salerno, 21–23 Novembre 2017. Accessed April 6, 2018, from http://atti.asita.it/ASITA2017/Pdf/191.pdf

Saaty, T. L. (1980). *The analytic hierarchy process* (pp. 1–11; 287 pages). New York: Education, McGraw Hill.

Saaty, T. L. (2004). Decision making – the analytic hierarchy and network processes (AHP/ANP). *Journal of Systems Science and Systems Engineering, 13*(1), 1–35. Retrieved September 14, 2017, from https://link.springer.com/content/pdf/10.1007%2Fs11518-006-0151-5.pdf

Saaty, T. L. (2008). Decision making with the analytic hierarchy process. *International Journal of Services Sciences, 1*(1), 83. https://doi.org/10.1504/IJSSCI.2008.017590

Sands, G. R., & Podmore, T. H. (2000). A generalized environmental sustainability index for agricultural systems. *Agriculture Ecosystems & Environment, 79*, 29–41. https://doi.org/10.1016/S0167-8809(99)00147-4

Stoate, C., Boatman, N. D., Borralho, R. J., Carvalho, C. R., Snoo, G. R. d., & Eden, P. (2001). Ecological impacts of arable intensification in Europe. *Journal of Environmental Management, 63*(4), 337–365.

Tallis, H., & Polasky, S. (2011). Assessing multiple ecosystem services: An integrated tool for the real world. In P. Kareiva, H. Tallis, T. H. Ricketts, G. C. Daily, & S. Polasky (Eds.), *Natural capital: theory and practice of mapping ecosystem services* (pp. 34–50). Oxford: OUP.

Van Wagner, C. (1987). *Development and structure of the Canadian forest fire weather index system, Ottawa*. Retrieved September 14, 2017, from http://cfs.nrcan.gc.ca/pubwarehouse/pdfs/19927.pdf

van Zanten, B. T., et al. (2013). European agricultural landscapes, common agricultural policy and ecosystem services: a review. *Agronomy for Sustainable Development, 34*(2), 309–325. https://doi.org/10.1007/s13593-013-0183-4

Villa, F., & McLeod, H. (2002). Environmental vulnerability indicators for environmental planning and decision-making: Guidelines and applications. *Environmental Management, 29*(3), 335–348. https://doi.org/10.1007/s00267-001-0030-2

Yang, A. L., et al. (2014). Spatial analysis of agri-environmental policy uptake and expenditure in Scotland. *Journal of Environmental Management, 133*, 104–115. https://doi.org/10.1016/j.jenvman.2013.11.038

You, W., et al. (2017). Geographical information system-based forest fire risk assessment integrating national forest inventory data and analysis of its spatiotemporal variability. *Ecological Indicators, 77*, 176–184. https://doi.org/10.1016/j.ecolind.2017.01.042

The Use of the Analytic Network Process for the Analysis of Public Goods Supply from Agricultural Systems: Advances and Challenges Ahead

Stefano Targetti, Anastasio J. Villanueva, Carlos Parra-López, and Davide Viaggi

Abstract This chapter aims to review and summarize potentials and limitations of the use of the Analytic Network Process in the context of the evaluation of public goods provision from agricultural systems. The chapter provides a description and a step-by-step explanation of the method, and presents insights from three recent papers using the Analytic Network Process to analyze public goods supply from agricultural systems. The papers were selected to show a range of diversified and complementary approaches to the problem and the possibility to integrate stakeholders in the different phases of the evaluation process. The first paper presents a comparison between three rural landscapes and provides a discussion of the role of different economic actors in supplying private- and public-type services. The second paper presents an integrated approach to support the policy-making aimed at a more efficient provision of public goods from a specific farming system. The third paper presents a farm level assessment of multifunctional performance considering a range of different farming practices and techniques. These studies provide evidence of the usefulness of the method to support policy-making and understand the relation between farmers' decision-making and the provision of public goods. The results are also discussed, pointing out the strengths and weaknesses of the method in this type of analysis as well as pathways for methodological refinements and integration

S. Targetti (✉)
Faculty of Environmental Sciences, Czech University of Life Sciences, Prague, Czech Republic

A. J. Villanueva
Water, Environmental and Agricultural Resources Economics (WEARE) Research Group, Universidad de Córdoba, Córdoba, Spain
e-mail: ajvillanueva@uco.es

C. Parra-López
Department of Food Chain Economics, Institute of Agricultural and Fisheries Research and Training, Centro Camino de Purchil, Granada, Spain
e-mail: carlos.parra@juntadeandalucia.es

D. Viaggi
Department of Agricultural and Food Sciences, University of Bologna, Bologna, Italy
e-mail: davide.viaggi@unibo.it

J. Berbel et al. (eds.), *Multicriteria Analysis in Agriculture*, Multiple Criteria Decision Making, https://doi.org/10.1007/978-3-319-76929-5_4

possibilities with other techniques with a particular attention toward the ex-ante and ex-post participation of stakeholders.

Keywords ANP · Public goods · Ecosystem services · Farm management · Rural development · Stakeholders · Agricultural policy

1 Introduction

Agricultural systems jointly provide a wide range of goods highly valued by society (Madureira et al. 2013). This provision has been studied following different strands such as the "multifunctionality" (e.g. OECD 2001), or the "ecosystem services" (e.g. Porter et al. 2009) frameworks. A common aspect concerns the differentiation between private (e.g. food and fiber) and public (e.g. amenities, regulation, and supporting functions) goods, which underlines the disconnection between marketed and non-marketed services and the need of a more balanced delivery of both good types from agricultural landscapes. That entails the necessity to design enhanced policies able to tackle several challenging issues, including the joint production of public and private goods and the socioeconomic context in which this provision takes place (Hanley et al. 2012). Therefore, the investigation of public goods provided by agricultural systems requires the use of integrated approaches, taking into account the different dimensions, from purely ecological and biophysical to socioeconomic, which play a role in their provision (Bennett 2017). In particular, the employment of a "portfolio" of methods (quantitative and qualitative, economic and bio-physical, etc.) is advocated to reflect different perspectives and cause-effect pathways for the assessment and identification of efficient policy solutions (Young et al. 2006). That is even more relevant considering the recent advances challenging a range of basic assumptions of economic theory (e.g. rational behavior) which suggest the opportunity to adopt more nuanced methodologies (Gowdy and Erickson 2005).

Multicriteria analysis (MCA) is considered among the effective approaches to investigate the provision of public and private goods of agricultural systems as it can overcome some inherent limitations of monetary evaluation (Spangenberg and Settele 2010; Parks and Gowdy 2013). Among the MCA techniques, the Analytic Network Process (ANP) stands out as a promising tool to analyze public goods (Parra-López et al. 2008a; Reig et al. 2010). In fact, there is a growing body of literature employing the ANP in agri-environmental policy evaluation and design. Recent studies particularly underline the potential of the ANP in the context of policy-making toward the supply of public goods from agricultural systems (e.g. Parra-López et al. 2008a; Nekhay et al. 2009; Reig et al. 2010; Villanueva et al. 2014, 2015; Schaller et al. 2015; Carmona-Torres et al. 2016). Indeed, the ANP enables to consider the multi-faceted interrelationships and feedbacks of the joint provision of private and public goods by using a systemic and trans-disciplinary approach (Carmona-Torres et al. 2014). Main advantages of the ANP approach are the possibility to analyze complex systems where only partial information is available, the potential to integrate quantitative and qualitative variables, and, therefore, the option to carry out cross-cutting assessments involving different expertise, methodologies and information sources (Villanueva et al. 2015). Alike other MCA approaches, the ANP can be integrated in a multi-step methodology including trans-disciplinary techniques (e.g. cognitive mapping, focus groups, Delphi, etc.). As such, the ANP is also fitting as a boundary object where stakeholders and researchers may discuss cause-effect chains involving multiple interests at stakes, and visualize the results to identify disconnections between different knowledge types.

This chapter aims to provide an overview of the ANP method and a review of latest advances, potentials and limitations of the use of the ANP in the context of the evaluation of public goods provision from agricultural systems. After a description of the method (Sect. 2), this chapter reports results from three selected papers applied to the analysis of public goods in different rural case studies (Sect. 3). Based on this, Sects. 4 and 5 provide a critical discussion and the conclusions highlighting main pros and cons in the analysis of public goods by agricultural systems and exploring the potential for methodological refinements and integration possibilities with other methods.

2 The Method

The ANP is based on pairwise comparisons between criteria, and matrix algebra calculations to produce a vector of scores (so called "priority vector") which outlines the importance of the criteria with respect to a specific goal (so called "control criterion"), and a measure of judgment consistency. The method was developed by T. L. Saaty as a generalization of the Analytic Hierarchy Process (AHP). The difference between the two methods is essentially in the network structure as the AHP strictly fits to hierarchical decision trees, and the ANP to problems involving feedbacks and loops between the system's constituents (Saaty 2005). Similarly, to other multicriteria techniques, the ANP was developed as a tool to outline the most important factors affecting a decision and to support decision-making processes involving multiple conflicting criteria in a wide variety of fields, such as finance, engineering processes, international policy, etc. (cfr. Hülle et al. 2013; Sipahi and Timor 2010, for reviews). More recently, the possibility to analyze problems even with incomplete or inconsistent inputs has raised the attention toward the ANP as a method able to address the relevance or influence of various factors in complex environmental evaluations (Huang et al. 2011). Indeed, the ANP is useful when data are lacking or difficult to collect, or in cases where feedback effects are not easily disentangled by the stakeholders/experts involved in the assessment. That feature is linked with the use of the eigenvector method for the calculation of the priority vector and the use of a specific psycho-cognitive rating scale for the pairwise comparisons.

An assessment based on the ANP can be developed in four subsequent phases: (a) identifying the network of elements of a system, (b) pairwise comparisons, (c) assessing the relative importance of the elements, and (d) result validation. This four-phase process is led by a narrowing-down rationale, from more abstract (i.e. the general elements at play within the system under study) to more specific (i.e. scoring the relative importance of the elements and estimating the final weights/ importance). Therefore, through this process the analyst can better approach the problem under analysis.

The first methodological phase consists of building the ANP network, which entails the identification of the most important elements contributing to the targeted system, as well as the clusters of elements (Saaty 2005). Usually, expert knowledge is required to identify the elements that have major relevance and organize the network of relations. This phase is particularly critical as the consistency of the whole analysis depends on considering all the elements that influence the system, but without going beyond an affordable number of elements for the subsequent evaluation. Indeed, failing to identify the elements influencing the system will bias the results and could induce to rank-reversal effects (Harker and Vargas 1987). On the other hand, including a wider number of elements implicates the need to manage long evaluation sessions and raises problems related to cognitive stress for the involved experts. The elements are organized in clusters following the rationale of the system functioning and each cluster should be homogenous under some specific characteristic that is considered relevant and comprehensible by the experts and/or stakeholders. Once the elements are identified and organized in clusters, the network can be finalized sketching the relations linking the different clusters with arrows on the basis of hypotheses about the system functioning. The arrows outline impacts or influences between the clusters of elements and can be uni- or bidirectional. Also, inner feedbacks can be included to outline the relationships between the elements belonging to a specific cluster.

The second phase concerns the pairwise comparisons among the elements of the network. The comparisons are organized in groups gathering the pairwise comparisons of the elements of one cluster with regard to one specific element of a linked cluster. The evaluation is based on the 1-to-9 ratio scale based on the empirical work of Saaty (1980, 2008a, b) which is aimed at capturing the individual's intensity of preference (Table 1).

Considering a simple network with two clusters, a typical question could be "Which element between "Element x1" and "Element x2" belonging to cluster X has more influence on "Element y2" of cluster Y and to what extent?". In the example (Fig. 1), the influence of element x1 is rated as "strongly more important" (i.e. 5-score) than element x2 on the element y2 (cell a_{12}). Reciprocal values in the pairwise comparison matrix are assigned following $a_{ij} = 1/x$, if $a_{ji} = x$ with $x \neq 0$. In the example, (1/5) is hence reported for the influence of element x2 in comparison to element x1 in influencing element y2 (cell a_{21}). The same pairwise comparison process is afterwards applied to assess the influence of each cluster (cluster weights) in the system.

Table 1 The 1-to-9 ANP rating scale

Rating	Definition	Explanation
1	Equal importance	The two compared elements contribute equally to the objective
3	Moderately more important	Experience and judgment slightly favor one element over the other
5	Strongly more important	Experience and judgment strongly favor one element over the other
7	Very strongly more important	Experience and judgment very strongly favor one element over the other
9	Extremely more important	One element is completely dominating the other
2, 4, 6, 8	Intermediate values	

Source: Adapted from Saaty (2005, 2008a, b)

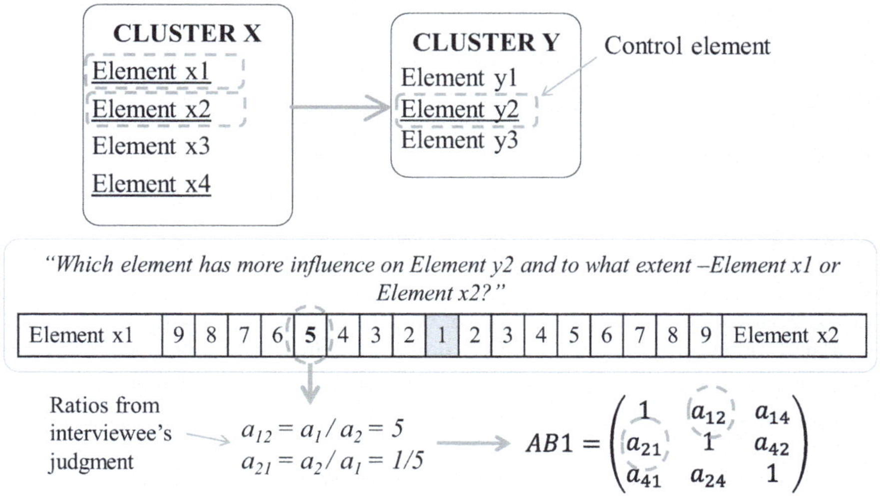

Fig. 1 A detailed pairwise comparison in a simplified ANP assessment. Source: Own elaboration

The third phase regards the calculation of the elements' overall importance in the system: The priority vector. This part can be performed with an *ad-hoc* software (www.superdecisions.com) specifically designed to aid the calculations required by the methodology. In the alternative, the calculation can be performed following the eigenvector techniques as in the sequential steps outlined below.

A matrix of pairwise comparisons and reciprocals is built for each question block. Considering a simple network composed by three elements and their relations as outlined in Fig. 2, the following matrix X of relations is built.

The matrix (Table 2) outlines that the influence on the network of element "A" is respectively two times and six times higher than the influence of elements "B" and

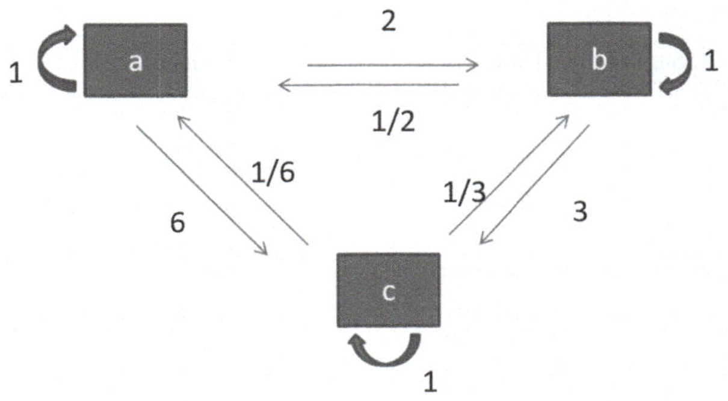

Fig. 2 Network of relations outlining feedbacks and loops between elements. Source: Own elaboration

Table 2 Matrix of relations as outlined in Fig. 2

Matrix X	A	B	C
A	1	2	6
B	1/2	1	3
C	1/6	1/3	1

Source: Own elaboration

Table 3 Matrix derived from inconsistent pairwise comparisons

Matrix A	A	B	C
A	1	2	4
B	1/2	1	3
C	1/4	1/3	1

Source: Own elaboration

"C". The vector could be written as (0.6; 0.3; 0.1) which is the normalized values calculated for an arbitrary column of the matrix. The matrix X is the case of perfect judgment consistency where If $a_{ij} = a_{ik}/a_{jk}$ for all i, j, k and the weights w_i, w_j, w_k from an arbitrary column of the matrix would be the vector of weights of the elements i, j, k.

As inconsistent judgements are usual especially in the analysis of complex systems, Saaty (2005) proposed the right eigenvector method and a related consistency index for the calculation of the priority vector of "near" consistent judgment matrices as AW = λmax W where λmax is the largest eigenvalue of the matrix A, and W is the eigenvector (Saaty 2013). The right eigenvector method is computed by raising and then normalizing the judgment matrix to an arbitrary large power *k* so that the (normalized) column vectors converge to the principal right eigenvector of the matrix (Harker and Vargas 1987). To explain the procedure, consider matrix A (Table 3) as derived from matrix X but with the inclusion of an inconsistent pairwise comparison which does not follow the axiom $a_{ij} = a_{ik}/a_{jk}$ for all i, j, k (i.e. A is two times more important than B, B is three times more important than C, but A is four times more important than C).

Normalizing the columns of the matrix A (with $n \rightarrow +\infty$) *n* and then taking the average of the corresponding entries in the columns the normalized values of the inconsistent matrix A converge to a unique vector that represents the influence

Table 4 Explanations of the right eigenvector methodology

Normalized vectors	Matrix A1	Matrix A2	Matrix A3	Matrix A4	Matrix An $\to +\infty$
A	0.517	0.554	0.5584	0.5584	0.558
B	0.286	0.323	0.3198	0.3196	0.320
C	0.143	0.123	0.1218	0.1220	0.122

Source: Own elaboration

weights of the single elements within each group of pairwise comparisons (Table 4). That vector is also converging to the normalized vector 0.6; 0.3; 0.1 derived by the consistent matrix X and explains why the eigenvector method is able to account for some judgment inconsistencies.

This calculation procedure of the eigenvector refers to the process of averaging over normalized columns (Meade and Sarkis 1999). Readers can consult Harker and Vargas (1987), and Saaty (2005, 2013) for a more extended presentation of the eigenvector method.

The eigenvectors from each question block are then arranged in a so-called "*unweighted supermatrix*", which is weighted according to the cluster weights to obtain the "*weighted supermatrix*". The latter is transformed in the "*limit supermatrix*" following the same eigenvector procedure described above. The normalized columns of the limit supermatrix converge to a same vector which is called "priority vector" which is the output of the ANP method summarizing the average coefficient or weight of each element in the network accounting for all the possible direct and indirect interactions (Harker and Vargas 1987; Saaty 2005).

The last steps aim to aggregate and validate the results and verify the extent to which the ANP output represents and is consistent with the studied system (Saaty 2005). Aggregation of weights consist of a procedure which synthesizes the opinions of different experts/stakeholders following a wide range of methods, including mathematical combinations of individual judgments (e.g. simple aggregation like arithmetic, geometric means, weighted-group means, or more complex rules such as the singular value decomposition, Gass and Rapcsák 1998; or Bayesian-based models, e.g. Lipscomb et al. 1998). Different weighting and aggregation techniques exist and a decisive solution for the selection of the most effective practice has not been attained (Bartolini et al. 2011). Nevertheless, the decision context, a clear articulation of the research structure framing the criteria at stake and the type of expert/stakeholder knowledge available are crucial factors that can help in the selection of the most appropriate elicitation process (Roy and Słowinski 2013).

The result validation can follow different procedures, but a stakeholder panel exercise involving the experts participating to the pairwise comparisons is usually recommended. The main reason is that the ANP results are not evident to the interviewee until the ANP procedure is completed and the priority vector calculated. Discussing the results with the same interviewees is therefore extremely useful for the interpretation of the results and the identification of limitations. The ANP method may involve multiple stakeholders in the process and therefore requires an attentive participatory assessment procedure. That goes beyond the scope of this chapter, but it is relevant to consider typical flaws and biases related to stakeholders' involvement like representativeness, anchoring, power balance, etc. (cfr. Kuhnert et al. 2010).

3 The Analytic Network Process Applied to the Analysis of Public Goods: A Summary of Selected Papers

3.1 Assessing the Role of Economic Actors in the Production of Private and Public Goods in Three EU Agricultural Landscapes (Villanueva et al. 2015)

3.1.1 Background and Rationale

Even though agriculture is still the main form of land-use in the EU, rural areas are undergoing changes in terms of economic structure and contribution of the different economic sectors to employment and added value (SEGIRA 2010). In many cases, the shift is from an agriculture-based economy to a more diversified contribution of tourism and tertiary sectors. It is therefore relevant to assess the impact of different economic sectors on ecosystem services (ES) as these are defined as the benefits for society generated by the environment (MEA 2005). Such assessments are timely in areas affected by huge changes like rural areas (van Zanten et al. 2014), but quantitative evaluations are hampered by the range of direct and indirect effects and trade-offs affecting ES (de Groot et al. 2010). The objective of the paper by Villanueva et al. (2015) is to present an application based on the ANP in three rural landscapes and provide a discussion of the potential of the method. The aim is to evaluate the influence of a range of local economic actors in three different agricultural landscapes covering distinct aspects and trends of agricultural regions in EU, and taking into consideration the feedback effects between ES and socioeconomic benefits.

The selected case study areas were: (a) a marginal area affected by out-migration and abandonment of agricultural activities, (b) intensive agriculture in a highly cultural-environmental value area, and (c) permanent groves shifting toward abandonment or intensive/industrial production. The first case study is located in an alpine valley in Austria (Mittleres Ennstal) where livestock-based agriculture and forestry are the traditional economic activities. Main land-use dynamic is the abandonment of these traditional activities linked to a consistent out-migration. The second case study is located on reclaimed lands in the Po River delta in Italy (Emilia-Romagna, Ferrara Province). Intensive agriculture occupies the major part of the area, but naturalistic stakes are noteworthy (around 30% of the area is part of the Regional Po Delta Park), and cultural heritage aspects are also relevant (the area is a UNESCO site). Main dynamics include urbanization concentrated on the coastal strip, the growing importance of the tourism sector, and a constant concentration of farm holdings. The third case study is located in a rural municipality in Spain

(Montoro, Andalusia). The most important activity is olive growing (with around 20,000 ha) but traditional land-uses such as the agro-forestry system "dehesa" and the "Sierra de Cardeña y Montoro Natural Park" are also relevant in the area. Traditional agricultural land-uses (e.g. mountain olive groves) are important for the conservation of endangered species, but profit reduction and competition from other sectors increase the rate of abandonment of these land-use types (Villanueva et al. 2017a).

3.1.2 Method Application

The application of the ANP was based on the analytical framework proposed by van Zanten et al. (2014) for the analysis of the connections between agricultural landscapes and regional competitiveness and the operational adaptation proposed by Schaller et al. (2015). Basically, the framework employs the ES cascade (Haines-Young and Potschin 2010) to connect the impact of different economic actors on local competitiveness. In this work, the focus was on differentiating the impacts of the main economic sectors on private- and public-type services and evaluate the perceived influence on socioeconomic benefits from local stakeholders. The ANP network consisted of two sub-networks composed by three clusters each: *Economic actors* (Agriculture & forestry, Tourism, Trade & services, Local population), *Private goods-type services* (Supply of food, Production of raw materials), *Public goods-type services* (Protection function, Natural processes, Biodiversity, Cultural services) and *Socioeconomic benefits* (Creation of jobs, Creation of added value, Stability of rural demography, Local investment) (Fig. 3). The structure of the ANP was developed on the basis of a focus group held in each case study and the terminology was defined to avoid misunderstandings between the different areas about the meaning of specific ecosystem services.

To obtain the influence of each economic actor on ES and socioeconomic benefits, the ANP was modified following a weighted matrix approach which consists of a matrix multiplication driven by the target of the analysis. While the ANP is based on the supermatrix to calculate the priority vector, the weighted matrix approach aims to a matrix that synthesizes the flow of influences among the elements of the network and, hence, provides an assessment of specific connections (Lee and Kim 2000; Karsak et al. 2003). In this work, the three matrices of influences of each sub-network (one matrix for each cluster) were multiplied to obtain two matrices representing respectively the influence of economic actors on private and public services. With reference to Fig. 3, the matrix outlining the influence of Economic actors on Private-type services was calculated as

$$PrS = [(w*A)*((1 - w)*S^t)]*P^t \quad (1)$$

where PrS is the matrix of actors' impact on private-type services mediated through the feedback between private-type services and socioeconomic benefits; w is the actor cluster weight; A is the matrix of direct impacts of economic actors on private-

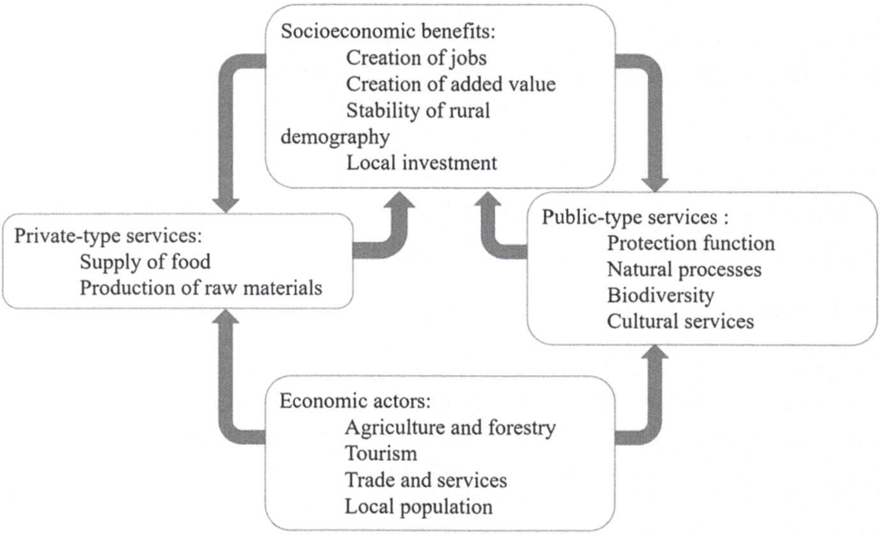

Fig. 3 Network for the analysis of economic actors' impact on private and public services in agricultural landscapes. Source: Adapted from Villanueva et al. (2015)

type services; S^t is the transposed matrix of socioeconomic benefits' impacts on private-type services; and P^t is the matrix of impacts of private-type services on socioeconomic benefits. The same procedure was followed to calculate the impact of economic actors on public-type services.

The two obtained matrices were subsequently normalized and weighted to create a final matrix outlining the contribution of the Economic actors on Socioeconomic benefits through their influence on Private-type and Public-type services.

With regards to data gathering, 28 interviews were performed in the three case studies (8–10 stakeholders from each case study) employing the Saaty (2005)'s ratio scale. The stakeholders were selected in order to cover a homogeneous range of relevant public institutions in each case study (agriculture, tourism, trade and services, local municipalities, and local experts). The control criteria concerned the "valorization of agricultural landscape" and therefore the elicited pairwise comparison focused on "impacts" as a positive influence (to avoid comparisons linked to negative impacts).

Regarding the aggregation method, the elicited priorities were aggregated in each case study to obtain a general assessment attaining to each specific agricultural landscape. As the stakeholders were interviewed separately, the aggregation was based on arithmetic means. Confidence intervals of the estimates were also calculated by means of bootstrapping to compare the results and test statistical differences between the case studies. A stakeholder workshop was finally organized in each case study to discuss the results against the specific features of the case study and provide keys for their interpretation.

3.1.3 Main Results

Slightly more than 50% of the influence on services was attributed to agriculture and forestry in the three case studies (Table 5). That was not a surprise, as the major land-use activity was perceived to have also significant importance for socioeconomic benefits and impacts on services. On the contrary, tourism influence was generally perceived as the lowest in comparison to the other economic actors and that could be related to the lower development of this sector in particular in the Spanish and Austrian areas. Trade and services and local population impacts denoted more evident differences between the case studies. Influence of the local population was the highest in M. Ennstal reflecting the greater involvement of the local population in contributing to ES provision, while trade and services impact were the highest in Montoro.

Whereas the overall impacts on services were related to land-use, the balance between influence on the private-type and public-type services outlined some differences linked to the specific socioeconomic features and agricultural system type of each case study area (Table 6). For instance, the impact of agriculture and forestry on private-type and public-type services was more balanced in M. Ennstal, whereas Montoro and the Po delta evidenced productive-based agricultural systems with lower influence on public services. The link to a more multifunctional landscape in the Austrian case study versus intensive agriculture in the Po delta was evident considering the impacts on natural processes and biodiversity. More nuanced was the Spanish case where a mix of traditional agriculture and intensive olive production was present. The local stakeholders attributed the results to the low adoption of environmentally friendly agricultural practices (despite close to the Natural Park) and in particular the related erosion risks in olive groves. Environmental concerns in the Montoro municipality were also stressed by the low contribution of the other actors to public-type services which characterized the area as a case strongly focused on delivering private-type services. Results were similar in the Po delta, but a sharp difference concerning the protection function was denoted in comparison to the Spanish case. The stronger influence on the protection function was related to the artificial nature of the reclaimed landscape in the Po delta and the vital role of all the actors in preserving the territory. However, the lower contribution to cultural

Table 5 Overall contribution of economic actors to services in the three case studies

	Case study			
Economic actor	M. Ennstal	Po delta	Montoro	Mean (%)
Agriculture and forestry	50[b]	54[a]	49[b]	51
Tourism	15[a]	11[b]	10[c]	12
Trade and services	11[c]	17[b]	24[a]	17
Local population	24[a]	18[b]	17[b]	20

Values are expressed in percentage; values with different letters in a row outline statistically significant differences with $p < 0.05$
Source: Villanueva et al. (2015)

Table 6 Economic actor contribution to private- and public-type services in the three case studies

		Private-good services		Public-good services			
				Environmental services			
Case study	Economic actor	Food	Raw-materials	Protection function	Natural processes	Biodiversity	Cultural services
Mittleres Ennstal	Agriculture and forestry	32^{c}	18^{b}	16^{a}	10^{a}	10^{a}	14^{a}
Po delta		53^{a}	13^{c}	13^{b}	7^{b}	5^{b}	9^{b}
Montoro		46^{b}	26^{a}	6^{c}	4^{c}	6^{b}	12^{a}
Mittleres Ennstal	Tourism	27^{b}	14^{b}	17^{b}	10^{a}	14^{a}	18^{b}
Po delta		43^{a}	9^{c}	18^{a}	9^{a}	7^{c}	14^{c}
Montoro		31^{b}	18^{a}	11^{b}	6^{b}	10^{b}	24^{a}
Mittleres Ennstal	Trade and services	35^{b}	21^{b}	13^{a}	8^{a}	10^{a}	13^{a}
Po delta		54^{a}	12^{c}	14^{a}	6^{b}	5^{b}	9^{b}
Montoro		55^{a}	31^{a}	3^{b}	2^{c}	3^{c}	6^{c}
Mittleres Ennstal	Local population	28^{b}	15^{b}	17^{b}	11^{a}	12^{a}	17^{a}
Po delta		42^{a}	12^{c}	20^{a}	8^{b}	6^{c}	12^{b}
Montoro		40^{a}	21^{a}	8^{c}	5^{c}	8^{b}	18^{a}

Values are expressed in percentage; values with different letters in a column for each economic actor outline statistically significant differences with $p < 0.05$
Source: Villanueva et al. (2015)

services in the Po delta in comparison to the other cases and in particular to M. Ennstal was evident. The result attributed to Cultural services was surprising given the inclusion of the Po delta among the UNESCO sites. Local stakeholders interpreted that result as an indicator of general lack of connections between the "young" agricultural landscape on reclaimed land and the traditional values of the territory which is clearly more connected with tourism activities. Indeed, both M. Ennstal and Montoro highlighted a higher contribution of agriculture and local populations to cultural services in comparison to the Po delta.

3.2 *Analysing the Provision of Agricultural Public Goods: The Case of Irrigated Olive Groves in Southern Spain (Villanueva et al. 2014)*

3.2.1 Background and Rationale

The joint production of private and public goods from farming activities is typically characterized by a high level of complexity derived from the intense relationship between the production of both kinds of outputs (OECD 2001). This complex nature of the joint production processes in agriculture calls for the use of *integrated approaches* to analyze them (Renting et al. 2009; Novo et al. 2015). This work presents a ANP application on a theoretical framework including a clear definition of each public good provided by the system and a priori set of relations explaining these production processes. This framework is based on the causality of producers' decision-making at farm level, allowing to build up a network enabling an ANP application.

The main objective of this work is the development of an integrated approach to analyze the production of public goods by agricultural systems to support public decision-making concerning the design and implementation of policies aiming at the governance of the farming sector. Specifically, it focuses on the analysis of what are the most relevant public goods to be provided by the agricultural system under study from the policy perspective and how farmers' decision-making on structural and management factors of the farm influence the provision of these goods.

The study is carried out in a specific farming system: The irrigated olive groves (IOG) of Andalusia, Southern Spain. Andalusia is the world's main olive oil production region producing roughly 35% of the world's output and IOG systems produce approximately half of it (EC 2012). IOG occupy half a million hectares, is located on low and moderate slopes and is characterized by low tree density (usually less than 140 trees ha^{-1}), the use of drip irrigation methods at low doses (around 1500 m^3 ha^{-1} per year) (Gómez-Limón et al. 2013). The analysis of IOG is highly pertinent due to its enormous expansion during the last two decades and the relevant environmental and socio-cultural impacts of this process. This expansion essentially results from the high profitability of the crop, which had a lot to do with high olive oil market prices and agricultural policy implementation, both boosting intensification processes, as well as water-management technological improvements and crop's low water needs (Gómez-Limón and Arriaza 2011). Nevertheless, this process of expansion and intensification of olive production involves important negative impacts both on the environment (including soil erosion, biodiversity decline, water resources overexploitation, and non-point water pollution, among others—see Gómez Calero 2009) and on sociocultural aspects such as cultural heritage and deterioration of traditional landscape (Guzmán Álvarez 2004). Also, positive impacts from the economic and social points of view may have resulted from the expansion of IOG, as these agricultural system is characterized by highly value-added production and intensive labour-use (Viladomiu and Rosell 2004). Therefore, this expansive trend together with its environmental and social impacts demands a deeper understanding from the perspective of its multifunctional performance.

3.2.2 Method Application

The application of the ANP method to the case study has been developed in three steps. The first step was the network design, which consists of three clusters: *Public Goods*, *Structural Factors* and *Management Factors* (see Fig. 4; Table 7). The elements of the last two clusters depended on farmer's decisions in the long and short term, respectively. In fact, the cluster of *Management Factors* groups the agricultural practices that are decided within a single season, including productive and non-productive decisions, while the cluster of *Structural Factors* includes farmer's decisions that can only be modified in the long term. Also, according to the theoretical framework (see Table 7), the cluster of *Public Goods* contains the most relevant public goods for our case study, considering a novel theoretical

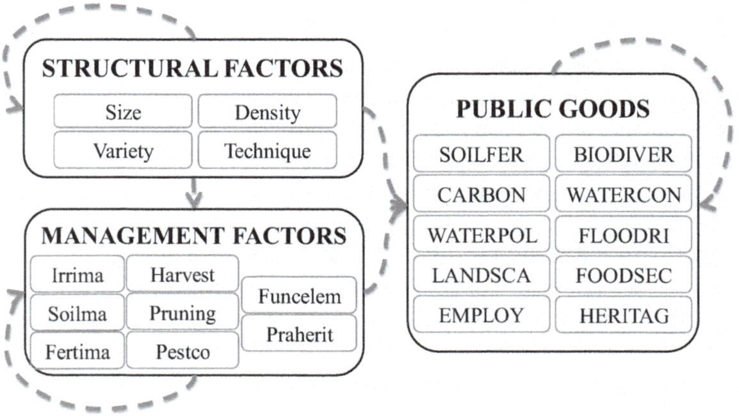

Fig. 4 ANP network for the analysis of public goods provided by IOG (see Table 7 for elements description). Source: Villanueva et al. (2014); reproduced with permission of the Editor

Table 7 Description of the ANP network elements

Cluster	Element	Brief description
Public Goods	*CARBON*	Carbon balance: GHG emissions and carbon sequestration (in trees and soil)
	WATERPOL	Water pollution (nutrients, pesticides and soil sediments)
	WATERCON	Irrigation water consumption
	FLOODRI	Flooding risk at the basin level (or sub-basin level)
	BIODIVER	Biodiversity associated to irrigated olive farmlands, excluding off-farm effects
	SOILFER	Soil physical, chemical and structural properties regarding its long-term fertility
	EMPLOY	Rural employment (limited to that produced inside the farm)
	FOODSEC	Contribution to food security (olive oil production in quantity and quality)
	HERITAG	Rural cultural heritage, including material (constructions, buildings, etc.) and immaterial (traditional food production, traditions, habits, etc.)
	LANDSCA	Visual quality of the rural landscape
Management Factors	*Fertima*	Fertilization management
	Irrima	Irrigation management
	Soilma	Soil management (including pruning residues and herbicides managements)
	Pruning	Pruning practices at olive groves
	Pestco	Pest & disease control (incl. mgmt. of phytosanitary products, except herbicides)
	Harvest	Harvesting practices
	Funcelem	Practices related to functional elements (hedges, riparian vegetation, terraces, etc.)
	Praherit	Practices related to management of material and immaterial cultural heritage
Structural Factors	*Technique*	Cultivation technique, considering only conventional and integrated
	Variety	Variety of the olive tree used
	Density	Tree density
	Size	Farm olive grove area

Source: Villanueva et al. (2014); reproduced with permission of the Editor

framework specifically developed for the analysis of the public goods provided by agricultural systems (see Villanueva et al. 2014, for further explanation).

The second step established feedback and dependency connections among them (i.e. the interaction matrix). This was done through a deliberative process that ended with the consensus among the authors and a significant proportion of the panel of experts consulted to support this research (see below). From the network already shown in Fig. 4, two interaction matrices were defined: (i) *matrix of received influences*; and (ii) *matrix of influences exerted*. This represents a novelty in comparison to the standard ANP applications. As Saaty (2008b) explains, there are two approaches of the interactions, namely "being influenced" and "influencing" (equivalent to our "received influences" and "influences exerted", respectively) and, depending on the way the user understands the problem, one approach should be chosen to ensure consistency of final results. In the IOG case, both results were relevant: the *received influences* approach shows the public goods productions that can be influenced by structural and management factors; whereas the *influences exerted* approach reveals which farmer's decisions are more influential. It is worth clarifying that Fig. 4 represents the ANP network for the *influences exerted* approach; the same figure would represent the ANP network for the *received influences* approach if opposite direction arrows were applied.

In the third step, 28 unweighted supermatrices were obtained from each questionnaire: 14 supermatrices of "received influences" and another 14 of "influences exerted". Matrix calculations were performed with the *SuperDecisions 2.2.3.0* software (for further details see Saaty 2005).

The data gathering consisted of interviewing the experts using two questionnaires (elicited from the two interaction matrices) including the following pairwise comparison question-types:

(a) An example of the *matrix of received influences*: "*Biodiversity of the olive groves*" (*BIODIVER*) and "*Soil fertility*" (*SOILFER*) are both influenced by farmer's decisions concerning "*Soil management*" (*Soilma*); which one is more influenced by those decisions and to what extent?
(b) An example of the *matrix of influences exerted*: Farmer's decisions concerning "*Pest and disease control*" (*Pestco*) and "*Soil management*" (*Soilma*) influence "*Biodiversity of the olive groves*" (*BIODIVER*); which one has more influence on it and to what extent?

The linear 9-point scale was used for answering the pairwise-comparisons (cfr. Sect. 2).

A pre-test was run to verify the network design, check the correct understanding of the questions, and to refine the number of connections among the elements, omitting the least relevant.

Regarding the composition of the panel of experts, 28 specialists were interviewed (14 for each questionnaire), including 6 olive production, researchers (AGR), 8 researchers in ecology and environmental sciences (ENV) 7 researchers in economics and other social sciences (ECO), as well as 7 agricultural training and extension specialists (TEC), all of them directly involved in irrigated olive growing systems. The interviews were carried out during January–March 2013.

As an aggregation method, the aggregation of individual priorities (AIP) (instead of judgments, AIJ) was used following Forman and Peniwati (1998), who recommend its use when it is considered that experts act as individuals (we interviewed them separately) instead of as a unit. They also recommend the use of arithmetic or geometric mean when experts are considered to be of equal importance (as is the case here). The former is used as it better fits our problem since extreme values (i.e. zeros) were frequent in the experts' priorities.

3.2.3 Main Results

In this section, the results concerning the public goods produced by the IOG from the *received influences* approach, and the structural and management factors from the *influences exerted* approach are discussed.

Public Goods (Received Influences Approach)

Table 8 shows the results related to the influence capacity of olive growers regarding the public goods provision. The final weights obtained show that productions of the public goods most modifiable by farmer's decisions at farm level in IOG are soil fertility (*SOILFER*), visual quality of agricultural landscapes (*LANDSCA*) and farmland biodiversity (*BIODIVER*), retaining 24.3%, 18.0% and 17.1%, respectively, of the total influence produced by olive growers' decision-making in this agricultural system. Other public goods whose provision can be affected by these farmers' decision-making are carbon balance (*CARBON*, 10.5%), irrigation water consumption (*WATERCON*, 9.9%) and the contribution to the food supply (*FOODSEC*, 7.6%). The olive growers' capacity to influence the production of the other public goods considered is rather limited, with less than 5% in each case.

It is worth noting that the four most influenced public goods by IOG producers are of "public good/bad type", that is, depending on what decisions the olive grower makes, a public good or bad will be produced. This can be explained by considering that these public goods are the ones with the widest range of possible production levels, as opposed to "strictly good" (e.g. *EMPLOY*) or "strictly bad" (e.g. *WATERPOL*) public goods, where the amount produced can vary between narrower thresholds.

Structural and Management Factors (Influences Exerted Approach)

As can be observed in Table 9, the *Structural Factors* are more influential than the *Management Factors* regarding the production of the public goods. In particular, 85% of the production of such goods depending on the olive grower decision-making in the long term (*Structural Factors*). Among these factors, it is worth highlighting farm size (*Size*), tree density (*Density*) and olive variety (*Variety*).

Density and *Size* are clearly the most influential factors, as these two factors influence the production of public goods both directly and indirectly through their influence on *Management Factors*. Regarding tree density, it must be pointed out that this is a typical indicator of extensification/intensification of olive growing (Viladomiu and Rosell 2004). Hence, there is a certain consensus among the panel of experts regarding the negative relationship between *Density* and environmental public goods production in IOG, except for *CARBON*, and a positive relationship

Table 8 Influence capacity on the production of Public goods provided by IOG in Andalusia (in %)

Scope	Public good	Experts														Mean (std. dev.)
		AGR1	AGR2	AGR3	ENV1	ENV2	ENV3	ENV4	ECO1	ECO2	ECO3	ECO4	TEC1	TEC2	TEC3	
Environmental	*SOILFER*	12.2	19.3	26.8	24.3	22.4	34.6	37.6	19.4	14.1	14.1	30.7	35.0	30.8	18.7	24.3 (8.4)
	BIODIVER	18.1	13.2	18.5	21.9	12.2	8.4	10.8	15.0	14.6	18.1	25.6	21.2	23.3	18.3	17.1 (5.0)
	CARBON	2.5	7.0	7.5	5.5	13.9	28.8	25.6	10.3	7.1	4.4	6.9	12.2	6.8	8.8	10.5 (7.7)
	WATERCON	5.2	11.2	15.6	6.5	11.2	8.2	10.0	12.1	14.2	10.1	10.7	6.9	8.8	8.0	9.9 (2.9)
	WATERPOL	2.8	7.4	3.0	2.8	5.0	3.3	6.0	5.7	3.3	5.3	6.0	0.9	2.7	3.8	4.1 (1.8)
	FLOODRI	2.0	12.7	2.6	3.4	4.5	1.9	2.8	3.1	2.2	4.6	1.8	6.9	4.4	1.7	3.9 (2.9)
Socio-cultural	*LANDSCA*	37.7	13.9	8.0	23.2	17.1	9.5	4.2	23.1	23.9	27.7	11.7	10.8	14.2	27.4	18.0 (9.4)
	FOODSEC	7.3	11.4	12.3	6.3	10.3	3.7	2.1	8.9	13.3	9.8	5.0	3.7	5.1	7.6	7.6 (3.5)
	EMPLOY	10.7	2.8	4.0	5.6	2.4	1.3	0.8	1.7	6.1	4.5	1.5	2.2	2.5	4.3	3.6 (2.6)
	HERITAG	1.6	1.2	1.8	0.4	0.9	0.2	0.1	0.7	1.2	1.4	0.2	0.3	1.4	1.5	0.9 (0.6)

Source: Villanueva et al. (2014); reproduced with permission of the Editor

Table 9 Influential factors regarding the production of Public goods provided by IOG in Andalusia (in %)

Cluster	Factor	Experts														Mean (std. dev.)
		AGR4	AGR5	AGR6	ENV5	ENV6	ENV7	ENV8	ECO6	ECO7	ECO8	TEC4	TEC5	TEC6	TEC7	
Management factors	*Fertima*	0.0	6.3	1.6	16.1	12.5	0.0	4.3	10.1	0.0	0.4	0.0	0.0	3.6	15.9	5.1 (6.1)
	Irrima	0.0	6.3	2.4	14.6	1.8	0.0	5.1	9.0	0.0	0.6	0.0	0.0	4.0	8.8	3.8 (4.5)
	Soilma	0.0	4.9	3.6	4.0	14.6	0.0	4.6	5.6	0.0	0.9	0.0	0.0	4.3	10.5	3.8 (4.4)
	Harvest	0.0	2.9	4.2	0.3	3.9	0.0	5.1	2.2	0.0	0.4	0.0	0.0	4.5	1.9	1.8 (2.0)
	Pruning	0.0	2.9	1.2	1.0	0.9	0.0	0.7	2.2	0.0	0.9	0.0	0.0	1.4	1.9	0.9 (0.9)
	Pestco	0.0	0.0	0.0	0.0	0.0	0.0	0.0	0.0	0.0	0.0	0.0	0.0	0.0	0.0	0.0 (0.0)
	Funcelem	0.0	0.0	0.0	0.0	0.0	0.0	0.0	0.0	0.0	0.0	0.0	0.0	0.0	0.0	0.0 (0.0)
	Praherit	0.0	0.0	0.0	0.0	0.0	0.0	0.0	0.0	0.0	0.0	0.0	0.0	0.0	0.0	0.0 (0.0)
Structural factors	*Density*	47.7	25.5	22.0	24.1	19.4	48.3	24.6	28.0	46.4	29.4	46.3	46.4	23.8	19.4	32.2 (11.8)
	Size	8.7	43.6	56.5	27.7	35.5	6.5	46.9	27.3	13.4	55.7	13.4	13.4	49.7	32.4	30.8 (17.7)
	Variety	43.6	6.4	6.5	8.6	3.5	45.2	7.9	14.5	40.2	11.4	40.3	40.2	5.8	3.6	19.8 (17.3)
	Technique	0.0	1.3	2.0	3.7	7.9	0.0	0.7	1.0	0.0	0.3	0.0	0.0	2.9	5.5	1.8 (2.4)

Source: Villanueva et al. (2014); reproduced with permission of the Editor with some socio-cultural public goods like *FOODSEC* and *EMPLOY*. In relation to *Size*, experts highlight that larger farms usually perform better than smaller ones regarding environmental public goods, given the former are more efficient in inputs use and are more prone to adopt conservation practices. According to the experts' opinion, this statement is also valid regarding socio-cultural public goods like *FOODSEC* and *HERITAG*, but not to *EMPLOY* and *LANDSCA*.

Regarding the *Management Factors*, the most influential short-term decisions made by the producer in terms of public goods production are: fertilization (*Fertima*), irrigation (*Irrima*) and soil managements (*Soilma*). These three factors influence the production of at least six public goods, coinciding in five of them, namely: *BIODIVER, CARBON*, *FOODSEC*, *WATERPOL* and *EMPLOY*. Other influential *Management Factors* are harvest (*Harvest*) and pruning (*Pruning*) practices, but with a relatively lower influence. Surprisingly, pest and disease control (*Pestco*) does not appear to be an influential factor. In part, due to its substantial influence on only one public good (*BIODIVER*) and the low variability in the pest control treatments that olive growers carry out.

As regards non-productive *Management Factors* (*Funcelem* and *Praherit*), no influence in the production of agricultural public goods is observed, mainly due to the absence (or notable scarcity) of the elements or components associated with each factor (buffer strips, margin vegetation, terraces, buildings, etc.) on irrigated olive farms (Gómez-Limón and Arriaza 2011).

3.3 Farm-Level Multifunctionality Associated with Farming Techniques in Olive Growing: An Integrated Modelling Approach (Carmona-Torres et al. 2014)

3.3.1 Background and Rationale

Evaluating multifunctionality of agriculture (MFA) poses problems of high complexity, uncertainty and risk. The complexity of the interactions between the functions of agriculture and agro-ecological processes, institutional conditions and technical restrictions (Zander et al. 2008) prevents from building excessively simple models of reality if it is wished to avoid considerable loss of information. The evaluation of MFA is a social decision-making process in which the conflicting values and interests of different groups and communities must be considered (Martinez-Alier et al. 1998). This process is affected by uncertainty due to the lack of hard data for many relevant processes and interactions. It also involves an element of risk because of what is at stake, since decisions usually affect not only the current generation but also those that follow, and many of these decisions, especially those affecting the environment, may be irreversible (Funtowicz and Ravetz 1993).

In this context, the objective of the work is to develop an integrated model for the evaluation of MFA based on the ANP. This integrated model evaluates the multifunctional performance at farm-level according to the farming techniques implemented by different farmers. The model is built on the basis of expert knowledge and an extensive review of the international literature, and draws from empirical data gathered from a survey of farmers in the main olive oil producing zones of Andalusia.

Olive agriculture is a strategic sector of the economy and the socio-territorial cohesion of Andalusia. The olive is the single most important crop in the region, covering 32% of the agricultural area (MARM 2010) and providing 28% of Andalusian fruit and vegetable production (CAP 2012). Moreover, it is a social cultivation that generates around one third of agricultural employment, of which approximately 47% is family-based (CAP 2009). In addition to its economic and social importance, olive agriculture has a high potential for affecting the environment in the region due to its wide territorial presence, influencing the welfare of Andalusian society significantly (Parra-López et al. 2008a). Most of the olive crop is cultivated in a traditional extensive manner, although an increasing surface area (ca. 15%) is cultivated more intensively with a massive use of productive inputs (see Sect. 3.2). Moreover, a growing surface area is now dedicated to alternative methods of agriculture, such as Integrated Production (Parra-López 2003), and certified quality systems such as Protected Designation of Origin (PDO). These alternatives to conventional techniques are adopted by around 16% of Andalusian olive growers (Hinojosa-Rodríguez et al. 2014).

3.3.2 Method Application

The application of the ANP method is based on a dynamic network definition involving three phases: (1) defining and clustering of the most relevant farming practices and technical alternatives, and their potential functions; (2) *a priori* definition of the potential relationships between alternatives based on literature; and (3) pre-testing and refining the proposed structure and relationships. The final structure of the model consists of: (a) Cluster of Functions (C_F), containing 11 relevant farm-level functions of agriculture relative to the economic, social and environmental dimensions; (b) Cluster of Farming practices (C_P), consisting of 22 olive farming practices, referring to 7 main groups of practices in olive growing ranging from planting to pruning; and (c) Cluster of Alternatives (for farming practices) (C_A), which consists of the technical alternatives for each farming practice. For instance, for practice P1, 'Olive variety', there are five technical alternatives, ranging from A1(P1) 'Picual' to A5(P1) 'Picudo'. A farming technique would be the use of the technical alternative 'Picual' for the farming practice 'olive variety'.

The relationships between the elements of the ANP model are represented as a super-matrix (Table 10) where $W_{A,F}$ accounts for the outer contribution of the alternatives for each farming practice (A) to achieve each function (F). $W_{F,F}$ represents the inner relationships between functions. Finally, I is a unity submatrix which shows that the alternatives for the farming practices, i.e. the farming techniques, are inner independent from a dominance/contribution point of view. Thus, these alternatives are technically outer restricted, meaning that a particular

Table 10 Supermatrix of the ANP model

			C_A: Alternatives (for farming practices)		
		C_F: Functions	P_1	...	Pp
C_F: Functions		$\mathbf{W_{F,F}}$	0		
C_A: Alternatives (for farming practices)	P1	$\mathbf{W_{A,F}}$	**I**		
	...				
	Pp				

combination of farming techniques to be jointly implemented, referred to henceforth as a 'farming pack', is conditioned by its technical feasibility. For instance, to irrigate or not to irrigate are independent alternatives. However, it is impossible not to irrigate and still apply fertilizers through irrigation water (fertirrigation).

The control submatrix consists of just one sub-matrix, $\mathbf{W_{P,F}}$, which reflects the different contribution of the practices to each function. It is needed to weight the column vectors of $\mathbf{W_{A,F}}$, since the practices may not all contribute equally to the achievement of a given function. For instance, olive variety may be more/equal/less important than soil management to achieve a high olive yield. To facilitate the evaluation of relationships between elements, some parts of the columns of the super- and control sub-matrix were set *a priori* at zero if the relations they represented were not possible based on previous literature or due to technical restrictions.

The assessment of the relationships between the elements was based on expert knowledge due to the technical nature of the information required and the lack of previous *ad hoc* hard data in most of the topics analyzed. 27 experts of Andalusian olive systems were selected to cover a wide range of specialization and profession fields (agricultural economics, olive soil management, olive pests and diseases, olive growing, fertilization, olive quality, and organic olive production). The experts individually filled in the parts of the supermatrix and the control submatrix for which they had knowledge and expertise. They were asked to evaluate the relationships, though 'direct rating' since the number of elements to compare exceeded the recommended limit for many items, e.g. 22 practices for each of the 11 functions. In particular, the rating scale was used to evaluate the strength of the relationships ranging from 1 (very weak) to 9 (very strong), reserving 0 for the absence of any relationship (Parra-López et al. 2008b). For instance, in the matrix $\mathbf{W_{A,F}}$, the alternative A1(P1) Picual could be related strongly to the function F1, lower production costs, according to an expert ($W_{A1(P1),F1}$ would be 9 in Table 10).

The arithmetic mean of individual priorities (AIP) method (Forman and Peniwati 1998; Gómez-Limón and Atance 2004) was selected as aggregation method because it is recommended for social problems among the diverse aggregation options (Ramanathan and Ganesh 1994). For the supermatrices and the control matrices the AIP is: $w_{i,j(aggr)} = \sum_{\forall i,j} w_{i,j(e)}/E$, where $w_{i,j}$ is an element of a submatrix; e is an expert; and E is the number of experts. Relationships not evaluated by individual experts due to their lack of knowledge were not included when calculating the AIP.

Finally, the submatrix was weighted and normalized thus obtaining the weighted supermatrix. The farm-level multifunctional performance associated with a given farming pack was calculated by synthesizing the information contained in the weighted supermatrix. The particular mathematical properties of reducibility, primitivity and cyclicity of this supermatrix (Kahraman et al. 2006; Karsak et al. 2003; Lee and Kim 2000; Parra-López et al. 2008b; Saaty and Takizawa 1986) make it advisable to carry out this synthesis as follows:

1. Calculate the matrix of interdependent relationships between the alternatives for practices and the functions ($\mathbf{W}_{\mathbf{A,F}}'$): This matrix represents the contribution of the farming techniques to each function once the inner relationships between functions have been taken into consideration. It is defined as: $\mathbf{W}_{\mathbf{A,F}}' = \mathbf{W}_{\mathbf{A,F(aggr,weigh)}} \cdot \mathbf{W}_{\mathbf{F,F(aggr)}}$.
2. Calculate the performance on the multiple functions associated with a farming pack ($\mathbf{pf}_{\mathbf{(pack)}}$): This is a row vector defined as: $\mathbf{pf}_{\mathbf{(pack)}} = \mathbf{pack.W}_{\mathbf{A,F}}'$, where **pack** is a row vector of the farming techniques that define the farming pack. For instance, in our model, **pack** = (1,0,0,0,0; 1,0,0,0; ...; 1,0) means the combined use of Picual as the variety, bare soil through constant tillage as the soil management technique, ... and traditional pruning.

Subsequently, the maximum (or minimum) performances achievable for each function F can be defined by implementing the optimal (or worst) technical alternatives for each farming practice that maximizes (or minimize) performance of this function and that constitute a technically feasible combination. In this way, 11 optimal (or worst) farming packs can be defined, one for each function. These maximum and minimum values define the range of potential performance achievable for each function in current technological conditions and according to our model. This allows evaluation of the farm-level performance in each function associated with a given farming pack in relative terms, by assigning 1 to the maximum performance and 0 to the minimum performance. In this way, the value of the performance associated with a farming pack in any function will range from 0 to 1. Furthermore, the economic performance of a given farming pack can be calculated as the average of its performances for the economic functions (F1–F3). The same process can be applied to social (F4–F5) and environmental (F6–F11) performances. Finally, the global performance associated with a farming pack can be calculated as the average of its economic, social and environmental performances.

As a novelty, to delimit the most sensitive farming practices, i.e. those with the highest potential to affect the multifunctional performance of olive growing at farm-level, a 'sensitivity index' $S_{(P,F)}$ is proposed for a given practice *P* over a function *F*, as follows:

$$S_{(P,F)} = \sum\nolimits_{\forall a_{(P,F)} \neq amax_{(P,F)}} \frac{\left[pf\left(amax_{(P,F)}\right) - pf\left(a_{(P,F)}\right)\right]}{(A_P - 1)} \quad (2)$$

where $pf(amax_{(P,F)})$ is the maximum performance achievable in function F by changing only the technical alternatives for farming practices P; this is achieved for this function when implementing the technical alternative $amax_{(P,F)}$ for this practice; $pf(a_{(P,F)})$ is the performance in function F when implementing the technical alternative $a_{(P,F)}$, other than $amax_{(P,F)}$; and A_P is the number of alternatives for practice P, including $amax_{(P,F)}$. The S index measures the variation of the performance in a given function relative to the maximum performance when changing the technical alternatives for a given practice. Furthermore, for a given farming practice it is possible to calculate its S index of the economic dimension as the average of its S indices for the economic functions. The same applies to the social and environmental S indices. The 'mean sensitivity' of a farming practice is the average of the economic, social and environmental S indices.

3.3.3 Results

Sensitive Farming Practices and Best Technical Alternatives

Soil management—P2—is by far the most sensitive farming practice with a mean sensitivity of 0.2618. Any change in the way soil management is carried out, i.e. in the technical alternative used, could substantially increase/decrease the performance of olive growing in almost all the functions analyzed, especially in the environmental ones. In this sense, the technical alternative 'soil covered by plants'—A4(P2)—produces, *ceteris paribus*, the maximum performance by far in almost all the functions compared to the three 'bare soil' options (Fig. 5). The only exception is product quality—F3—which is not influenced by the soil management technique (S index = 0.0000). Quality is defined on the basis of oil yield and the organoleptic, physical, chemical and nutritional attributes of the olive oil subsequently produced. Of the bare soil techniques, little or shallow tillage—A3(P2)—is the best alternative, especially for the environmental functions. Conventional tillage—A1(P2)—, i.e. constant tillage, is the worst technical alternative (Fig. 5). Irrigating—P3—, the group of fertilization practices—GP4—, phytosanitation—P11—are also very sensitive farming practices (Fig. 5).

Average and Optimal Farming Techniques and Their Multifunctional Performance

Using the results from the ANP application, impacts of different olive growing farming techniques can be assessed. Particularly, current average techniques are compared to past average techniques—implemented 10 years ago—to analyze the evolution of multifunctionality over time. To determine the current average farming techniques for the year 2011, a survey of 400 olive farmers of the main olive oil producing zones—Jaen, Cordoba and Granada—was carried out. For the past average techniques, data from Parra-López and Calatrava-Requena (2006) was used. Moreover, the optimal farming techniques, one optimal farming pack for each function, were defined according to the standardization procedure proposed.

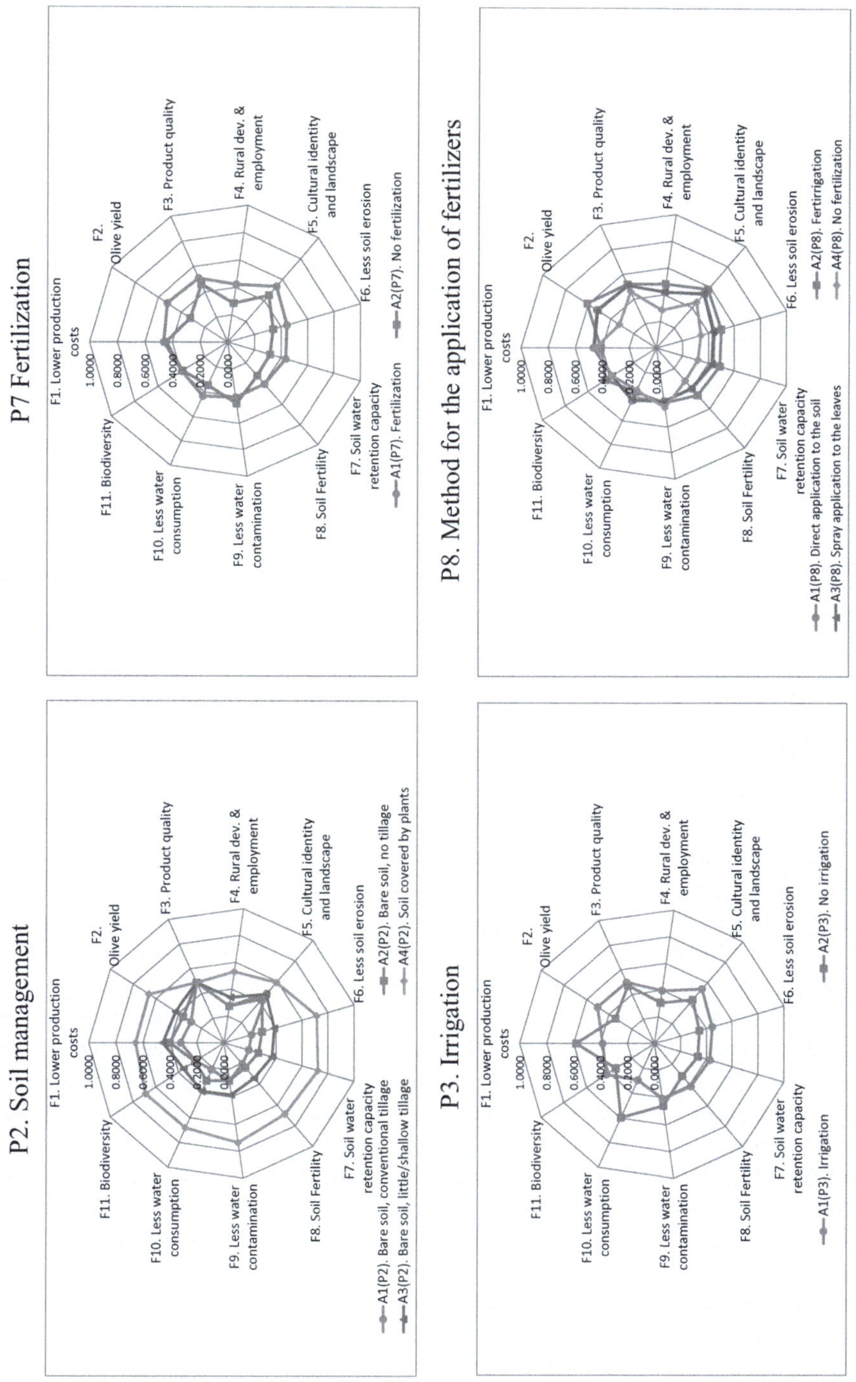

Fig. 5 Multifunctional performance associated with the highly sensitive farming practices. Source: Carmona-Torres et al. (2014)

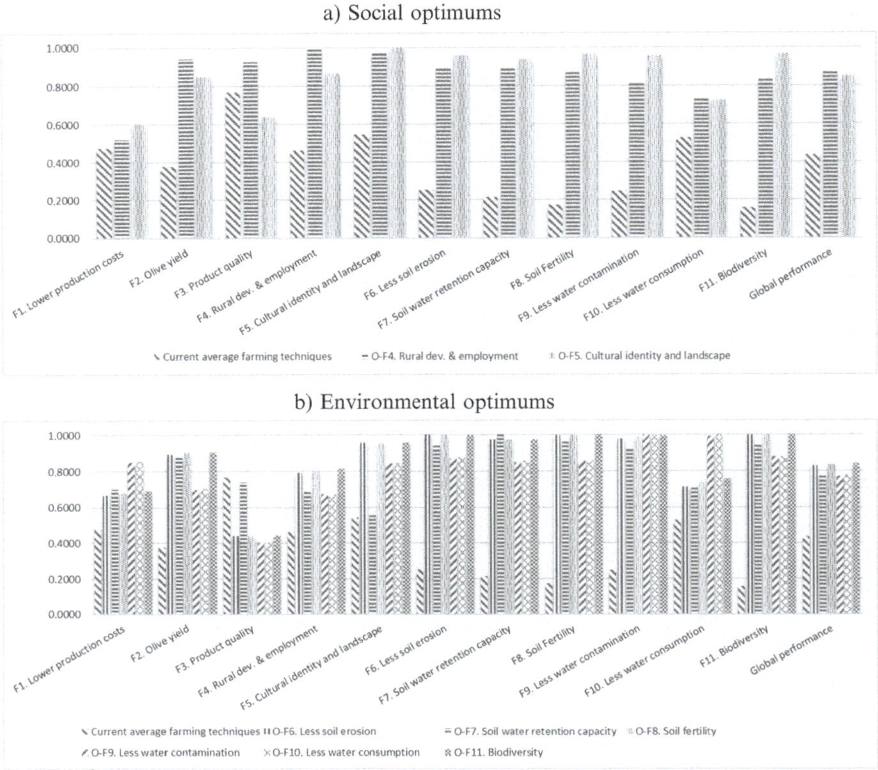

Fig. 6 Multifunctional performance associated with the average and optimal farming techniques. Source: Carmona-Torres et al. (2014)

The results indicate that there is much room for improvement in current farming techniques (Fig. 6). Especially, farming techniques today are far from being socially and environmentally optimal. The environmental functions are currently, on average, far from being fully achieved. The optimization of 'less soil erosion'—O-F6—would require these changes with respect to the current situation in the highly sensitive practices: use of soil cover instead of bare soil, irrigation, fertirrigation instead of spraying leaves, use of organic fertilizers instead of inorganic ones, no phytosanitation, and hand-pole beating the olives from trees instead of using vibrators. The environmental functions seem to be highly correlated: if one of them is optimized, the rest also achieves very high performances, close to maximum performance levels (Fig. 6b). This relationship is especially notable for less soil erosion—F6—, soil fertility—F8—and biodiversity—F11—. Therefore, if less soil erosion is maximized, as for the proposed optimal pack O-F6, high soil fertility and biodiversity are also achieved. The positive effects on the environmental functions are due to using soil cover, avoiding phytosanitary treatments, using organic fertilizers and irrigation and fertirrigation. On the other hand, the maximization of the environmental functions entails a reduction in olive quality—F3—due to the lack of phytosanitation, and lower costs performance—F1—due to the need for irrigation and hand-pole beating.

The optimal pack for maximizing the social function of rural development and employment—O-F4—seems to be one of the best optimal packs, since it achieves a relatively high performance (over 0.8000) in most of the functions (Fig. 6a). This optimal farming pack would entail these changes in the highly sensitive practices currently implemented: use of soil cover instead of bare soil, irrigating, fertirrigation instead of spraying the leaves with fertilizers, use of organic fertilizers instead of inorganic ones, and not using machinery for collecting olives from the trees. An especially superior performance was detected for O-F4 in the social functions due basically to soil cover, irrigation and the method of collecting olives from the trees, and in olive yield due to soil cover, irrigation and fertirrigation with organic fertilizers. On the other hand, the production costs for this optimum O-F4 are far from the best score (Fig. 6a) due to the handpicking of olives from the trees and irrigation, despite the improvement brought by the use of soil cover.

4 Discussion

The selected papers present applications of the ANP to the relation between agriculture and public goods at different scale of analysis.

The first paper compares the role of agriculture in three different agricultural landscapes and outlines that farming activity is still perceived as the most relevant actor in those case study areas. The influence of agriculture on private and public services was however diverse across the case studies and connected to the specific features of the agricultural and socioeconomic context. For instance, the multifunctional role of extensive-traditional agriculture in M. Ennstal underlined a balanced, positive impact on private and public services including cultural services, whereas in more intensive systems private services were the main “target” of agriculture. That supports the necessity to consider that the impact on public goods related to changes of the agricultural sector will strongly depend on the local socioeconomic context. The difficulties to understand the role of environmental services could for example hamper the generation of socioeconomic benefits in more agriculture-intensive areas.

The second paper presents a study at farming system level aiming at enhancing the policy efficiency concerning the provision of public goods from irrigated olive groves in Andalusia. The study pointed out the most influenced public goods that should be focused to increase policy efficiency (i.e. *SOILFER*, *LANDSCA*, *BIODIVER* and *CARBON*). The study also showed the little room for effective incentives for other less influenced public goods (e.g. *WATERPOL* or *EMPLOY*) which are focused by society’s demand and that would incur in non-efficient public intervention. The results suggest that policies oriented to influence *structural* and *management factors* should consider their impacts on public goods and in particular the complementarity between the supply of the most influenced public goods (i.e. *SOILFER*, *LANDSCA*, *BIODIVER* and *CARBON*).

The third paper contributes to modelling and evaluation of multifunctionality at farm-level with an integrated assessment of the functions of alternative farming techniques. The results outlined changes to improve the farming techniques implemented in the Andalusian olive growing sector: using soil cover instead of bare soil, irrigating, fertirrigation instead of spraying the leaves with fertilizers, and using organic fertilizers instead of inorganic ones. Furthermore, some of the identified optimal farming packs highlighted the possibility to improve economic performance together with social objectives and environmental protection. Therefore, the study concluded that agricultural policies favoring rural development may indirectly be reinforcing the diffusion of a multifunctional agriculture compatible with rural development, employment, and conservation of soil, water and biodiversity.

The selected papers underline the usefulness of the ANP method to the study of public goods provision of agricultural systems thanks to the prospect to integrate the tangible and non-tangible issues at play in a common evaluation framework. A particular advantage of the ANP is the possibility to account for feedbacks and a wide range of cause-effect interrelations characterizing public goods and agricultural activities. Compared to other expert/stakeholder-based assessment, the ANP is able to elicit knowledge in contexts of scarce information and complexity with a lower risk to generate a high cognitive stress and consequently reduce the quality of results. Indeed, the pairwise comparison procedure combined with the 1–9 measurement scale allows to cope with complex problems reducing the judgment efforts. As shown by these papers, ANP is therefore helpful to channel the knowledge of experts and stakeholders on complex questions, and allows to deal with analyses requiring the integration of different knowledge sources such as stakeholders' and "hard" data. Indeed, the selected papers prove the flexibility of the ANP to carry out assessments at different conceptual levels, from more abstract (e.g. Villanueva et al. 2015) to more in-detail (e.g. Carmona-Torres et al. 2014). In the same direction, it is also possible to appreciate the flexibility in the balance between economic or value dimensions, and more technical representations of cause-effect relationships. Indeed, the method fits to the combination with participatory approaches, also showing good potential for combination with other methods, such as mechanistic modelling and stated preferences methods, by dealing with qualitative, subjective, and intangible information, which is typically beyond their scope.

In this regard, it is important to notice how an intensive process characterized by a strong stakeholder participation and framework and network development was a relevant issue with all the presented papers, as this is of special importance with regards to the analysis of public goods provided by agricultural systems (Villanueva et al. 2017b; Novo et al. 2017). Relevant aspects concern the role and engagement of stakeholders from the beginning of the analysis. That entails the need to perform a careful selection of experts/stakeholders committed and interested to participate in the assessment like for other participatory methods. It may be also very relevant to include the right balance between stakeholders able to give a value judgement and experts able to give a technical insight about functional connections among items. The validation step was also relevant in the presented works to provide keys for interpretation by the stakeholders and obtain consistent and coherent results from the analyses. The validation was important in particular to understand potential limitations and biases, get suggestions for e.g. sensitivity analyses, and represented an opportunity to stimulate discussion between the stakeholders. Inputs from stakeholders were also of primary importance to build a network and make a balance between depth and length of the questionnaire. In this context, the possibility to build more complex networks and ask the stakeholders to answer to specific sections of questionnaires (instead of full questionnaires) could be an interesting development which deserves further research and a deeper understanding of whether partial answering is appropriate. That also relates to the need of enhanced techniques for aggregating the elicited judgments, for example by incorporating the level of specialization each respondent has on the specific field s/he is giving her/his judgments for.

However, according to the results of the selected papers, there are some limitations of the ANP method which are worth noting. For instance, the ANP results usually reflect only one fixed point in time for 'average' conditions and therefore may be affected by circumstances and context. In this sense, as for any other static approach, the fact that the work by Villanueva et al. (2015) was developed in a period characterized by economic crisis could have affected perceptions and opinions of the stakeholders and hence their final results. Also, results from Villanueva et al. (2014) and Carmona-Torres et al. (2014) refer to the average situation (e.g. average farm), so they fail to provide information on the heterogeneity of situations. However, to partly overcome this limitation, the innovative sensitivity analysis used in Carmona-Torres et al. (2014) may represent an option to show some information about deviations from that average results. Also, the use of fuzzy approaches (Vinodh et al. 2016) and combined approaches [such as ANP with agent-based models, e.g. Knoeri et al. (2011), and with spatial analysis, e.g. Cervelli et al. (2016)] would help to deal with heterogeneity in ANP applications. Clearly, further research is needed to develop the method to analyze the provision of public goods by agricultural systems, which is heterogeneous by definition.

In addition, compared to purely economic models, ANP can be insufficient to get to straightforward policy recommendations in terms of incentive structure. The second case study is a nice example of how some policy related discussion can be necessary to complement ANP (as most multicriteria tools) but at the same time can highly benefit from insights from ANP itself.

5 Conclusions

The ANP is effective in providing an approach to the analysis of the public goods provided by agricultural systems as it is (a) holistic (i.e. involving complete consideration of the main factors at play), dealing with complexity and including relevant aspects and facets to manage uncertainties, contradictions and inconsistencies; (b) systemic, i.e. incorporating the underlying mechanisms that link the multiple functions of agriculture to the characteristics of farming systems, and to forecast results for new scenarios; (c) integrative, dealing with models of different levels of complexity ranging from statistical to expert-based; and (d) transdisciplinary, incorporating knowledge and facilitating the interchange of information between diverse scientific disciplines, and between scientists and non-academic stakeholders, such as managers, administrators, and the general public. This has been evidenced here by critically discussing three recently published papers using the ANP to investigate the agricultural provision of public goods, with each of them providing insights based on distinct approaches and representing good proofs of the adaptive capacity of the method to different contexts and theoretical frameworks.

However, this method also shows some limitations for the analysis of public goods, especially with regards to excessive simplification of complex systems, including the assumption of average conditions and the use of static approach, among others. Therefore, the integration of the ANP with other methods is highly recommended to provide a more complete assessment and help to better understand processes and dynamics of public good provision from agricultural systems. Indeed, the use of the ANP as a complementary tool together with methods focusing on specific disciplinary aspects may disclose the full potential of the ANP as a "descriptive" tool able to improve the understanding of a system and enhance the range of results of an integrated assessment.

Acknowledgements This chapter is derived in part from articles published in: the Journal of Environmental Planning and Management 2015 copyright Taylor & Francis, available online: http://www.tandfonline.com/doi/abs/10.1080/09640568.2014.1001022; Land Use Policy 2014 copyright Elsevier available online: https://doi.org/10.1016/j.landusepol.2013.11.018; Agricultural Systems 2014 copyright Elsevier available online: https://doi.org/10.1016/j.agsy.2014.02.001.

References

Bartolini, F., Gallerani, V., & Viaggi, D. (2011). What do agro-environmental measures actually promote? An investigation on AES objectives for the EU 2000–2006 rural development program. *Spanish Journal of Agricultural Research, 9*(1), 7–21.

Bennett, E. M. (2017). Research frontiers in ecosystem service science. *Ecosystems, 20*, 31–37.

CAP. (2009). *Documento de reflexión: Ley del Olivar*. Seville: Consejería de Agricultura y Pesca (CAP). Available at: http://www.infaoliva.es/documentos/documentos/Documento%20de%20Reflexion%20Ley%20del%20Olivar%20Produccion%20Olivar.pdf

CAP. (2012). *Evolución macromagnitudes agrarias provinciales. 2005–2010*. Seville: Consejería de Agricultura y Pesca (CAP). Available at: http://www.cap.junta-andalucia.es/agriculturaypesca/portal/servicios/estadisticas/estadisticas/agrarias/macromagnitudes-y-economia-agraria.html

Carmona-Torres, C., Parra-López, C., Hinojosa-Rodríguez, A., & Sayadi, S. (2014). Farm-level multifunctionality associated with farming techniques in olive growing: An integrated modeling approach. *Agricultural Systems, 127*, 97–114.

Carmona-Torres, C., Parra-López, C., Sayadi, S., & Chirosa-Ríos, M. (2016). A public/private benefits framework for the design of polices oriented to sustainability in agriculture: An application to olive growing. *Land Use Policy, 58*, 54–69.

Cervelli, E., Pindozzi, S., Capolupo, A., et al. (2016). Ecosystem services and bioremediation of polluted areas. *Ecological Engineering, 87*, 139–149.

de Groot, R., Alkemade, S. R., Braat, L., et al. (2010). Challenges in integrating the concept of ecosystem services and values in landscape planning, management and decision making. *Ecological Complexity, 7*(3), 260–272.

EC (European Commission). (2012). *Economic analysis of the olive sector*. Brussels: European Commission.

FAO. (2011). FAOSTAT webpage. http://faostat.fao.org

Forman, E., & Peniwati, K. (1998). Aggregating individual judgments and priorities with the analytic hierarchy process. *European Journal of Operational Research, 108*, 165–169.

Funtowicz, S., & Ravetz, J. R. (1993). *Espistemología política Ciencia con la gente*. Buenos Aires: Centro Editor de América Latina.

Gass, S. I., & Rapcsák, T. (1998). A note on synthesizing group decisions. *Decision Support Systems, 1*(22), 59–63.

Gómez Calero, J. A. (2009). *Sostenibilidad de la producción de olivar en Andalucía*. Seville: Consejería de Agricultura y Pesca.

Gómez-Limón, J. A., & Arriaza, M. (2011). *Evaluación de la sostenibilidad de las explotaciones de olivar en Andalucía*. Málaga: Analistas Económicos de Andalucía.

Gómez-Limón, J. A., & Atance, I. (2004). Identification of public objectives related to agricultural sector support. *Journal of Policy Modeling, 26*, 1045–1071.

Gómez-Limón, J. A., Arriaza, M., & Villanueva, A. J. (2013). Typifying irrigated areas to support policy design and implementation: The case of the Guadalquivir river basin. *Irrigation and Drainage, 62*(3), 322–329.

Gowdy, J., & Erickson, J. D. (2005). The approach of ecological economics. *Cambridge Journal of Economics, 29*, 207–222.

Guzmán Álvarez, J. R. (2004). *El palimpsesto cultivado. Historia de los paisajes del olivar andaluz*. Seville: Consejería de Agricultura y Pesca.

Haines-Young, R. H., & Potschin, M. B. (2010). The links between biodiversity, ecosystem services and human well-being. In D. G. Raffaelli & C. L. J. Frid (Eds.), *Ecosystem ecology: A new synthesis* (pp. 110–139). Cambridge: Cambridge University Press.

Hanley, N., Banerjee, S., Lennox, G. D., & Armsworth, P. R. (2012). How should we incentivize private landowners to 'produce' more biodiversity? *Oxford Review of Economic Policy, 28*(1), 93–113.

Harker, P. T., & Vargas, L. G. (1987). The theory of ratio scale estimation: Saaty's analytic hierarchy process. *Management Science, 33*(11), 1383–1403.

Hart, K., Baldock, D., Weingarten, P., et al. (2011). *What tools for the European agricultural policy to encourage the provision of public goods?* Brussels: European Parliament.

Hinojosa-Rodríguez, A., Parra-López, C., Carmona-Torres, C., et al. (2014). Certified quality systems and farming practices in olive growing: The case of integrated production in Andalusia. *Renewable Agriculture and Food Systems, 29*, 291–309.

Huang, I. B., Keisler, J., & Linkov, I. (2011). Multi-criteria decision analysis in environmental sciences: Ten years of applications and trends. *Science of the Total Environment, 409*, 3578–3594.

Hülle, J., Kaspar, R., & Möller, K. (2013). Analytic network process – An overview of applications in research and practice. *International Journal of Operational Research, 16*(2), 172–213.
Kahraman, C., Ertay, T., & Buyukozkan, G. (2006). A fuzzy optimization model for QFD planning process using analytic network approach. *European Journal of Operational Research, 171*, 390–411.
Karsak, E. E., Sozer, S., & Alptekin, S. E. (2003). Product planning in quality function deployment using a combined analytic network process and goal programming approach. *Computers & Industrial Engineering, 44*, 171–190.
Knoeri, K., Binder, C. R., & Althaus, H. J. (2011). An agent operationalization approach for context specific agent-based modelling. *Journal of Artificial Societies and Social Simulation, 14*(2), 4.
Kuhnert, P. M., Martin, T. G., & Griffiths, S. P. (2010). A guide to eliciting and using expert knowledge in Bayesian ecological models. *Ecology Letters, 13*, 900–914.
Lee, J. W., & Kim, S. H. (2000). Using analytic network process and goal programming for interdependent information system project selection. *Computers & Operations Research, 27*, 367–382.
Lipscomb, J., Parmigiani, G., & Hasselblad, V. (1998). Combining expert judgment by hierarchical modeling: An approach to physician staffing. *Management Science, 44*(2), 149–161.
Madureira, L., Santos, J. L., Ferreira, A., & Guimarães, H. (2013). *Feasibility study on the valuation of public goods and externalities in EU agriculture*. Luxembourg: Publications Office of the European Union.
MARM. (2010). *Anuario de Estadística 2010* [Statistics Yearbook 2010]. Madrid: Ministerio de Medio Ambiente y Medio Rural y Marino (MARM). Available at: http://www.magrama.gob.es/estadistica/pags/anuario/2010/ae_2010_avance.pdf
Martinez-Alier, J., Munda, G., & O'Neill, J. (1998). Weak comparability of values as a foundation for ecological economics. *Ecological Economics, 26*, 277–286.
MEA, Millennium Ecosystem Assessment. (2005). *Ecosystems and human well-being: Synthesis*. Washington, DC: Island Press.
Meade, L. M., & Sarkis, J. (1999). Analyzing organizational project alternatives for agile manufacturing processes: An analytical network approach. *International Journal of Production Research, 37*(2), 241–261.
Nekhay, O., Arriaza, M., & Boerboom, L. (2009). Evaluation of soil erosion risk using analytic network process and GIS: A case study from Spanish mountain olive plantations. *Journal of Environmental Management, 90*(10), 3091–3104.
Novo, P., Faccioli, M., & Byg, A., et al. (2017). *Guidelines and report on initial interviews and workshops* (Deliverable 2.3). PROVIDE Project (No. 633838), Brussels.
Novo, P., Slee, B., & Byg, A., et al. (2015) *Conceptual paper on the 'unpacked' notion of public goods* (Deliverable 2.2). PROVIDE Project (No. 633838), Brussels.
OECD, Organisation for Economic Co-operation and Development. (2001). *Multifunctionality: Towards an analytical framework*. Paris: OECD Publications Service.
OECD, Organisation for Economic Co-operation Development. (2003). *Multifunctionality in agriculture: The policy implications*. Paris: OECD Publishing.
Parks, S., & Gowdy, J. (2013). What have economists learned about valuing nature? A review essay. *Ecosystem Services, 3*, e1–e10.
Parra-López, C. (2003). *Sistemas de producción ecológica, integrada y convencional en olivar: Estudio de difusión de innovaciones y evaluación multifuncional* [Organic, integrated and conventional agricultural systems in olive grove: A diffusion of innovations analysis and multifunctional assessment]. PhD Thesis University of Córdoba, Spain.
Parra-López, C., & Calatrava-Requena, J. (2006). Comparison of farming techniques actually implemented and their rationality in organic and conventional olive groves in Andalusia. *Biological Agriculture & Horticulture, 24*, 35–59.
Parra-López, C., Calatrava-Requena, J., & de-Haro-Giménez, T. (2008a). A systemic comparative assessment of the multifunctional performance of alternative olive systems in Spain within an AHP-extended framework. *Ecological Economics, 64*, 820–834.

Parra-López, C., Groot, J. C. J., Carmona-Torres, C., & Rossing, W. A. H. (2008b). Integrating public demands into model-based design for multifunctional agriculture: An application to intensive Dutch dairy landscapes. *Ecological Economics, 67*, 538–551.
Porter, J., Costanza, R., Sandhu, H., et al. (2009). The value of producing food, energy, and ecosystem services within an agro-ecosystem. *Ambio, 38*(4), 186–193.
Ramanathan, R., & Ganesh, L. S. (1994). Group preference aggregation methods employed in AHP: An evaluation and an intrinsic process for deriving members' weightages. *European Journal of Operational Research, 79*, 249–265.
Reig, E., Aznar, J., & Estruch, V. (2010). A comparative analysis of the sustainability of rice cultivation technologies using the analytic network process. *Spanish Journal of Agricultural Research, 8*(2), 273–284.
Renting, H., Rossing, W. A. H., Groot, J. C. J., et al. (2009). Exploring multifunctional agriculture. A review of conceptual approaches and prospects for an integrative transitional framework. *Journal of Environmental Management, 90*(2), S112–S123.
Roy, B., & Słowinski, R. (2013). Questions guiding the choice of a multicriteria decision aiding method. *EURO Journal on Decision Processes, 1*, 69–97.
SEGIRA. (2010). *Study on employment, growth and innovation in rural areas: Main report*. Report Prepared for DG Agriculture and Rural Development. Available at: http://ec.europa.eu/agriculture/analysis/external/employment/
Saaty, T. L. (1980). *The analytic hierarchy process*. New York: McGraw-Hill.
Saaty, T. L. (2005). *Theory and applications of the analytic network process: Decision making with benefits, opportunities, costs, and risks*. Pittsburgh: RWS Publications.
Saaty, T. L. (2008a). Decision making with the analytic hierarchy process. *International Journal of Services Sciences, 1*(1), 83–98.
Saaty, T. L. (2008b). Relative measurement and its generalization in decision making why pairwise comparisons are central in mathematics for the measurement of intangible factors the analytic hierarchy/network process. *Revista de la Real Academia de Ciencias Exactas, Fisicas y Naturales – Serie A: Matematicas, 102*(2), 251–318.
Saaty, T. L. (2013). The modern science of multi-criteria decision making and its practical applications: The AHP/ANP approach. *Operations Research, 61*(5), 1101–1118.
Saaty, T. L., & Takizawa, M. (1986). Dependence and independence: From linear hierarchies to nonlinear networks. *European Journal of Operational Research, 26*, 229–237.
Schaller L, Kantelhardt J, Viaggi D, et al. (2015) Agricultural landscapes as a driver for socio-economic benefits in rural regions – Empirical results of the EU project 'CLAIM'. *Paper presented at the 29th International Congress of Agricultural Economists*, University of Milan, Milan, 8–14 August 2015.
Sipahi, S., & Timor, M. (2010). The analytic hierarchy process and analytic network process: An overview of applications. *Management Decision, 48*(5), 775–808.
Spangenberg, J. H., & Settele, J. (2010). Precisely incorrect? Monetising the value of ecosystem services. *Ecological Complexity, 7*, 327–337.
van Zanten, B. T., Verburg, P. H., Espinosa, M., et al. (2014). European agricultural landscapes, common agricultural policy and ecosystem services: A review. *Agronomy for Sustainable Development, 34*(2), 309–325.
Viladomiu, L., & Rosell, J. (2004). Olive oil production and the rural economy of Spain. In F. Brouwer (Ed.), *Sustaining agriculture and the rural economy: Governance, policy and multifunctionality*. Cheltenham: Edward Elgar Publishing.
Villanueva, A. J., Gómez-Limón, J. A., Arriaza, M., & Nekhay, O. (2014). Analysing the provision of agricultural public goods: The case of irrigated olive groves in southern Spain. *Land Use Policy, 38*, 300–313.
Villanueva, A. J., Targetti, S., Schaller, L., et al. (2015). Assessing the role of economic actors in the production of private and public goods in three EU agricultural landscapes. *Journal of Environmental Planning and Management, 58*(12), 2113–2136.

Villanueva, A. J., Gómez-Limón, J. A., & Rodríguez-Entrena, M. (2017a). Valoración de la oferta de bienes públicos por parte de los sistemas agrarios: El caso del olivar de montaña en Andalucía. *Economía Agraria y Recursos Naturales, 17*(1), 25–57.

Villanueva, A. J., Rodríguez-Entrena, M., & Gómez-Limón, J. A., et al. (2017b). *Report on selection of valuation tools and guidelines* (Deliverable 4.1). PROVIDE Project (No. 633838), Brussels.

Vinodh, S., Sai Balagi, T. S., & Patil, A. (2016). A hybrid MCDM approach for agile concept selection using fuzzy DEMATEL, fuzzy ANP and fuzzy TOPSIS. *The International Journal of Advanced Manufacturing Technology, 83*(9), 1979–1987.

Young, O. R., Lambin, E. F., Alcock, F., et al. (2006). A portfolio approach to analyzing complex human-environment interactions: Institutions and land change. *Ecology and Society, 11*(2), 31.

Zander, P., Groot, J. C. J., Josien, E., et al. (2008). Farm models and economic valuation in the context of multifunctionality: A review of approaches from France, Germany, The Netherlands and Portugal. *International Journal of Agricultural Resources, Governance and Ecology, 7*, 339–360.

Allocating Shadow Prices in a Multi-objective Chance Constrained Problem of Biodiesel Blending

Carla Caldeira, Luis Dias, Fausto Freire, Dimitris Kremmydas, and Stelios Rozakis

Abstract Biodiesel can be produced from different vegetable oils and the choice of the blend (mix of oils) to be used for biodiesel production has an important impact on its cost and environmental performance. This chapter presents a model that determines the optimal blend that minimizes production costs and GHG emissions and assesses the influence of technical constraints on the decision objectives. For this purpose, an algorithm for the allocation of shadow prices to the constituent parts of the composite objective function was implemented. The technical constraints in the model control biodiesel properties based on the feedstock's chemical composition, taking into account inherent compositional uncertainty. The information obtained from the shadow prices allowed the identification of which technical constraint limits GHG reduction and cost effectiveness. Thus, the model can be used for evaluating the effects of technical progress or policy mandatory measures relatively to the cost and GHG emissions of the biodiesel production process.

C. Caldeira (✉) · F. Freire
ADAI-LAETA Department of Mechanical Engineering, Faculty of Sciences and Technology, University of Coimbra, Coimbra, Portugal
e-mail: carla.caldeira@dem.uc.pt; fausto.freire@dem.uc.pt

L. Dias
INESC Coimbra and CeBER, Faculty of Economics, University of Coimbra, Coimbra, Portugal
e-mail: lmcdias@fe.uc.pt

D. Kremmydas
Department of Agricultural Economics and Rural Development, Agricultural University of Athens, Athens, Greece
e-mail: Kremmydas@aua.gr

S. Rozakis
School of Environmental Engineering, Technical University of Crete, Chania, Greece

Department of Bioeconomy and Systems Analysis, Institute of Soil Science and Plant Cultivation, Pulawy, Poland
e-mail: srozakis@isc.tuc.gr

J. Berbel et al. (eds.), *Multicriteria Analysis in Agriculture*, Multiple Criteria Decision Making, https://doi.org/10.1007/978-3-319-76929-5_5

Keywords Biodiesel blends · Uncertainty · Chance constrained programming · Shadow price decomposition · Multi-objective programming · GHG emissions · Costs

1 Introduction

Biodiesel is typically produced from vegetable oils as feedstocks that represent between 80% and 85% of the total production costs (Gülşen et al. 2014). Many types of vegetable oils can be used in biodiesel production and diversification in feedstock blending may reduce costs while maintaining biodiesel quality. Gülşen et al. (2014) provided evidence that a skillful selection of a diversified portfolio of feedstocks at the conversion phase can provide a significant financial advantage and stabilize overall costs, reducing financial risk.

The implementation of policies such as the European Directive on Renewable Energy (RED) (European Comission 2009) led to a significant increase of biofuels production so that the share of renewable energy to be used in transportation imposed by the Directive could be achieved. However, the controversy raised about GHG emissions of biofuels, mainly due to direct and indirect land use change (Soimakallio and Koponen 2011), has forced the biodiesel industry to take into account not just the costs but also GHG emissions (Buratti et al. 2012; Tomaschek et al. 2012). Models that accommodate GHG constraints and cost of biofuel chains have been used to analyze trade-offs between costs and GHG emissions (You et al. 2012; Akgul et al. 2012; Thomas et al. 2013; Bairamzadeh et al. 2016). Focusing on policy analysis, Palak et al. (2014) used a model to assess the impacts of carbon regulatory mechanisms on emission and cost performance of biofuel supply chains. Caldeira et al. (2014) developed a bi-objective mathematical programming model to determine the blend of virgin oils for biodiesel production minimizing costs and life-cycle GHG emissions to analyze the trade-offs between these two dimensions.

However, none of these studies considered the uncertainty of the feedstock's chemical composition, which influences the biodiesel quality. Indeed, biodiesel properties are highly influenced by the compositional uncertainty of the feedstock (Caldeira et al. 2017) and, for this reason, the cost and GHG dimensions should be examined taking into account the feedstock composition uncertainty. The latter was considered in the work developed by Gülşen et al. (2014) and Olivetti et al. (2014) using Chance Constrained Programming (CCP) to assess, respectively, cost effectiveness and GHG emissions uncertainty. However, this was not done considering a multi-objective approach.

To ensure that the biodiesel has the requisite quality to be used as automotive diesel fuel, standard specifications for biodiesel have been established worldwide (Hoekman et al. 2012). However, some specifications vary from country to country or can change with the technical evolution of engines. Therefore, it is important that policy makers are aware of the opportunity costs associated with technical specifications imposed. In mathematical programming blending models this piece of information is embedded in the dual or shadow prices associated to the constraints.

The goal of this chapter is to present an approach to determine optimal blends (minimizing biodiesel production costs and GHG emissions) considering technical uncertainties and to provide detailed shadow price information useful for policy analysis. To determine optimal blends we have used a multi-objective programming model. To inform decision makers about the opportunity costs of biodiesel technical specifications, we have implemented a special approach for assigning the sensitivity on marginal changes of each different objective caused by the right hand side value of the imposed technological constraints.

The multi-objective formulation of the blending problem is presented in Sect. 2.1 and the shadow price decomposition method for multi-objective problems in Sect. 2.2. To incorporate the feedstock compositional uncertainty within the required technical specifications we used CPP. The methodological issues concerning the joint application of multi-objective programming and CCP to the blending problem are presented in Sect. 2.3. The model is illustrated with a case study for a two-objective (costs and GHG emissions) biodiesel blending problem in Portugal (Sect. 3). Results are presented in Sect. 4 along with discussion, analyzing the constraints on fuel specifications currently enacted in the EU and the US. Conclusive comments in Sect. 5 complete the chapter.

2 Material and Methods

2.1 The Deterministic Multi-objective Blending Problem

Blending problems consist in determining the combination of raw materials that leads to the optimal value of the objective function. The general mathematical formulation for a multi-objective blending problem can be written as:

$$\begin{aligned}
&\min\left\{\sum_{i\in I}(c_{ki}\,x_i) : k \in K\right\}\\
&\text{st.} \sum_{j\in J}\sum_{i\in I}\left(q_{ji}\,x_i\right) \le b_p \quad \forall\ p \in P\\
&\sum_{i\in I} x_i = D\\
&x_i \ge 0\ \forall\ i \in I
\end{aligned} \tag{1}$$

where K is the set of objectives, I is the set of raw materials, J is the set of ingredients, P is the set of regulated properties of the final blend which are functions of its ingredient composition, c_{ki} are the individual objective coefficients, x_i is the raw material quantity (decision variable), q_{ji} is the concentration of j-ingredient in i-raw material, b_p is the limit of p-property, and D is the demand.

The main characteristic of multi-objective problems is that the concept of optimal solution gives place to the one of non-dominated (Pareto efficient) solutions, i.e., those solutions that cannot be improved in one objective without worsening at least one of the other objectives. Multi-objective methods can be classified into three categories: the *a priori* methods, the interactive methods and the generation or *a posteriori* methods (Hwang and Masud 1979). In *a priori* methods the decision maker expresses his or her preferences before the solution process (e.g. setting goals or weights for the objective functions) so that a most preferred solution is identified without no further involvement of the decision maker. In interactive methods, phases of dialogue with the decision maker alternate with computation phases, iteratively computing new solutions until the most preferred solution is identified. In generation methods the efficient solutions of the problem (all of them or a sufficient representation) are generated and then the decision maker may compare them in order to select the most preferred one, or may simply explore the trade-offs involved thus supporting the decision process.

In this work we opted for an *a posteriori* approach implementing the "weighting method" to generate the Pareto efficient solutions, minimizing a weighted sum of the K objectives for several weight vectors. Although the weighting method's caveats are known and more sophisticated algorithms are proposed in the literature (Mavrotas 2009), it is appropriate for this case study since a large number of alternative solutions is provided so that the stylized blending problem adequately illustrates the shadow price decomposition. For the blending problem presented in (1), the mathematical formulation of the weighting method corresponds to a single-objective optimization model as follows:

$$\begin{aligned} &\min \sum_{k \in K} \left(w_k \sum_{i \in I} (c_{ki} x_i) \right) \\ &\text{st.} \sum_{j \in J} \sum_{i \in I} \left(q_{ji} x_i \right) \geq b_p \ \forall \ p \in P \\ &\sum_{i \in I} x_i = D \\ &x_i \geq 0 \ \forall \ i \in I \end{aligned} \tag{2}$$

The w_k represents the weight of the individual objectives and problem (2) is solved for various weight combinations (such that $w_k > 0$ and $\sum_{k \in K} w_k = 1$) assigned to the objectives in order to obtain the Pareto efficient set of solutions.

2.2 Decomposing Shadow Price for the Various Objective Function Components

The shadow prices generated by linear programming models represent the objective value change for a unit change on the Right Hand Side (RHS) value of a certain constraint (Cohon 1978). In resource allocation problems, the shadow price of a resource constraint can be interpreted as the maximum value the decision maker is willing to pay for obtaining an additional unit of that resource. In blending problems, the shadow price represents the improvement in the objective function for relaxing a requirement of the final blend.

The allocation of the shadow price information has been applied in joint production: Nejad M. (2007) proposed a two-stage methodology based on the marginal contribution of oil products and the production elasticity of unit processes to provide an additive CO_2 allocation scheme in joint product industries; Moghaddam and Michelot (2009) presented a methodology to use the shadow price information for joint cost allocation; and revenue loss from decreasing nitrogen pollution was estimated by Shaik et al. (2002).

In multi-objective problems the interpretation of shadow prices can be useful for decisions in policy and industry. For this type of problems, the shadow prices resulting from solving Model (2) give the marginal change in the weighted objective function for a unit change on the RHS value of a constraint. This type of information, an aggregate measure difficult to interpret, is not particularly useful for decision makers since they are rather more interested in the distinct effect on the individual objectives that compose the multi-objective function. To overcome this issue, McCarl et al. (1996) presented a technique to decompose the dual values of binding constraints in multi-objective problems that allows the allocation of the shadow price information to each specific objective. We applied this technique to the blending problem. A description of the technique is presented in the following paragraphs.

The weighting form of the problem given in Model (2) in matrix notation is transcribed in vector form in (3) to illustrate the decomposition process:

$$\begin{gathered}\min\left(w_1\,\overrightarrow{C_1}+w_2\,\overrightarrow{C_2}+\ldots+w_k\,\overrightarrow{C_k}\right)\cdot\overrightarrow{x}\\ I_{PxJ}Q\,\overrightarrow{x}\geq\overrightarrow{b}\\ \overrightarrow{x}\geq 0\end{gathered}\tag{3}$$

where $\overrightarrow{C_k}$ is a $1\times I$ vector containing the objective coefficients for the k-objective, $\overrightarrow{x}$ is the $I\times 1$ vector of the decision variables, I_{PxJ} is an $P\times J$ unity matrix, Q is a $J\times I$ matrix containing the q_{ji} elements, $\overrightarrow{b}$ is a $P\times 1$ vector containing the property limits.

The decomposed form of the objective function of the problem that is given in Eq. (3) is equal to:

$$\min\overrightarrow{C_f}\cdot\overrightarrow{x},\tag{4}$$

where $\overrightarrow{C_f}=w_1\cdot\overrightarrow{C_1}+w_2\cdot\overrightarrow{C_2}+\ldots+w_k\cdot\overrightarrow{C_k}$.

We know that the shadow prices are given by

$$\overrightarrow{U_k}=\overrightarrow{C_{fB}}\cdot B^{-1}\tag{5}$$

where $\overrightarrow{C_{fB}}$ is a vector that contains the objective functions coefficients for the basic variables of the optimal solution and B^{-1} is the basis inverse. From Eq. (4) we obtain:

$$\overrightarrow{C_{fB}} = w_1 \overrightarrow{C_{1B}} + w_2 \overrightarrow{C_{2B}} + \ldots + w_k \overrightarrow{C_{kB}} \tag{6}$$

where $\overrightarrow{C_{1B}}, \overrightarrow{C_{2B}}, \ldots, \overrightarrow{C_{kB}}$ are the coefficients of the basic variables in the individual objectives context.

Thus, the shadow prices are equivalent to

$$\overrightarrow{U_k} = \overrightarrow{C_{fB}} \cdot B^{-1} = w_1 \cdot \overrightarrow{C_{1B}} \cdot B^{-1} + w_2 \cdot \overrightarrow{C_{2B}} \cdot B^{-1} + \ldots + w_k \cdot \overrightarrow{C_{kB}} \cdot B^{-1} \tag{7}$$

The $\overrightarrow{C_{kB}} \cdot B^{-1}$ component indicates the decomposed shadow prices corresponding to the k-th objective. It is a $1 \times P$ dimension vector, expressing the extent at which one unit of increasing the p-th constraint will affect the k-objective's value, considering that we are already at the optimal solution.

So we need to contrive a way to compute each $\overrightarrow{C_{kB}} \cdot B^{-1}$ in order to complete the decomposition of the shadow prices. However, as McCarl et al. (1996) states: "linear programming solvers do not generally yield the basis inverse". Furthermore, computing the basis inverse from scratch would be computationally equivalent to solving again the LP problem. The algorithm proposed consists of the following steps: Solve the composite problem and save the basis. For each k-th objective set $w_k = 1$ and all other weights equal to zero. Load the saved basis of the composite problem and startup the problem (but make no iterations). The reported shadow prices are given by the $\overrightarrow{C_{kB}} \cdot B^{-1}$ product.

This method cannot be directly applied to problems that exhibit degeneracy. Degenerate problems are a special case as far as sensitivity analysis is regarded (Koltai and Terlaky 2000). Although McCarl et al. (1996) gives a technical solution for computing a consistent decomposition of the shadow prices, degenerate problems are expected to have different positive and negative shadow prices (Gal 1986) representing a diverse effect on the optimal price of an increase versus a decrease on the right hand side of a constraint. In this case different approaches would be more appropriate as discussed in Ho (2000).

2.3 *Introducing Uncertainty in the Constraints: Chance Constraint Programming (CPP)*

A critical aspect in blending problems is the stochastic nature of the composition of the raw materials. The consideration of chemical composition uncertainty in blending processes using CCP has been considered by several authors (Kumral 2003; Rong et al. 2008; Sakallı and Baykoç 2013). CCP is a stochastic programming technique that was first presented by Charnes and Cooper (1959) to address system feasibility in an uncertain environment, which is expressed as a requirement on the minimum probability of satisfying constraints (Sahinidis 2004). By controlling the probability that a constraint may be violated, it adds flexibility to the model reflecting the reality under consideration (Kampempe 2012). The advantages of the CCP approach in the development of blending models in relation to deterministic ones are presented by Olivetti et al. (2011). According to the authors, the CCP model formulation always performs better or equal than Linear Risk formulation (LR). The CCP formulation allows increasing the variation while still meeting technical specifications because it identifies portfolios of raw materials whose uncertainty characteristics are better than that of any individual raw material. The creation of these portfolios of raw materials allows to manage risk and cost simultaneously (Olivetti et al. 2011). In this approach, the deterministic constraints are replaced by non-deterministic ones. First, the decision maker specifies a minimum probability of $1 - \alpha$ that each constraint should satisfy:

$$P\left(\sum_{i=1}^{N} a_i x_i \leq b\right) \geq 1 - \alpha, \qquad x_i \geq 0 \text{ and } 0 < \alpha < 1 \tag{8}$$

If a_i is normally distributed parameter, $a_i \sim N(\mu_i, \sigma_i^2)$ and all a_i are independent, the constraint is converted as follows:

$$P\left(\frac{\sum_{i=1}^{N} a_i x_i - \sum_{i=1}^{N} \mu_i x_i}{\sqrt{\sum_{i=1}^{N} \sigma_i^2 x_i^2}} \leq \frac{b - \sum_{i=1}^{N} \mu_i x_i}{\sqrt{\sum_{i=1}^{N} \sigma_i^2 x_i^2}}\right) \geq 1 - \alpha, \tag{9}$$

Where $\frac{\sum_{i=1}^{N} a_i x_i - \sum_{i=1}^{N} \mu_i x_i}{\sqrt{\sum_{i=1}^{N} \sigma_i^2 x_i^2}}$ represents a standard normal variate with a mean of zero and a variance of one. Then, the stochastic chance-constraint is transformed into the following inequality:

$$\varphi\left(\frac{b - \sum_{i=1}^{N} \mu_i x_i}{\sqrt{\sum_{i=1}^{N} \sigma_i^2 x_i^2}}\right) \geq \varphi\left(K_{1-\alpha}\right) \tag{10}$$

Where $K_{1-\alpha} = 1 - \alpha$ and $\varphi(\cdot)$ represents the standard normal cumulative distribution function (Sakallı et al. 2011) This yields the following nonlinear deterministic constraint:

$$\sum_{i=1}^{N} \mu_i x_i + K_{1-\alpha} \sqrt{\sum_{i=1}^{N} \sigma_i^2 x_i^2} \leq b \tag{11}$$

Segarra et al. (1985) demonstrate a linearized, more conservative, substitute for Eq. (11) given by:

$$\sum_{i=1}^{N} \mu_i x_i + K_{1-\alpha} \sum_{i=1}^{N} \sigma_i x_i \leq b \tag{12}$$

3 Application to Biodiesel Produced in Portugal

The proposed approach was implemented for biodiesel produced in Portugal. The feedstocks considered for the application of the model were the main feedstocks used in Portugal for biodiesel production: palm, canola and soya. According to information provided by the Portuguese Energy Agency (DGEG), in 2012, 49% of the feedstocks used for biodiesel production in Portugal were soya, 34% canola and 14% palm.

Costs were calculated by multiplying the quantity of each one of the three feedstocks (palm, canola and soybean oil) by its market price. The prices of the feedstock oils used in the model are the average prices between November 2008 and November 2013, provided by IndexMundi (2014). GHG emissions were calculated by multiplying the quantity of each feedstock by its life-cycle emissions per quantity unit. GHG emissions were drawn from Olivetti et al. (2014) for soybean and palm, and were drawn from Malça et al. (2014) for canola. The price and GHG coefficients are presented in Table 1. In order to scale the objectives, the price and GHG coefficients were divided by the largest value in each row resulting in the relative price and GHG emissions that is given in parentheses in Table 1.

Furthermore, the model is subject to technical specification constraints that the biodiesel must comply with. These constraints consider biodiesel properties derived from prediction models based on the chemical composition (fatty acids, FA) of the vegetable oils. Prediction models were used for the following biodiesel properties: density (Den), cetane number (CN), cold filter plugging point (CFPP), iodine value (IV) and oxidative stability (OS) (Bamgboye and Hansen 2008; CEN 2008; Park et al. 2008; Ramos et al. 2009; Refaat 2009; Giakoumis 2013). These prediction models are presented and discussed in Caldeira et al. (2014), who demonstrated that the derived results were in agreement with values found in the literature. The targets (constraint

Table 1 Price and GHG coefficients used in the model (relative value in parenthesis)

	Feedstock Oil		
	Palm	Canola	Soybean
Price (€/t)	629 (0.761)	826 (1.000)	753 (0.911)
GHG emission (g CO_2eq/MJ)	67 (1.000)	48 (0.716)	58 (0.856)

right hand side levels) were established according the European Standard EN 14214 (CEN 2008) that defines the biodiesel (Fatty Acid Methyl Esters—FAME) requirements for diesel engines. For CFPP, EN 14214 climate-dependent requirements are given to allow for seasonal grades to be defined for each country. There are six CFPP grades for temperate climates and five different classes for arctic climates. Level B, with a maximum of 0 °C was selected for this work. To address the compositional uncertainty, the technical constraints were formulated according to the CCP technique described previously. The chemical composition information (average and standard deviation) used in the model was adopted from Hoekman et al. (2012).

To analyze the proportions of each feedstock in the blend the demand (D) is set equal to one. We implicitly consider that the biodiesel produced is fully consumed by the oil refinery industry and that there are no feedstock supply limitations.

The non-linear model is presented in a minimization under constraints form given by Model (13). To apply the shadow price decomposition we proceed with the linearized version of the chance constraint model. The linearized version of the non-linear chance constraints is given by Model (14). Table 2 presents the notation of the biodiesel blend problem

Non-linear Version

$$\begin{aligned}
&\min\left\{\sum_{i\epsilon I}\left(C_{Pr,i}\,x_i\right),\sum_{i\epsilon I}\left(CC_{GHG,i}\,x_i\right)\right\}\\
&\text{Subject to :}\\
&PropConst_p+\sum_{j\in J}\left(PropCoef_{p,\ j}\sum_{i\in I}x_i\overline{q_{j,i}}\right)\\
&\qquad-K_{1-\alpha}\sqrt{\sum_{j\in J}\left(PropCoef_{p,j}{}^2\sum_{i\in I}x_i{}^2\sigma_{j,i}{}^2\right)}\\
&\qquad\geq PropGTgt_p\quad\forall p\in Plb\\
&PropConst_p+\sum_{j\in J}\left(PropCoef_{p,\ j}\sum_{i\in I}x_i\overline{q_{j,i}}\right)\\
&\qquad+K_{1-\alpha}\sqrt{\sum_{j\in J}\left(PropCoef_{p,j}{}^2\sum_{i\in I}x_i{}^2\sigma_{j,i}{}^2\right)}\\
&\qquad\leq PropLTgt_p\quad\forall p\in Pub\\
&\sum_{i\epsilon I}x_i=1\\
&x_i\geq 0\qquad\forall\ i\ \epsilon\ I
\end{aligned}\tag{13}$$

Table 2 Notation used along the manuscript

Variables	
x_i	Quantity of each raw material i to be used in the blend
$\vec{x}$	I × 1 vector of the decision variables
Sets/Indices	
K	Set of objectives
I	Set of raw materials
J	Set of ingredients
P	Set of properties
$k \in K$	K = {cost, GHG}, objectives
$i \in I$	I = {soya, canola, palm}, feedstock oils
$j \in J$	J = {1, 2,. . ., 18}, Fatty Acids index
$p \in P$	P = {DenLB, DenUB, IV, CN, OS, CFPP}, set of properties
Subsets	Plb = {DenLB, CN, OS}, set of properties with lower bound
	Pub = {DenUB, IV, CFPP}, set of properties with upper bound
Parameters	
$c_{k,i}$	Coefficient of objective k for raw material i
$\vec{C_k}$	$\vec{C_k}$ is a 1 × I vector containing the objective coefficients for the k-objective
w_k	Weight of the individual objectives
$q_{j,i}$	Concentration of ingredient j in raw material i
b_p	Limit value for property p
$\vec{b}$	P × 1 vector containing the property limits
I_{PxJ}	P × J unity matrix
Q	Q is a J × I matrix containing the q_{ji} elements
$\vec{C_{fB}}$	Objective functions coefficients vector for the basic variables of the optimal solution
$\vec{C_{kB}}$	Coefficients of the basic variables in the individual objective
B^{-1}	The basis inverse
U_k	Shadow prices of objective k
$\vec{C_f}$	Composed form of the objective function
a_i	Uncertain parameter in the CCP formulation
μ_i	Average value of parameter a_i
σ_i	Standard deviation of parameter a_i
$\overline{q_{j,i}}$	Average value of ingredient j in raw material i
$\sigma_{j,\,i}$	Standard deviation of the concentration of ingredient j in raw material i
α	Confidence level
$K_{1-\alpha}$	Test coefficient for normal distribution, one tailed: z-value corresponding to the chosen confidence value level
$C_{Pr,\,i}$	Ratio of the price of feedstock i to the most expensive feedstock
$CC_{GHG,\,i}$	Ratio of the GHG emission of feedstock i to the feedstock with the highest GHG emissions
$PropCoef_{p,\,j}$	Coefficient of FA-j in the prediction model for property p
$PropConst_p$	Constant in the prediction model for property p
$PropGTgt_p$	Target for properties with lower bound
$PropLTgt_p$	Target for properties with upper bound

Linearized Version of the Constraints

$$\begin{aligned}
&\mathrm{PropConst}_p + \sum_{j \epsilon J}\left(\mathrm{PropCoef}_{p,\ j}\sum_{i \epsilon I}\left(x_i \overline{q_{j,i}}\right)\right)\\
&\qquad - K_{1-\alpha}\sum_{j \epsilon J}\left(\mathrm{PropCoef}_{p,j}\sum_{i \epsilon I}\left(x_i \sigma_{j,i}\right)\right)\\
&\qquad \geq \mathrm{PropGTgt}_p \ \forall p \in \mathrm{Plb}\\
&\mathrm{PropConst}_p + \sum_{j \epsilon J}\left(\mathrm{PropCoef}_{p,\ j}\sum_{i \epsilon I}\left(x_i \overline{q_{j,i}}\right)\right)\\
&\qquad + K_{1-\alpha}\sum_{j \epsilon J}\left(\mathrm{PropCoef}_{p,j}\sum_{i \epsilon I}\left(x_i \sigma_{j,i}\right)\right)\\
&\qquad \leq \mathrm{PropLTgt}_p \quad \forall p \in \mathrm{Pub}
\end{aligned} \tag{14}$$

Both approaches, non-linear and linear, were implemented in GAMS version 23.7.3 (GAMS 2011). The non-linear problem was solved using the CONOPT solver and the linearized version using CPLEX. The Pareto frontier was obtained using the weighted method. The weight combination of the two objectives was calculated in 0.01 steps, in a total of 101 points. For a 95% confidence level (z-value = 1.96) the model was infeasible. This can be attributed to the fact that the linearized model is a much more restricting transformation of the probabilistic constraint, since $\sqrt{\sum_{i=1}^{N} \sigma_i^2 x_i^2} < \sum_{i=1}^{N} \sigma_i x_i$. For this reason we reduced the confidence interval to 90% (z-value = 1.645).

4 Results and Discussion

Figure 1 depicts the Pareto frontier obtained for the non-linear chance constraint problem (blue points, triangles) along with those of the linearized version (red points, crosses).

The linearized version provides just two possible solutions: one when more relevance is given to the environmental objective (Cost weight ∈ [0.00, 0.54]), and the other when the economic objective is more relevant for the decision maker (Cost weight ∈ [0.55, 1.00]). The solutions are presented in Table 3.

Figure 2 shows the blend composition for the mentioned weight combinations for the linearized version. The X-axis gives the weight combination and the Y-axis indicates the proportion of each input feedstock in the final blend. The solutions are associated with blends composed only by palm and canola. When we shift from an environmental to an economic "preference" we can observe an increase of the quantity of palm and a reduction of the quantity of canola in the blend. As palm is the feedstock with the lowest cost (Table 1), this change in the blend leads to a 2.4%

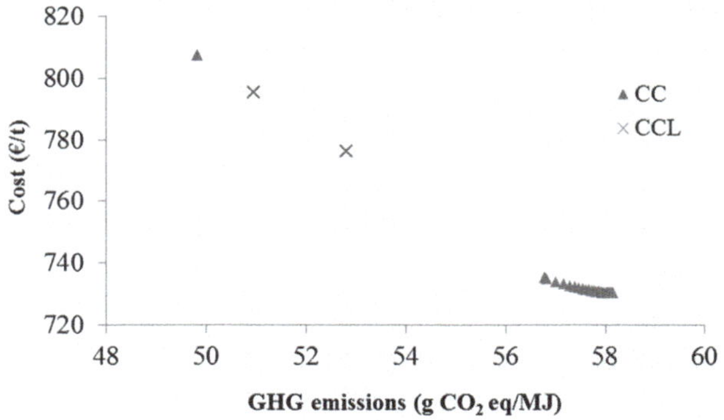

Fig. 1 Pareto frontier of non-linear (CC) and linear (CCL) chance constraint models obtained for a 90% confidence level

Table 3 Pareto frontier solutions of the linearized chance constraint model

Weights range	Cost (€/t)	GHG (g CO_2 eq/MJ)
Cost weight ∈ [0.00, 0.54]	795.44	50.94
Cost weight ∈ [0.55, 1.00]	776.24	52.79

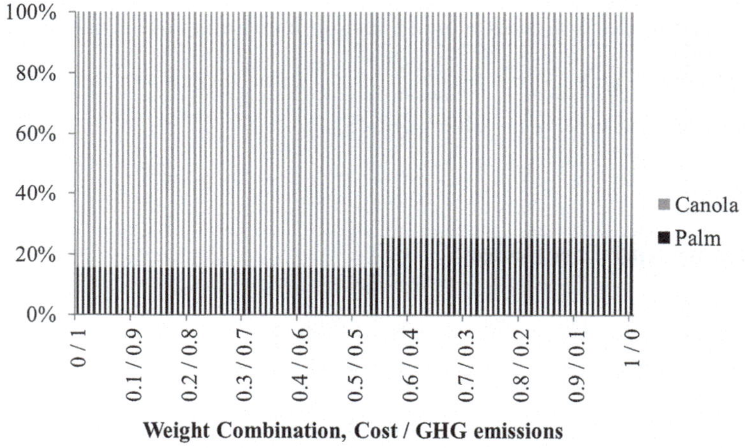

Fig. 2 Blend composition for all weight combinations of the linearized chance constraint model

cost reduction. However, since palm presents higher GHG emissions (Table 1), the improvement in the cost objective due to the increase of palm in the blend, worsens the environmental performance by 5%.

The non-linear version presents more solutions than the linearized one due to the fact that the technical constraints are not as conservative as in the linear version. For this reason, when more weight is given to the environmental objective (Cost weight ∈ [0.00, 0.54]), the algorithm suggests a blend with less palm and

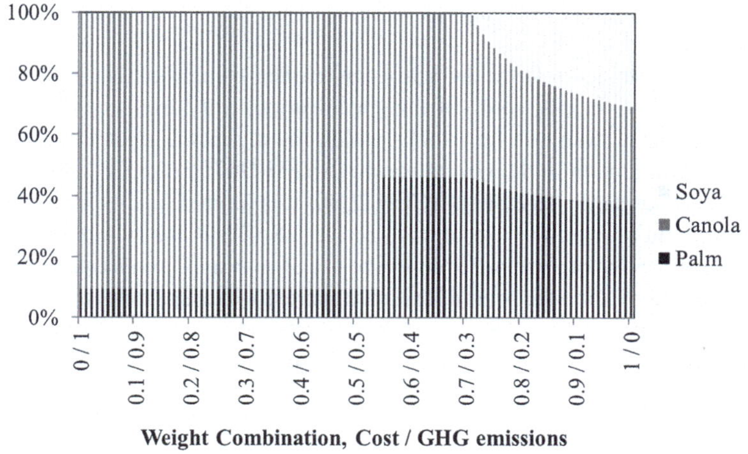

Fig. 3 Blend composition for all weight combinations of the non-linear chance constraint model

Table 4 Shadow price decomposition results for the linearized version

Weight combination	Binding constraint	Cost (€/t)	GHG (g CO_2 eq/MJ)
Cost weight ∈[0.00, 0.54]	Upper bound of iodine value ($IV \leq IV_0$)	+0.0034	−0.041
Cost weight ∈[0.55, 1.00]	Upper bound of cold filter plugging point ($CFPP \leq CFPP_0$)	−0.0179	+0.021

more canola. The proposed solution corresponds to lower GHG emissions than the one obtained for the same preference weights in the linearized approach. When the cost objective becomes more important, the blend increases the use of soya at the same time reducing canola quantity (which is the feedstock with higher cost) leading to solutions with lower cost than the ones obtained for the linearized version. The blend composition for the mentioned weight combinations in the non-linear version is shown in Fig. 3.

The shadow price decomposition discussed in the methodology section was applied to the linearized version and the results are presented in Table 4.

For both solutions the binding constraint was identified and shadow prices were allocated to both objectives. In the case where GHG emissions are evaluated as more important (Cost weight ∈[0.00, 0.54]) the binding constraint is iodine value (IV). If the upper bound of IV (IV_0) is increased by one unit, then the cost component of the objective value would increase by 0.0034 (2.74 €/t) while the GHG emissions would decrease by 0.041 (2.09 g CO_2 eq/MJ).

When the cost is more important than GHG emissions (Cost weight ∈[0.55, 1.00]), CFPP replaces IV as the binding technical constraint. In this case, an increase of one unit (1 °C) in the upper bound of CFPP ($CFPP_0$) would result in a decrease (−0.0179, i.e. 13.89 €/t) in the cost component (which has a positive context since the target is to minimize cost) and an increase in the GHG emissions (+0.021, i.e., 1.12 g CO_2 eq/MJ).

The results obtained refer to the European market, using the EN 14214 standard limit values. Nevertheless, some of the specifications vary from region to region, favoring the use of domestic feedstocks within the regions. This is the case for IV that has no requirement in the US standard—ASTM D6751 (ASTM 2008) while in the European standard—EN 14214 (CEN 2008) it is limited to a maximum of 120. In one hand, this favors the use of soya in the US (US is the world leading soya producer) and, on the other hand, it limits the use of soya and favors canola in the EU as it is observed by the blends composition obtained with our model (Fig. 2). As IV is the limiting property for GHG emission reduction and the US market has no limit for IV, it favors, according to our results, the reduction of GHG emissions.

5 Conclusions

Worldwide biofuel policies have been implemented leading the biodiesel industry to take into account not just costs but also GHG emissions, together with biodiesel technical performance. Some biodiesel technical specifications are "soft"and vary (e.g. between regions, climate conditions). It is therefore important that policy makers learn what the opportunity costs of technical specifications are. This article presents an approach to provide this information using an algorithm for the allocation of shadow prices to the constituent parts of the composite objective function articulated in a multi-objective chance constrained formulation. The information obtained from the shadow prices allowed the identification of the limiting technical properties for GHG reduction and cost effectiveness in the Portuguese policy context with three feedstocks entering the blend: CFPP (cold filter plugging point) is the limiting factor for cost effectiveness and IV (Iodine Value) is the limiting factor for reducing GHG emissions.

The biodiesel cost effectiveness can be increased when the biodiesel is commercialized in warmer countries. CFPP, the binding constraint for cost effectiveness, was limited to a maximum of 0 °C in this article. This property is determined regionally based on climate conditions. In warmer countries, this constraint could be limited to a maximum of 5 °C (Grade A in EN 14214) favoring biodiesel cost reduction. As both IV and CFPP have different limits in regional standards, biodiesel producers can adapt their production (choice of feedstocks blends) according to specific environmental and economic goals and to the destination market.

To illustrate the method we focused on technical specifications that are directly related to the chemical composition of the three feedstocks considered in the case study. However, this model can be further used with alternative feedstocks in order to analyze "regional blends" that would favor cost or GHG reduction targets. The shadow prices obtained provide information to the decision maker about the changes in each objective function that result from relaxing a requirement of the final blend. Thus it can be used as a guideline for evaluating the effectiveness of technical specifications relatively to the cost and GHG emissions and other objectives related to the biodiesel production process. That could be the case when new technologies with the potential to alter technological specifications of the input biodiesel oil are under consideration.

Further research should focus on the decomposition of shadow prices in non-linear multi-objective chance constrained formulations of the blending problem. As it is shown in the aforementioned example, the nonlinear version outperforms the linearized one in numbers of efficient blends. Although efficient blends obtained with the nonlinear formulation lie in the same frontier as those by the linear formulation, they represent more detailed information that could result in smoother dual values enriching policy options regarding the technical standard levels.

Acknowledgements Carla Caldeira acknowledges financial support from the Portuguese Science and Technology Foundation (FCT) through grant SFRH/BD/51952/2012. This work has also been supported by FCT project FEDER/FCTIPTDC/AAG-MAA/6234/2014 (POCI-01-0145-FEDER-016765). The research presented in this article has been developed under the framework of Energy for Sustainability Initiative of the University of Coimbra and the MIT Portugal Program. Stelios Rozakis acknowledges financial support from BioEcon project (ID: 669062) financed from the EU H2020-WIDESPREAD-2014-2 program.

References

Akgul, O., Shah, N., & Papageorgiou, L. G. (2012). An optimisation framework for a hybrid first/second generation bioethanol supply chain. *Computers and Chemical Engineering, 42*, 101–114. https://doi.org/10.1016/j.compchemeng.2012.01.012

ASTM. (2008). *Standard specification for biodiesel fuel blend stock (B100) for middle distillate fuels*. Report no. D6751–08.

Bairamzadeh, S., Pishvaee, M. S., & Saidi-Mehrabad, M. (2016). Multiobjective robust possibilistic programming approach to sustainable bioethanol supply chain design under multiple uncertainties. *Industrial and Engineering Chemistry Research, 55*, 237–256. https://doi.org/10.1021/acs.iecr.5b02875

Bamgboye, A. I., & Hansen, A. C. (2008). Prediction of cetane number of biodiesel fuel from the fatty acid methyl ester (FAME) composition. *International Agrophysics, 22*, 21–29.

Buratti, C., Barbanera, M., & Fantozzi, F. (2012). A comparison of the European renewable energy directive default emission values with actual values from operating biodiesel facilities for sunflower, rape and soya oil seeds in Italy. *Biomass and Bioenergy, 47*, 26–36. https://doi.org/10.1016/j.biombioe.2012.10.008

Caldeira, C., Gülsen, E., Olivetti, E. A., et al. (2014). A multiobjective model for biodiesel blends minimizing cost and greenhouse gas emissions. In B. Murgante, et al. (Eds), Computational science and its applications – ICCSA 2014. *Lecture Notes in Computer Science, 8581*, 653–666. https://doi.org/10.1007/978-3-319-09150-1_48

Caldeira, C., Freire, F., Olivetti, E. A., & Kirchain, R. (2017). Fatty acid based prediction models for biodiesel properties incorporating compositional uncertainty. *Fuel, 196*, 13–20. https://doi.org/10.1016/j.fuel.2017.01.074

CEN. (2008). *EN 14214: Automotive fuels – fatty acid methyl esters (FAME) for diesel engines – requirements and test methods*.

Charnes, A., & Cooper, W. (1959). Chance-constrained programming. *Management Science, 6*, 73–79.

Cohon, J. L. (1978). *Multiobjective programming and planning* (1st ed.). New York: Academic Press.

European Comission. (2009). Directive 2009/28/EC of the European Parliament and the council of 23 April 2009 on the promotion of the use of energy from renewable sources, pp. 16–62.

Gal, T. (1986). Shadow prices and sensitivity analysis in linear programming under degeneracy. *State-of-the-Art-Survey, Operational Research Spectrum, 8*, 59–71.

GAMS. (2011). GAMS Development Corporation: General Algebraic Modeling System (GAMS) Release 23.7.3. Washington, DC, USA.

Giakoumis, E. G. (2013). A statistical investigation of biodiesel physical and chemical properties, and their correlation with the degree of unsaturation. *Renewable Energy, 50*, 858–878. https://doi.org/10.1016/j.renene.2012.07.040

Gülşen, E., Olivetti, E., Freire, F., et al. (2014). Impact of feedstock diversification on the cost-effectiveness of biodiesel. *Applied Energy, 126*, 281–296. https://doi.org/10.1016/j.apenergy.2014.03.063

Ho, J. (2000). Computing true shadow prices in linear programming. *Informatica, 11*(4), 421–434.

Hoekman, S. K., Broch, A., Robbins, C., et al. (2012). Review of biodiesel composition, properties and specifications. *Renewable and Sustainable Energy Reviews, 16*, 143–169. https://doi.org/10.1016/j.rser.2011.07.143

Hwang, C.-L., & Masud, A. S. M. (1979). Multiple objective decision making — Methods and applications: A state-of-the-art survey. *Systems, Lecture Notes in Economics and Mathematical Volume 164*.

IndexMundi. (2014). Retrieved May 19, 2014, from http://www.indexmundi.com/. http://www.indexmundi.com/

Kampempe, J. D. B. (2012). Chance-constrained approaches for multiobjective stochastic linear programming problems. *American Journal of Operations Research, 2*, 519–526. https://doi.org/10.4236/ajor.2012.24061

Koltai, T., & Terlaky, T. (2000). The difference between the managerial and mathematical interpretation of sensitivity analysis results in linear programming. *International Journal of Production Economics, 65*, 257–274.

Kumral, M. (2003). Application of chance-constrained programming based on multi-objective simulated annealing to solve a mineral blending problem. *Engineering Optimization, 35*, 661–673. https://doi.org/10.1080/03052150310001614837

Malça, J., Coelho, A., & Freire, F. (2014). Environmental life-cycle assessment of rapeseed-based biodiesel: Alternative cultivation systems and locations. *Applied Energy, 114*, 837–844. https://doi.org/10.1016/j.apenergy.2013.06.048

Mavrotas, G. (2009). Effective implementation of the ε-constraint method in Multi-Objective Mathematical Programming problems. *Applied Mathematics and Computation, 213*, 455–465.

McCarl, B., Rister, M. E., Stokes, J., & Ziari, H. (1996). Improving shadow price information: Obtaining relevant shadow prices from problems with decomposable objectives. *American Journal of Agricultural Economics, 78*, 599–705.

Moghaddam, A. T. N., & Michelot, C. (2009). A contribution to the linear programming approach to joint cost allocation: Methodology and application. *European Journal of Operational Research, 197*, 999–1011. https://doi.org/10.1016/j.ejor.2007.12.043

Nejad, M. A. T. (2007). Allocation of CO2 emissions in petroleum refineries to petroleum joint products: A linear programming model for practical application. *Energy Economics, 29*, 974–997. https://doi.org/10.1016/j.eneco.2006.11.005

Olivetti, E., Gaustad, G. G., Fiels, F. R., & Kirchain, R. E. (2011). Increasing secondary and renewable material use: A chance constrained modeling approach to manage feedstock quality variation. *Environmental Science & Technology, 45*, 4118–4126.

Olivetti, E., Gülşen, E., Malça, J., et al. (2014). Impact of policy on greenhouse gas emissions and economics of biodiesel production. *Environmental Science & Technology, 48*, 7642–7650. https://doi.org/10.1021/es405410u

Palak, G., Ekşioğlu, S. D., & Geunes, J. (2014). Analyzing the impacts of carbon regulatory mechanisms on supplier and mode selection decisions: An application to a biofuel supply chain. *International Journal of Production Economics, 154*, 198–216. https://doi.org/10.1016/j.ijpe.2014.04.019

Park, J.-Y., Kim, D.-K., Lee, J.-P., et al. (2008). Blending effects of biodiesels on oxidation stability and low temperature flow properties. *Bioresource Technology, 99*, 1196–1203. https://doi.org/10.1016/j.biortech.2007.02.017

Ramos, M. J., Fernández, C. M., Casas, A., et al. (2009). Influence of fatty acid composition of raw materials on biodiesel properties. *Bioresource Technology, 100*, 261–268. https://doi.org/10.1016/j.biortech.2008.06.039

Refaat, A. A. (2009). Correlation between the chemical structure of biodiesel and its physical properties. *International journal of Environmental Science and Technology, 6*, 677–694. https://doi.org/10.1007/BF03326109

Rong, A., Lahdelma, R., Rong, A., & Lahdelma, R. (2008). Fuzzy chance constrained linear programming model for optimizing the scrap charge in steel production. *European Journal of Operational Research, 186*, 953–964. https://doi.org/10.1016/j.ejor.2007.02.017

Sahinidis, N. V. (2004). Optimization under uncertainty: State-of-the-art and opportunities. *Computers and Chemical Engineering, 28*, 971–983. https://doi.org/10.1016/j.compchemeng.2003.09.017

Sakallı, Ü. S., & Baykoç, Ö. F. (2013). Strong guidance on mitigating the effects of uncertainties in the brass casting blending problem: A hybrid optimization approach. *The Journal of the Operational Research Society, 64*, 562–576. https://doi.org/10.1057/jors.2012.50

Sakallı, Ü. S., Baykoç, Ö. F., & Birgören, B. (2011). Stochastic optimization for blending problem in brass casting industry. *Annals of Operations Research, 186*, 141–157. https://doi.org/10.1007/s10479-011-0851-1

Segarra, E., Kramer, R. A., & Taylor, D. B. (1985). Stochastic programming analysis of the farm level implications of soil erosion control. *Southern Journal of Agricultural Economics, 17*, 147–154.

Shaik, S., Helmers, G. A., & Langemeier, M. R. (2002). *Direct and indirect shadow price and cost estimates of nitrogen pollution abatement direct and indirect shadow price pollution abatement.*

Soimakallio, S., & Koponen, K. (2011). How to ensure greenhouse gas emission reductions by increasing the use of biofuels? – Suitability of the European Union sustainability criteria. *Biomass and Bioenergy, 35*, 3504–3513.

Thomas, A. R. C., Bond, A. J., & Hiscock, K. M. (2013). A multi-criteria based review of models that predict environmental impacts of land use-change for perennial energy crops on water, carbon and nitrogen cycling. *GCB Bioenergy, 5*, 227–242. https://doi.org/10.1111/j.1757-1707.2012.01198.x

Tomaschek, J., Özdemir, E. D., Fahl, U., & Eltrop, L. (2012). Greenhouse gas emissions and abatement costs of biofuel production in South Africa. *GCB Bioenergy, 4*, 799–810. https://doi.org/10.1111/j.1757-1707.2011.01154.x

You, F., Graziano, D. J., & Snyder, S. W. (2012). Optimal design of sustainable cellulosic biofuel supply chains: Multiobjective optimization coupled with life cycle assessment and input – output analysis. *AIChE Journal, 58*, 1157–1180. https://doi.org/10.1002/aic

Promoting Small-Scale Biofuel Production: A Qualitative GIS-OWA Methodology for Land Suitability Analysis of Winter Rapeseed

Mauro Viccaro, Mario Cozzi, Antonella Vastola, and Severino Romano

Abstract Biofuels could be a possible solution to promote agricultural development in rural areas by increasing farm income. Different studies suggest that all problems linked to large-scale biofuel production can be overcome by promoting small-scale production, particularly of rapeseed straight vegetable oil (SVO) used as self-supply agricultural biofuel, specially if the rapeseed is cultivated in crop rotation systems with minimum tillage practices. However, an ex-ante analysis would be very important to explore the feasibility of rapeseed production, via the evaluation of land use suitability.

As land planning issues are complex problems with multiple decision makers and criteria, we propose a spatial multi-criteria analysis model for supporting decision makers in the site selection process for winter rapeseed production. The methodology applied is the Ordered Weighted Averaging (OWA) extended by means of fuzzy linguistic quantifiers. The results have shown as the proposed methodology is more flexible compared to the other MCA methods, in particular for the possibility to make the choice in qualitative rather than quantitative terms, enabling the decision-maker to explore different decision strategies or scenarios, thus facilitating a better understanding of alternative land use suitability models.

Keywords Biofuel · SVO · Land suitability · GIS · Ordered Weighted Averaging · Linguistic quantifiers

This research was carried out in the framework of the project 'Smart Basilicata' (Contract n. 6386 – 3, 20 July 2016).
Smart Basilicata was approved by the Italian Ministry of Education, University and Research (Notice MIUR n.84/Ric 2012, PON 2007–2013 of 2 March 2012) and was funded with the Cohesion Fund 2007–2013 of the Basilicata Regional authority.

M. Viccaro (✉) · M. Cozzi · A. Vastola · S. Romano
School of Agricultural, Forestry, Food and Environmental Sciences, University of Basilicata, Potenza, Italy
e-mail: mauro.viccaro@unibas.it

J. Berbel et al. (eds.), *Multicriteria Analysis in Agriculture*, Multiple Criteria Decision Making, https://doi.org/10.1007/978-3-319-76929-5_6

1 Introduction

Agricultural sector plays an important role in the rural economy, and it is considered one of the most important elements to take into account in the rural development processes (Sánchez-Zamora et al. 2014). However, farmers face different shocks, mainly related to the climate change, global market instability and political decisions that frequently make them vulnerable (Eakin 2005). In several rural regions, like Basilicata (in Southern Italy), marginal farmland areas are being increasingly abandoned due to their low productivity and as a result of the reforms of the EU Common Agricultural Policy (Romano and Cozzi 2008). Different studies suggest that farm diversification could be a viable solution to reduce risk management and increase farm income (Meert et al. 2005; Barbieri and Mahoney 2009; Gautam and Andersen 2016).

In this scenario, the development of bioenergy–as a new business model integrated with environmental and social dimensions of a region–is a valuable tool with positive impacts both in socio-economic and environmental terms. In particular, biofuels are not only the main alternative for fossil fuels to reduce global greenhouse gas emissions, but they can provide local and regional benefits such as energy security, rural development, positive impacts on regional gross domestic product, and mitigation of local pollutant emissions (Franke et al. 2012).

All problems linked to large-scale biofuel production (land grabbing, land-use change, competition with the main agricultural products) can be overcome by promoting small-scale production of rapeseed straight vegetable oil (SVO) used as self-supply agricultural biofuel (Baquero et al. 2010). Rapeseed can be cultivated in crop rotation generating many economic and environmental benefits, *primarly* the non-competition between fuel and food production (Zegada-Lizarazu and Monti 2011).

However, given the economic relevance of investment related to SVO production (Baquero et al. 2011), an ex-ante analysis would be very important to explore the feasibility of rapeseed cultivation in a given area, via the evaluation of land use suitability (Cozzi et al. 2015). Land evaluation is the process of predicting the potential use of land on the basis of its attributes (Rossiter 1996), and in particular it is considered the basic tool for the consideration of agriculture in rural development plans (Hafif et al. 2013).

Several studies suggest that crop selection based on land suitability analysis, using a Multi-Criteria Analysis (MCA) and Geographical Information Systems (GIS) approach, is the most efficient low-cost method to determine the optimal cropping system as a function of biophysical variables. Pirbalouti (2009), Grassano et al. (2011) and Kamkar et al. (2014) use the weighted linear combination method (WLC) to evaluate land use suitability for rapeseed cultivation. However, there are some major limitations associated with the use of conventional MCA procedures (as well as WLC method) in a decision process, especially in situations that involve a high number of assessment criteria (Malczewski 2004). The main difficulty is to combine the criterion maps in a way that the results reflect decisions-makers' preferences. In these circumstances, the key issues of decision-making might be specified in terms of some linguistic quantifiers such as, for example, *"most criteria should be satisfied"* or *"at least 80% of criteria should be satisfied"*, etc. (Malczewski 2006; Mokarram and Hojati 2017; Romano et al. 2013). This necessitates extending the conventional MCA procedure so as to include situations that involve *qualitative* statements in the form of fuzzy linguistic quantifiers (Yager 1996).

This work is aimed to propose a *qualitative* GIS-OWA methodology for land use suitability analysis in order to identify the target investment areas for the cultivation of rapeseed to be use as self-supply agricultural biofuel at regional scale. The *qualitative* GIS-OWA enables the decision-maker to explore different decision strategies or scenarios, thus facilitating a better understanding of alternative land use suitability models (Malczewski 2006; Mokarram and Hojati 2017; Romano et al. 2013).

The Case Study is introduced in Sect. 2, and data and methods are presented in Sect. 3. Section 4 provides the results deriving from a set of alternative land use suitability maps, and the paper ends with a discussion section containing final remarks.

2 Case Study Area

The study area was carried out in Basilicata region, a rural region of Southern Italy (Fig. 1). The study area, typically Mediterranean, is located between latitude 39°54' N and 41°12' N and longitude 15°21' E and 16°51' E. The approximate surface area of the region is 9995 km^2 with a population of 570,365 inhabitants (ISTAT 2017), a mostly rural territory with the population being concentrated by the two thirds in the few large urban towns.

In geomorphological terms, the region is characterised by mountainous and hilly areas of the Apennine range (in the NW-SE direction), limited by the limestone base of the Murge hills and the Bradano depression in the north-east and by the Ionian coastal plains in the east.

In terms of climate there are differences specifically due to the complex orography of the region and its geographical position. The elevation varies between the sea level and 2200 m so, while a large portion of the territory shows typically Mediterranean features (Ionian coast, Bradano depression and Murge hills), the areas above 800 m asl are characterized by a temperate-cool climate with quite dry summers. Average annual precipitation ranges from 529 till about 2000 mm, concentrated in the South-Western area of the region, as the Apennine range intercepts most of the Atlantic weather perturbations into the Mediterranean. The most rainy months are November and December, the driest are July and August, when severe droughts are frequent. The temperature is characterised by wide variations, with very hot summers and very cold winters. The coldest month is usually January (with an average temperature between −4 and 7 °C).

How it's possible to observe in Fig. 1, agricultural land covers about 67% of the regional surface area. According with the last agricultural census (ISTAT 2010), the utilised agricultural area (UAA) is equal to 519,127 ha (52% of the total regional area), mostly dedicate to cereal cultivation on non-irrigated arable land (158,851 ha), followed by olive groves (31,351 ha), vegetable and orchards on permanently

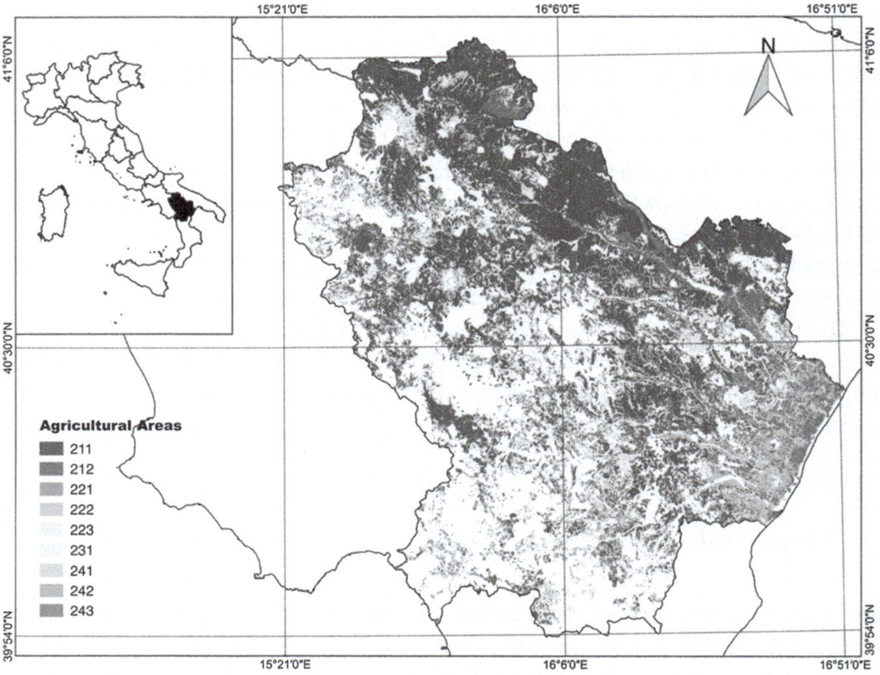

Fig. 1 Location and agricultural land use map of the study area (CLC code: 211 = Non-irrigated arable land; 212 = Permanently irrigated land; 221 = Vineyards; 222 = Fruit trees and berry plantations; 223 = Olive groves; 231 = Pastures; 241 = Annual crops associated with permanent crops; 242 = Complex cultivation; 243 = Land principally occupied by agriculture, with significant areas of natural vegetation)

irrigated land (about 16,000 ha), and vineyards (5361 ha). In this context, there exist good conditions to cultivate rapeseed in crop rotation with cereals on non-irrigated arable land. During the last decade, in Basilicata there was a reduction of land for the production of rapeseed, from 2700 ha (in 2000) to 343 ha (in 2010) (ISTAT 2010); but in the near future, rapeseed cultivation for straight vegetable oil (SVO) production could be a sustainable solution to diversify farm income, especially now that the traditional cultivations (e.g. durum wheat cultivation) in Basilicata region show a high risk management for farmers (Vastola et al. 2017).

3 Materials and Methods

3.1 Qualitative GIS-OWA Methodology

Since 2006, when Malczewski proposed, for the first time, the OWA approach with linguistic quantifiers in GIS environment, the method was widely used in different fields of study. Among others, Romano et al. (2013) and Mokarram and Hojati

(2017) have used it for land evaluation in agriculture showing its flexibility and easyness in land use analysis.

There are three main input components of GIS-OWA: (i) criterion maps (with associated standardization procedures); (ii) criterion weights (with associated procedures for defining preferences regarding the relative importance of criteria); and (iii) order weights (with associated ORness degree of the OWA operator) (Malczewski and Liu 2014).

The extension of conventional GIS-OWA approach with linguistic quantifiers–as in the case of others GIS-based approaches for land use suitability analysis–can be considered as a combination of purely MCA methods and Artificial Intelligence (AI) techniques (Malczewski 2004). After criterion map selection, we use *fuzzy logic* techniques, as standardization procedure of criterion maps (see Sect. 3.1.1), and AHP method, to calculate the relative criterion weights (see Sect. 3.1.2), classified as AI technique and MCA method respectively. At last, we use the OWA operator (MCA method) to aggregate the criterion maps after calculating the order weights through the linguistic quantifiers (AI technique) (see Sect. 3.1.3).

3.1.1 Criterion Maps

In order to assess the land suitability in agriculture for any crop type, all possible suitability criteria and their characteristics should be collected (Mendas and Delali 2012).

In our study, the criterion maps used in the analysis are related to the agro-ecological needs of rapeseed; topographic characteristics such as slope were not included, as the analysis was carried out only in non-irrigated arable land, where rapeseed can be cultivated in crop rotation systems.

The agro-ecological factors (climatic and soil factors) were selected from those proposed by Grassano et al. (2011), after an experts' panel evaluation. For each single factor under investigation was generated a geo-referenced *raster* layer (100 × 100 m cell size), by using Gauss Boaga East, on datum Monte Mario–Roma 1940 as geographic reference system.

Regarding climatic factors, we used the Crop-specific Thermal Index (CTI), but we modified the formula to calculate Seasonal Rainfall Deficit (SRD).

CTI was calculated on the basis of thermal requirements of rapeseed as average of the Monthly Thermal Indices (MTI) calculated for each month of the crop cycle (Eq. 1).

$$MTI = (x-B)(x-L)\,[(B+L-2T)(x-T)+(T-B)(T-L)] \,/ (T-B)^2(T-L)^2 \quad (1)$$

where x = average monthly temperature of the site; B = base temperature (0 °C); L = heat stress temperature (30 °C); T = optimum temperature (18 °C).

SRD can be assimilated to the irrigation water requirement (IWR), i.e. the amount of water that has to be applied in addition to rainfall to meet crop water requirements. It is calculated by difference between crop evapotranspiration (ET_c)

and that part of rainfall which is effectively used by plants (Pe) (Brower and Heibloem 1986).

The ET_c is calculated by multiplying the reference crop evapotranspiration (ET_0) by a crop coefficient (K_c) (Allen et al. 1998). In the study, the monthly ET_c was calculated using raster images representing the monthly ET_0, and the K_c values of each growth stage were derived from FAO paper n. 56 (Allen et al. 1998).

Effective rainfall (Pe) was calculated by the formula proposed by the Soil Conservation Service of the United States Department of Agriculture (Martin and Gilley 1993), adjusted for units converted from inches to mm:

$$Pe = fc\ (1.253 \times P^{0.824} - 2.935) \times 10^{0.001ET_c}$$

where *fc* is the correction factor depending on the soil available moisture; for the present work it is assumed to equal 1 (standard soil condition); *P* is the total monthly rainfall. In this way SRD values for rapeseed were calculated for the critical period of plant life cycle, from March to May.

Concerning the soil factors, we considered soil physical and chemical characteristics, such as texture, percentage of gravel, pH, soil depth, total carbonate content, and drainage. As suggested by the experts' panel, we did not take into account salinity due to the great adaptability of rapeseed and to the negligible influence of this factor on regional agriculture. Moreover, the map of the organic matter content was replaced with the map of land use capability.

As CTI is the only factor that ranges between 0 and 1, with 0 unsuitable and 1 suitable, the next stage involved the use of *fuzzy logic* technique (Zadeh 1965): given a fuzzy set (membership functions) is possible to standardize criterion maps defining the suitability degree within a range from 0 to 1. The fuzzy functions were chosen on the basis of the type of processed data and the uncertainty associated with it (Caniani et al. 2011, 2016; Eastman 2012) from those proposed by Cozzi et al. (2014), after an experts' panel evaluation. Criterion maps and fuzzy functions used in the analysis are shown in Table 1.

3.1.2 Criterion Weights

Because not all criteria affecting land suitability have equal levels of significance, the Analytical Hierarchy Process (AHP) method (Akıncı et al. 2013; Saaty 1977) was used for defining preferences regarding the relative importance of criteria and calculating the criterion weights necessary for the OWA aggregation procedure.

The AHP approach is one of the most widely known and used multi-criteria analysis approaches in GIS environment (especially for raster data models), allowing users to determine the weights associated with suitability maps. After the suitability maps (criteria) are set on a hierarchical structure, the weights can be derived by taking the principal eigenvector of a square reciprocal matrix of pairwise comparisons between the criteria (Eastman 2012). The comparisons concern the relative

Table 1 Criterion maps and related fuzzy function for land use suitability of rapeseed

Fuzzy function	Criterion maps	Criterion value	Fuzzy value
Null[a]	Crop-specific Thermal Index	–	–
Decreasing sigmoidal	Seasonal Rainfall Deficit (mm)	0	1
		50	0
User defined	Carbonates (% $CaCO_3$)	<0.5	1
		0.5–1	1
		1–5	1
		5–10	1
		10–25	1
		25–40	0.93
		>40	0.84
	Soil depth (cm)	<25	0.58
		25–50	0.70
		50–100	0.90
		100–150	1
		>150	1
	Gravel (%)	0	1
		1–5	0.90
		5–15	0.85
		15–35	0.65
		35–70	0.50
		>70	0.20
	Land use capability	Without limitations	1
		Moderate limitations	0.95
		Severe limitations	0.90
		Very severe limitations	0.80
		Accurate management	0.70
		Forestry and pasture use	0.50
		Very strong limitations	0.45
	Soil reaction (pH)	<4.5	0.75
		4.5–5.5	0.85
		5.6–6.5	0.92
		6.6–7.3	1
		7.4–7.8	0.95
		7.9–8.4	0.95
		8.5–9.0	0.90
	Soil texture	Coarse	0.65
		Moderately coarse	0.88
		Medium	0.88
		Moderately fine	0.95
		Fine	0.91
	Drainage	Rapid	0.70
		Good	0.93
		Mediocre	0.80
		Slow	0.70
		Very slow	0.50
		Prevented	0.30

[a]CTI (Crop-specific Thermal Index) range between 0 and 1 so it's no necessary standardize it

Table 2 Criterion weights resulting from the AHP approach (CR = 0.03)

Criterion map (j)	Criterion weights (u_j)
Carbonates	0.0192
Soil depth	0.0788
Soil reaction	0.0192
Soil texture	0.0378
Seasonal Rainfall Deficit	0.2865
Gravel	0.0378
Drainage	0.1554
Land Use Capability	0.0788
Crop-specific Thermal Index	0.2865

importance of the two criteria involved in determining suitability for the stated objective and it is made by using the preference scale suggested by Saaty.

Since performing pairwise comparisons of criteria in the AHP method a certain level of inconsistency may occur, Saaty proposes also a procedure to calculate an index of consistency, known as a consistency ratio (CR), indicating that, in case the CR of a matrix is above 0.10, the matrix of pairwise comparisons should be revaluated (Akıncı et al. 2013).

In our study, the criterion weights resulting from the AHP approach, calculated according to the estimation of the criterion influence on the rapeseed cultivation suitability, are shows in Table 2.

3.1.3 Order Weights

The order weights are relevant for the GIS-OWA combination procedures (Malczewski 2004). From different sets of order weights a wide range of OWA operators may be generated, including the most common map combination procedures: the weighted linear combination (WLC) and Boolean overlay operations, like the intersection (AND) and union (OR). In the conventional OWA approach, the OWA operators are defined by two parameters: the measures of trade-off and ORness (Yager 1996; Malczewski 2006). The trade-off is a compensation measure (substitutability criterion) ranging between 0 and 1, so that 0 indicates the lack of compromise between criteria, whereas 1 indicates a full compromise. The measure of ORness indicates the degree to which an OWA operator is similar to the logical connective OR in terms of its combinations behaviour. In this case as well the degree of OR required goes from 0 (*risk-averse*, operator MIN, AND) to 1 (*risk-taking*, operator MAX, OR).

However, in a complex spatial decision situation decisions-makers might be expected to find difficulties (or even impossible, notably for the problems that involve a number of criteria) to formulate accurate numerical information in relation to the OWA parameters.

In these situations, the key issue of decision-making might be specified in *qualitative* terms through the use of fuzzy linguistic quantifiers. A linguistic

Table 3 Regular increasing monotone quantifiers and their proprieties

Quantifier (Q)	α	GIS combination procedures	Position in the decision-strategy space (see Fig. 2)
All	$\alpha \to \infty$	OWA (AND, MIN)	1
Almost all	$\alpha = 10$	OWA	–
Most	$\alpha = 2$	OWA	–
Half (identity)	$\alpha = 1$	OWA (WLC)	2
A few	$\alpha = 0.5$	OWA	–
At least a few	$\alpha = 0.1$	OWA	–
At least one	$\alpha \to 0$	OWA (OR, MAX)	3

Source: Malczewsky (2006)

quantifier, used for computer–human interaction, enables decision makers to formulate OWA procedure in a simple way (Romano et al. 2013). Malczewsky (2006) proposes a set of linguistic quantifiers known as Regular Increasing Monotone (RIM) (Table 3), so that, given a set of standardized criterion map ($j = 1, 2, \ldots, n$) and criterion weight, the qualitative GIS-OWA for each i-th location (cell) is defined as follows:

$$OWA_i = \sum_{j=1}^{n} v_j z_{ij}$$

with the order weight

$$v_j = \left(\sum_{k=1}^{j} u_k\right)^{\alpha} - \left(\sum_{k=1}^{j-1} u_k\right)^{\alpha}, \text{ such that } v_j \in [0, 1], \ \sum_{j=1}^{n} v_j = 1$$

where $z_{i1} \geq z_{i2} \geq \ldots \geq z_{in}$ is the sequence obtained by reordering the standardized criterion values $a_{i1}, a_{i2}, \ldots, a_{in}$, u_j is the criterion weight reordered according to the value of z_{ij} and α is the parameter associated with RIM.

By specifying an appropriate linguistic quantifier, in the continuum that goes from the quantifier "*All*" (position 1) to the quantifier "*At least one*" (position 3), it's possible to generate a wide range of decision-making strategies (alternative models of land use suitability) with different degrees of ORness and trade-off (Fig. 2).

It's important to point out that in land use analysis, the linguistic quantifier to be adopted changes case by case. In the case of land use suitability analysis for agricultural crops, Romano et al. (2013) argue that the success of crop depends on the species finding the best climatic and edaphic conditions; it is evident that higher is the number of criteria considered, more reliable is the result.

In our case study, the linguistic quantifiers that best express this concept and that have contributed to the calculation of order weights are: "all criteria should be satisfied" ("*All*" quantifier, AND operator) and "almost all" ("*Almost all*" quantifier). All considered quantifiers are associated with a low ORness (low risk) and low trade-off (low compromise) degree. However, in order to evaluate the differences with the approach used by Pirbalouti (2009), Grassano et al. (2011) and Kamkar et al. (2014) in rapeseed land use analysis, we have chosen also the quantifier "*Half*" representing the WLC operator.

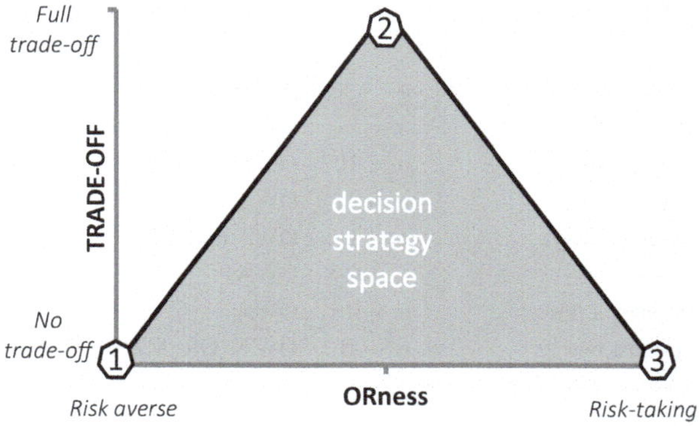

Fig. 2 Space of decision-making strategy in MCA and position of the main OWA operators (see Table 3)

4 Results and Discussions

Different scenarios representing suitability map of rapeseed cultivation, obtained with the multi-criteria analysis model, are described in Fig. 3.

The maps obtained with the "*All*" and "*Almost All*" quantifiers (All and Almost all quantifier scenario) look quite similar, showing some variability across the region with suitability values ranging between a minimum value of 0.32 and 0.33 and a maximum value of 0.68 and 0.70, for All and Almost all quantifier scenario respectively. About 75% of analysed arable lands (over the first quartile) shows suitability values higher than 0.50, despite the wide range (see box-plots). Most of the areas with higher values are mainly concentrated in the North-East, in the flat part of the region. The factor that has mainly influenced this distribution is the CTI. In fact, while other factors generally show a high suitability value, the CTI shows a mean value around 0.55, due to the presence of the Apennine ridge extending from North-West to South-East, where the mean annual temperatures are lower and not useful for heat requirements for rapeseed production.

Conversely, the map derived from the "*Half*" quantifier (WLC operator, Half quantifier scenario) shows a restricted range of suitability values (between 0.66 and 0.88). However, the resulting scenario is more optimistic: 50% of investigated area record suitability values range between 0.79 and 0.88. In this case, there is not the same variability across the region as described before. In the WLC approach, characterized by an ORness degree of 0.5 and full trade-off (position 2 in the decision strategy space, see Fig. 2), the low values of the CTI criterion (its relevance in the analysis, see Table 2) are compensated by the high values of the all other criteria.

To facilitate the reading of the results obtained using the OWA method, the non-irrigated arable lands have been classified into suitability classes for rapeseed cultivation using Chen-Hwang method (Chen and Hwang 1992). This method is a well-established tool to convert cardinal values to quality attributes, as it provides the mathematical representation of a linguistic term. Chen and Hwang identify 8 scales of linguistic terms. By the use of scale 4, four suitability classes were obtained (null, low, medium, high) (Table 4).

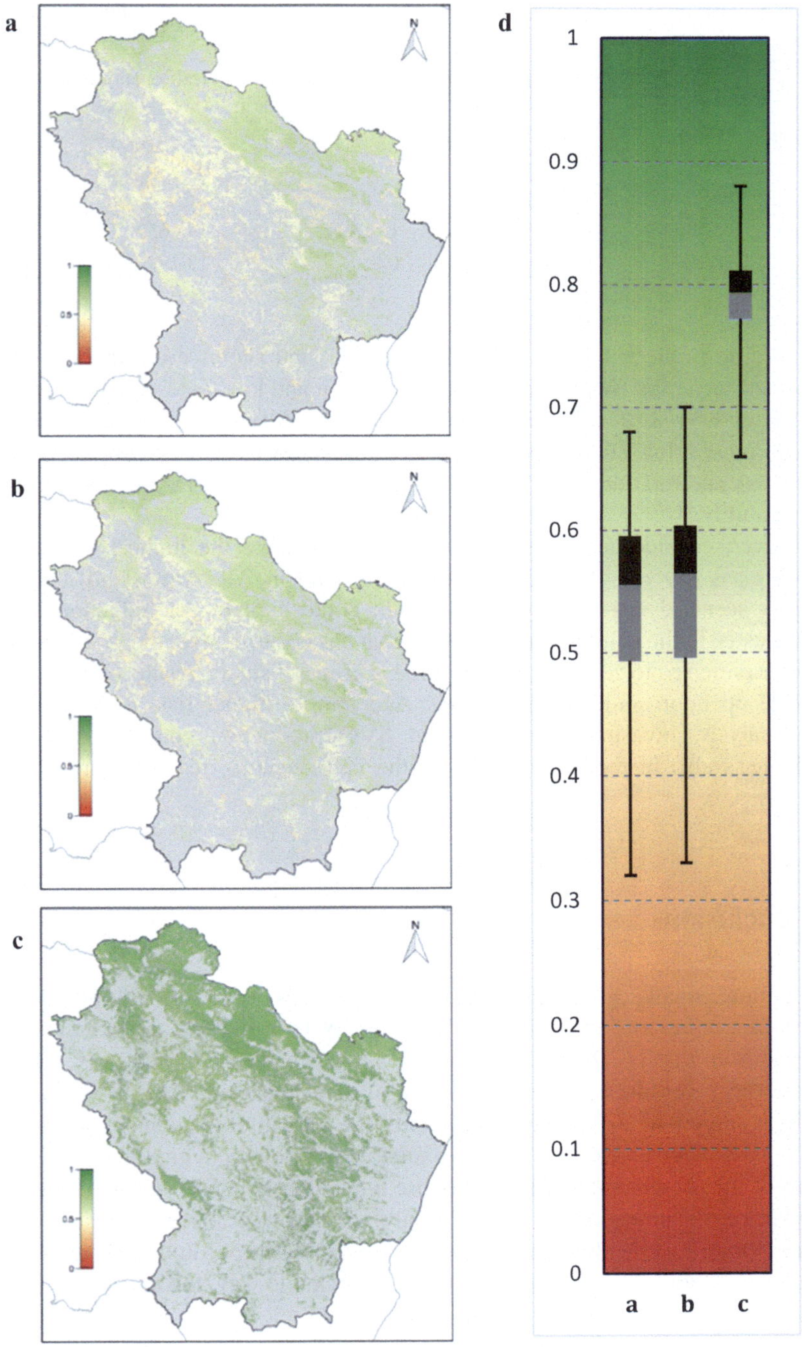

Fig. 3 Land suitability maps [(**a**) All quantifier scenario, (**b**) Almost all quantifier scenario, (**c**) Half quantifier scenario] for rapeseed cultivation and distribution of suitability values [(**d**) box-plot] for different scenarios

Table 4 Suitability classes of non-irrigated arable lands for rapeseed cultivation in different scenarios (ha)

Description	Range	All quantifier scenario	Almost all quantifier scenario	Half quantifier scenario (WLC operator)
		Surface (ha)		
Null	0–0.17	0	0	0
Low	0.17–0.5	33,259	25,048	0
Medium	0.5–0.83	125,592	133,803	145,652
High	0.83–1	0	0	13,199

According to the results in Table 4, no arable land shows non-suitability (null class) in all scenarios. With a non and low risk (non and low trade-off) associated to the "*All*" and "*Almost all*" quantifiers respectively, All and Almost all quantifier scenarios present low and medium classes; no high class has been recorded. With an average risk and full trade-off (WLC operator), arable lands show only medium and high suitability.

However, considering the surface, all scenarios have the highest area in the medium class (79%, 84% and 92% of arable land in All, Almost all and Half quantifier scenario respectively). Such result could explain the reduction of the area dedicated to the cultivation of rapeseed that has occurred over the years: farmers may have preferred to cultivate more profitable crops. If so, the results obtained with the WLC operator could be misleading. Accepting a higher risk associated with MCA analysis, investments in rapeseed SVO production could be not cost-effectiveness also in those areas that, in the half scenario, result to have a high suitability.

5 Conclusions

In agriculture, small-scale production of rapeseed SVO used as self-supply agricultural biofuel represents an opportunity to diversify farm income and achieve independence from fossil fuels.

However, given the economic relevance of investment related to SVO production, it is important to acquire instruments for agricultural planning to address investments towards areas that are more suitable for the crop's growing. Therefore, the aim of the present study was to propose a qualitative GIS-OWA methodology, applied to Basilicata region (in Southern Italy), helpful to produce land use suitability maps for rapeseed cultivation. The qualitative OWA procedure, through the use of linguistic quantifiers, enables to translate, in a simple way, the decision-maker's preferences in MCA combination procedures.

In order to produce land suitability maps, firstly the criterion maps were standardized by the use of fuzzy functions and then the relative criterion weights were calculated using the AHP method. Lastly, in order to aggregate the criteria with

OWA operators, the most suitable linguistic quantifiers were chosen. Since the success of rapeseed production depends on the specie finding the best climatic and edaphic conditions, it is evident that higher is the number of criteria considered, more reliable is the result. In our case study, the linguistic quantifiers that better express this concept and that have contributed to the calculation of order weights are "*All*" and "*Almost all*" quantifiers. The WLC operator was also applied in order to make a comparison with the most used approach in land use suitability analysis for rapeseed cultivation. Results showed that, in Basilicata region, the highest area has medium suitability values in all scenarios and this may explain the contraction of the surface dedicated to rapeseed cultivation over the years. The rest of the area in All and Almost all quantifier scenarios shows low suitability. Areas with a high level of suitability are recorded only in WLC scenario. This scenario is certainly more optimistic but unrealistic: the WLC operator corresponding to "Half quantifier, i.e. "half criteria should be satisfied". This expression is in disagreement with the aim of the analysis and the preferences of the decision maker for which all criteria or almost all must be met. Therefore, in a context of high-risk investments in agriculture, the WLC operator could be not appropriate in land use suitability analysis.

The proposed methodology is more flexible compared to the WLC methods, in particular for the possibility to make the choice in qualitative rather than quantitative terms, enabling the decision-maker to explore different decision strategies or scenarios, thus facilitating a better understanding of alternative land use suitability models.

References

Akıncı, H., Özalp, A. Y., & Turgut, B. (2013). Agricultural land use suitability analysis using GIS and AHP technique. *Computers and Electronics in Agriculture, 97*, 71–82.

Allen, R. G., Pereira, L. S., Raes, D., & Smith, M. (1998). Crop evapotranspiration-guidelines for computing crop water requirements-FAO irrigation and drainage paper 56. *FAO, Rome, 300*(9), D05109.

Baquero, G., Esteban, B., Rius, A., Riba, J. R., & Puig, R. (2010). Small-scale production of straight vegetable oil from rapeseed and its use as biofuel in the Spanish territory. *Energy Policy, 38*(1), 189–196.

Baquero, G., Esteban, B., Riba, J. R., Rius, A., & Puig, R. (2011). An evaluation of the life cycle cost of rapeseed oil as a straight vegetable oil fuel to replace petroleum diesel in agriculture. *Biomass and Bioenergy, 35*(8), 3687–3697.

Barbieri, C., & Mahoney, E. (2009). Why is diversification an attractive farm adjustment strategy? Insights from Texas farmers and ranchers. *Journal of Rural Studies, 25*(1), 58–66.

Brower, C., & Heibloem, M. (1986). Irrigation Water Management Training Manual No. 3: Irrigation water management: Irrigation water needs. Rome: FAO, International Support Program for Irrigation water Management, Land and Water Development Division.

Caniani, D., Lioi, D. S., Mancini, I. M., & Masi, S. (2011). Application of fuzzy logic and sensitivity analysis for soil contamination hazard classification. *Waste Management, 31*(3), 583–594.

Caniani, D., Labella, A., Lioi, D. S., Mancini, I. M., & Masi, S. (2016). Habitat ecological integrity and environmental impact assessment of anthropic activities: A GIS-based fuzzy logic model for sites of high biodiversity conservation interest. *Ecological Indicators, 67*, 238–249.

Chen, S. J., & Hwang, C. L. (1992). *Fuzzy multiple attribute decision making: Methods and applications*. Berlin: Springer.

Cozzi, M., Napoli, F. D., Viccaro, M., Fagarazzi, C., & Romano, S. (2014). Ordered weight averaging multicriteria procedure and cost-effectiveness analysis for short rotation forestry in the Basilicata Region, Italy. *International Journal of Global Energy, 37*(5/6), 282–303.

Cozzi, M., Viccaro, M., Di Napoli, F., Fagarazzi, C., Tirinnanzi, A., & Romano, S. (2015). A spatial analysis model to assess the feasibility of short rotation forestry fertigated with urban wastewater: Basilicata region case study. *Agricultural Water Management, 159*, 185–196.

Eakin, H. (2005). Institutional change, climate risk, and rural vulnerability: Cases from Central Mexico. *World Development, 33*(11), 1923–1938.

Eastman, J. R. (2012). IDRISI Selva Manual, Manual Version 17.01, Clark University, Worcester, MA.

Franke, B., Reinhardt, G., Malavelle, J., Faaij, A., & Fritsche, U. (2012, February 29). Global assessments and guidelines for sustainable liquid biofuels. A GEF Targeted Research Project organized by UNEP, FAO, UNIDO. Heidelberg/Paris/Utrecht/Darmstadt.

Gautam, Y., & Andersen, P. (2016). Rural livelihood diversification and household well-being: Insights from Humla, Nepal. *Journal of Rural Studies, 44*, 239–249.

Grassano, N., Tedone, L., Verdini, L., & De Mastro, G. (2011). Evaluation of rapeseed cultivation suitability in Apulia with GIS-multicriteria analysis. *Italian Journal of Agronomy, 6*(2), 101–105.

Hafif, B., Barus, J., & Masganti, M. (2013). Zoning rural area for the development of annual plants. *International Journal of Advanced Science, Engineering and Information Technology, 3*(1), 33–37.

ISTAT. (2010). Censimento Agricoltura 2010. Retrieved July 12, 2017, from http://dati-censimentoagricoltura.istat.it/Index.aspx?lang=it

ISTAT. (2017). I.Stat, Il tuo accesso diretto alla statistica italiana. Retrieved July 12, 2017, from http://dati.istat.it/Index.aspx

Kamkar, B., Dorri, M. A., & da Silva, J. A. T. (2014). Assessment of land suitability and the possibility and performance of a canola (*Brassica napus* L.)–soybean (*Glycine max* L.) rotation in four basins of Golestan province, Iran. *Egyptian Journal of Remote Sensing and Space Science, 17*(1), 95–104.

Malczewski, J. (2004). GIS-based land-use suitability analysis: A critical overview. *Progress in Planning, 62*(1), 3–65.

Malczewski, J. (2006). Ordered weighted averaging with fuzzy quantifiers: GIS-based multicriteria evaluation for land-use suitability analysis. *International Journal of Applied Earth Observation, 8*(4), 270–277.

Malczewski, J., & Liu, X. (2014). Local ordered weighted averaging in GIS-based multicriteria analysis. *Annals of GIS, 20*(2), 117–129.

Martin, D. L., & Gilley, J. (1993). Irrigation water requirements, Chapter 2, Part 623, National Engineering Handbook. Washington, DC: USDA, Soil Conservation Service.

Meert, H., Van Huylenbroeck, G., Vernimmen, T., Bourgeois, M., & Van Hecke, E. (2005). Farm household survival strategies and diversification on marginal farms. *Journal of Rural Studies, 21* (1), 81–97.

Mendas, A., & Delali, A. (2012). Integration of MultiCriteria Decision Analysis in GIS to develop land suitability for agriculture: Application to durum wheat cultivation in the region of Mleta in Algeria. *Computers and Electronics in Agriculture, 83*, 117–126.

Mokarram, M., & Hojati, M. (2017). Using ordered weight averaging (OWA) aggregation for multi-criteria soil fertility evaluation by GIS (case study: Southeast Iran). *Computers and Electronics in Agriculture, 132*, 1–13.

Pirbalouti, A. G. (2009). GIS-based land suitability evaluation for rapeseed oil crop. *Journal of Food, Agriculture and Environment, 7*, 837–840.

Romano, S., & Cozzi, M. (2008). Valutazione delle trasformazioni del suolo e delle dinamiche in atto mediante analisi territoriale e metriche statistiche: il caso dell'hinterland potentino, Basilicata. *Agribusiness Paesaggio Ambiente, 11*(2), 98–108.

Romano, S., Cozzi, M., Viccaro, M., & Di Napoli, F. (2013). The green economy for sustainable development: A spatial multi-criteria analysis-ordered weighted averaging approach in the siting process for short rotation forestry in the Basilicata Region, Italy. *Italian Journal of Agronomy, 8*(3), 158–167.

Rossiter, D. G. (1996). A theoretical framework for land evaluation. *Geoderma, 72*(3), 165–190.

Saaty, T. L. (1977). A scaling method for priorities in hierarchical structures. *Journal of Mathematical Psychology, 15*(3), 234–281.

Sánchez-Zamora, P., Gallardo-Cobos, R., & Ceña-Delgado, F. (2014). Rural areas face the economic crisis: Analyzing the determinants of successful territorial dynamics. *Journal of Rural Studies, 35*, 11–25.

Vastola, A., Zdruli, P., D'Amico, M., Pappalardo, G., Viccaro, M., Di Napoli, F., Cozzi, M., & Romano, S. (2017). A comparative multidimensional evaluation of conservation agriculture systems: A case study from a Mediterranean area of Southern Italy. *Land Use Policy, 68*, 326–333.

Yager, R. R. (1996). Quantifier guided aggregation using OWA operators. *International Journal of Intelligence Systems, 11*(1), 49–73.

Zadeh, L. A. (1965). Fuzzy sets. *Information and Control, 8*(3), 338–353.

Zegada-Lizarazu, W., & Monti, A. (2011). Energy crops in rotation. A review. *Biomass and Bioenergy, 35*(1), 12–25.

Multi-criteria Decision Analysis: Linear and Non-linear Optimization of Aqueous Herbal Extracts

J. Gajdoš Kljusurić, A. Jurinjak Tušek, D. Valinger, M. Benković, and T. Jurina

Abstract This chapter is aimed to present the multi-criteria decision analysis: Linear and non-linear optimization of aqueous herbal extracts. Modelling is an indispensable part of food production, from "farm to fork", where it is used to optimize the initial production of food and feed as well as in the food and feed processing. Different particle sizes of olive leaves were used in extraction of biologically active components using water as a solvent. Experiment conditions varied in mixing times (5, 10, 15 min), heating treatments (40 °C, 60 °C, 80 °C), Revolutions per minute: rpm (250, 500, 750 min^{-1}) and particle sizes (100, 300, 500 μm). Based on the measured bioactive compounds (pH, total dissolved solids, conductivity, dry matter, total polyphenols and the antioxidant capacity by ABTS method, DPPH method and FRAP method). Aim was to develop models to support the optimization of this decision-making process—find the best experiment conditions for extraction of a certain bioactive compound. Two approaches; linear and nonlinear approaches were investigated. Linear optimization is presented with two models: Response Surface Methodology and using linear programing based on the Simplex method while the nonlinear approach is presented by developing membership functions using fuzzy logic approach. Final results showed that, simple or complex, i.e. linear or nonlinear approach(es) in the search for optimal experiment conditions in extraction of bioactive compounds from olive leaves, will lead to an optimal solution, but the engineer will decide which approach is suitable for further application. Linear optimization and application of fuzzy logic resulted with the best possible offer per set limitations. But each approach resulted with other optimal extraction conditions. However, the application of fuzziness allowed the extension of the set of acceptable experiment combinations to achieve the best extraction of a bioactive component.

Keywords Olive leaves extracts · Multi-criteria decision analysis · Linear optimization · Non-linear optimization · Process optimization

---

J. Gajdoš Kljusurić (✉) · A. Jurinjak Tušek · D. Valinger · M. Benković · T. Jurina
Faculty of Food Technology and Biotechnology, Department of Process Engineering, University of Zagreb, Zagreb, Croatia
e-mail: jasenka.gajdos@pbf.hr

J. Berbel et al. (eds.), *Multicriteria Analysis in Agriculture*, Multiple Criteria Decision Making, https://doi.org/10.1007/978-3-319-76929-5_7

1 Introduction

Growing world population and consequential rise in food consumption is the main concern and challenge for agriculture and food production technology and science. Today, the food industry faces many different challenges: from satisfying the original nutritive minimum requirements in some countries and on the other hand, in more developed countries, food has an additional function as the settlement of a special enjoyment or the promotion of health (Klatt et al. 2016). Food consumption can be directly out of nature, out of the characteristic agriculture or as semi-processed or processed foods which dominate in the industrialized countries. Savings and profit encouraged the industry to slowly accept mathematical modelling, automatization and control as useful investments. Mathematical modelling with the goal of automatization and process control has been developed and applied in almost every industry: it is also widespread in agricultural sector and food production. Main impulses for such changes pointed out are: (i) increased competition that forces manufacturers to produce a wider variety of products more quickly, (ii) escalating labor costs and raw material costs, (iii) increasingly stringent regulations that have resulted from increasing consumer demands for standardized, safe foods and international harmonization of legislation and standards (Mogale et al. 2017). Mathematical modelling, automatization and process control is an indispensable part of food production where it is used to from the initial production (in the field and barn) as well as in the further treatment where the main goal is food processing/production aimed for direct consumption or partial preservation before preparation and consumption.

2 Application in Agricultural Science

Multi-Criteria Decision Analysis (MCDA) is a valuable tool that can be applied in many complex decisions such as finding optimal conditions for extraction, production etc. It can be evaluated by application of linear optimization and, in most cases, by fuzzy logic approach.

Although plants were used in the human diet since ancient times, their bioactive properties have not yet been sufficiently explored. Herbs are used in traditional medicine as teas or alcoholic tinctures. In recent years, the focus of agricultural studies has been turned towards optimization of growth soil and cultivation with an aim to maximize their nutritive characteristics and the yield of bioactive compounds with antioxidant activity (e.g. polyphenols).

The aim of this study was to apply linear and non-linear optimization to present advantages or potential disadvantages of the MCDA in practice. Aqueous extracts of olive leaves were used as an example, according to the extraction procedure previously described by Jurinjak Tušek et al. (2016). Extraction of milled leaves of different particle sizes (100, 300, 500 μm) different diameters, using water as solvent at different temperatures (40 °C, 60 °C and 80 °C), mixed for 5, 10 or 15 min at different rpm (250, 500 and 700 min^{-1}).

Physical properties of aqueous plant extracts (pH, TDS, conductivity and dry matter), content of total polyphenols (*TP*) and antioxidant activity (AOA) were determined and used as parameters for linear optimization and the fuzzy logic approach. Content of *TP* and AOA determined by three methods (*DPPH* radical scavenging essay, *ABTS* and *FRAP*) were used as the chemical properties of the prepared extracts.

Linear and fuzzy logic approaches both showed advantage(s) and disadvantage (s): linear programming was concentrated on the goal function that was subjected to different limitations, while in the fuzzy logic approach, each membership curve was observed as an independent goal function. The fuzzy logic approach could be observed as an optimization procedure in the Pareto sense. Multivariate models that showed the interrelationship of the mentioned parameters are also presented.

Multi-Criteria Decision Analysis is a helpful tool for deciding which treatment to use if a special parameter is set as the goal function. As Dehe and Bamford emphasized (2015), in planning of a new development the decisions have always been a major issue and different modelling methods are used or developed to support the optimization of this decision-making process. Although it seems that different herbs should be treated separately in the optimization sense, application of MCDA shows no necessity for such approach. Linear optimization shows the simplicity of the optimization process, while the fuzzy logic approach shows the MCDA in the best light, awarding the optimization process with a set of solutions that allows the user to see what to expect in which case. In this study, real experimental data obtained by aquatic olive leaves analysis was used to show the potential and usefulness of linear and nonlinear modelling for optimization of the extraction process.

3 Collection of Experimental Data Used for Optimization

To obtain different particle sizes of olive leaves, milling of previously dried herbal material was carried out at rotational speed of rpm = 15,000 min^{-1} using IKA Tube Mill (IKA, Germany). The milling time was adjusted according to the desired particle size ($t = 10–40$ s). Sieving was carried out to separate different particle sizes using standardized DIN sieves (Fritsch, Germany) with pore openings of $d = 1000$, 800, 500, 355, 250 and 100 μm. For further investigation, fractions with particle sizes of $d = 500$, 300 and 100 μm were taken. Classical extraction of biologically active components was conducted using water as a solvent. Extraction of the biologically active component is based on denaturation of the cell membrane and release of molecules with biological activity (Kaushik et al. 2010). In a glass cup volume of $V = 200$ mL, $m = 1$ g of milled dried olive leaf was mixed with $V = 50$ mL of distilled water heated to a predetermined temperature.

Aim in this study was to develop models to support the optimization of this decision-making process. The main goal is to find the best experiment conditions for extraction of a certain bioactive compound (that was measured in this study).

Linear optimization is presented with two models: Response Surface Methodology and using linear programing based on the Simplex method (Orešković et al. 2015). The nonlinear approach is presented by developing membership functions using fuzzy logic approach (Gajdoš Kljusurić et al. 2014).

Extraction experiments were performed according to the conditions defined by the experiment plan using the response surface method (Table 1).

Table 1 Experimental plan based on three mixing times (5, 10 and 15 min), three heating treatments (40 °C, 60 °C and 80 °C), per three rpm (250, 500 and 750 min^{-1}), for fractions with three particle sizes (100, 300 and 500 μm)

Experiment no.	Mixing time (*t*/min)	Temperature (*T*/°C)	Revolutions per minute (*rpm*/min^{-1})	Particle size (*d*/μm)
1	5	40	500	300
2	15	40	500	300
3	5	80	500	300
4	15	80	500	300
5	10	60	250	100
6	10	60	750	100
7	10	60	250	500
8	10	60	750	500
9	10	60	500	300
10	5	60	500	100
11	15	60	500	100
12	5	60	500	500
13	15	60	500	500
14	10	40	250	300
15	10	80	250	300
16	10	40	750	300
17	10	80	750	300
18	10	60	500	300
19	5	60	250	300
20	15	60	250	300
21	5	60	750	300
22	15	60	750	300
23	10	40	500	100
24	10	80	500	100
25	10	40	500	500
26	10	80	500	500
27	10	60	500	300

4 Modeling and Optimization Results

The Response Surface Methodology (RSM) analyzes the relationships between several input variables and one or more response variables. The basic motive for using the RSM method is to find optimal conditions for obtaining an optimal response. The experimental plan for optimizing the conditions for extraction of biologically active molecules from olive leaves was made using the Box Behnken experimental design in the Statistica 13.0 (StatSoft, USA) program package. The Box Behnken experimental design requires at least three parameters. The effect of temperature, extraction time, mixing speed and particle size on the extraction efficacy of biologically active components using a classical extraction process was investigated. Each of the variables was analyzed on three levels. For 4 factors on 3 levels, the algorithm suggests 27 experiments. The conditions under which the experiments were performed are shown in Table 1. The basic idea of response surface methodology is to get the relationship of influencing factors to the dependent variable (response) through the response function. The response surface method also allows the discerning of the effects of individual model members as major effects or interactions. In most RSM problems, the form of connection between responses and independent variables is unknown, so the first step in the RSM method is to find the appropriate approximation link. Usually, the lower-order polynomial is used in the appropriate independent variable domains. For example, first order model function (Eq. 1) or second order model function (Eq. 2) in case of nonlinearity (Anderson-Cook et al. 2009):

$$y = \beta_0 + \beta_1 \cdot x_1 + \beta_2 \cdot x_2 + \ldots + \beta_k \cdot x_k + \varepsilon \tag{1}$$

$$y = \beta_0 + \sum_{i=1}^{k} \beta_i \cdot x_i + \sum_{i=1}^{k} \beta_{ii} \cdot x_i^2 + \sum_{i<j} \sum_{j=2} \beta_{ij} \cdot x_i \cdot x_j + \varepsilon \tag{2}$$

where y represents the dependent variable, x independent variables, β model coefficients, and ε the error that occurs in the response. The model coefficients are determined by minimizing the sum of the squares of the deviation, and then validating the response surface on an independent data set. Application of this particular method in the case of our study will result in optimal conditions for extraction of biologically active components from olive leaf.

The optimization process is related to the physicochemical parameters determined for the aquatic extracts of olive leaves. The following physical properties of aqueous extracts were determined and further used as parameters in the optimization process: pH, conductivity (G), total dissolved solids (TDS), dry matter by drying method (AOAC 1995). Determination of chemical properties vs based on the content of polyphenols as well as the antioxidative capacity, measured spectrophotometrically: total polyphenols (*TP*) by the method previously described by Pinelo et al. (2005) and the antioxidant capacity by ABTS method (Re et al. 1999), DPPH method (Brand-Williams et al. 1995) and FRAP method (Benzie and Strain 1996).

Experimental data shown in Table 2 was further used for RSM (StatSoft Inc., 2008). RSM results for optimization conditions of extraction conditions of biologically active components from olive leaf are presented in Fig. 1.

Influence of four variables [temperature (*T*), time (*t*), particle size (*d*) and mixing rate (*rpm*)] on the extractability of total polyphenols (*TP*) and antioxidant activity was analyzed at three levels (extraction temperature: $T = 40$ °C, 60 °C and 80 °C; extraction time: $t = 5$, 10 and 15 min; particle size: $d = 100$, 300 and 500 μm & mixing rate: $rpm = 250$, 500 and 750 min^{-1}). Figure 1 shows two-dimensional representations of the dependence of output variables and process conditions and the estimated optimal process conditions. From the results it can be seen that in the prepared aqueous extracts, the proportion of total polyphenols ranged from $TP = 60$ $\text{mg}_{\text{GAE}}\,\text{g}_{\text{d.m.}}^{-1}$ to $TP = 240$ $\text{mg}_{\text{GAE}}\,\text{g}_{\text{d.m.}}^{-1}$, whereas, according to the model, the optimal proportion of total polyphenols was $TPF = 152.12$ $\text{mg}_{\text{GAE}}\,\text{g}_{\text{d.m.}}^{-1}$. It was also visible that the increase in extraction temperature, particle diameter and mixing rate did not affect the proportion of total polyphenols, whereas the share of total polyphenols increased with increasing extraction time. Also, it can be seen that the antioxidant activity measured by the *ABTS* method was in the range from $ABTS = 0.10$ $\text{mmol}_{\text{Trolox}}\,\text{g}_{\text{d.m.}}^{-1}$ to $ABTS = 0.80$ $\text{mmol}_{\text{Trolox}}\,\text{g}_{\text{d.m.}}^{-1}$ and that the optimal predicted value of was $ABTS = 0.37$ $\text{mmol}_{\text{Trolox}}\,\text{g}_{\text{d.m.}}^{-1}$. It was also noticeable that the minimum response function was achieved for temperature $T = 60$ °C.

As far as the extraction time was concerned, the increase of extraction time did not significantly affect the antioxidant activity measured by the ABTS method. Also, an increase in particle diameter and mixing rate did not affect the antioxidant activity significantly neither. The antioxidant activity measured by the *DPPH* method was in the range of $DPPH = 0.25$ $\text{mmol}_{\text{Trolox}}\,\text{g}_{\text{d.m.}}^{-1}$ to $DPPH = 0.55$ $\text{mmol}_{\text{Trolox}}\,\text{g}_{\text{d.m.}}^{-1}$ with the optimal value of $DPPH = 0.38$ $\text{mmol}_{\text{Trolox}}\,\text{g}_{\text{d.m.}}^{-1}$. Increase of temperature and stirring speed lead to a decrease of the antioxidant activity measured by the *DPPH* method ant the *TP* values while the particle diameter has no significant influence on the content of *TP* and AOA measured by use of the DPHH method. The optimum antioxidant activity measured by the *FRAP* method was $FRAP = 1.07$ $\text{mmol}_{\text{FeSO4x7H2O}}\,\text{g}_{\text{d.m.}}^{-1}$. It can also be seen that at $T = 40$ °C and $T = 80$ °C there was a decrease in antioxidant activity. Duration of the extraction and mixing rate did not influence antioxidant activity measured by FRAP method.

Based on the RSM optimization, it can be seen that the optimal conditions for extraction of total polyphenols and antioxidant activity measured by ABTS, FRAP and DPPH methods are: temperature of 60 °C, 10 min of extraction time, particle diameter of 300 μm and the stirring speed of 500 min^{-1}. Derrien et al. (2017) also used effectively the response surface methodology in optimizing the green process for the extraction of lutein and chlorophyll from spinach by-products. However, still remains the question if the optimal solution i.e. whether the proposed conditions of extraction lead to the best utilization of the observed bioactive components.

Defined optimal ranges of physicochemical parameters were further used for linear and fuzzy optimization.

Table 2 Physical and chemical properties of prepared olive leave aqueous extracts

Exp	*pH*	*TDS* (mg L −1)	*G* (μS cm −1)	Dry matter (*d.m.*) %	*TP* (mg_{GAE} $g_{d.m.}^{-1}$)	*ABTS* ($mmol_{Trolox}$ $g_{d.m.}^{-1}$)	*DPPH* ($mmol_{Trolox}$ $g_{d.m.}^{-1}$)	*FRAP* ($mmol_{FeSO4x7H2O}$ $g_{d.m.}^{-1}$)
1	5.58	320	639.5	0.4886	150.812	0.464	0.422	0.927
2	5.47	375.5	756.0	0.5069	126.997	0.390	0.378	0.773
3	5.40	374.5	760.5	0.5734	160.980	0.452	0.386	1.015
4	5.38	430.5	861.0	0.5405	194.272	0.442	0.502	1.141
5	5.46	387.5	776.5	0.5516	145.489	0.470	0.367	0.898
6	5.47	382.0	766.0	0.5430	152.053	0.471	0.383	0.929
7	5.46	389.0	778.5	0.5209	151.431	0.481	0.347	0.897
8	5.36	403.0	810.0	0.6144	152.191	0.412	0.303	0.908
9	5.45	395.0	789.0	0.5422	142.116	0.441	0.382	1.025
10	5.48	391.5	782.0	0.5612	153.458	0.445	0.3990	0.797
11	5.45	407.0	809.0	0.5953	134.918	0.453	0.352	0.810
12	5.41	358.0	712.5	0.4906	154.137	0.571	0.429	0.818
13	5.45	377.5	753.5	0.5771	133.099	0.460	0.332	0.778
14	5.43	341.0	687.5	0.5254	139.406	0.507	0.397	0.874
15	5.41	405.0	811.5	0.5907	166.428	0.391	0.419	0.791
16	5.48	395.0	790.0	0.5322	150.795	0.439	0.369	0.845
17	5.46	388.5	775.0	0.5730	160.221	0.495	0.433	1.027
18	5.42	422.5	843.5	0.6476	146.213	0.353	0.376	0.973
19	5.40	385.5	771.0	0.5572	155.354	0.423	0.495	0.994
20	5.41	394.5	790.0	0.5919	145.618	0.449	0.422	1.073
21	5.48	382.5	767.0	0.5568	153.217	0.389	0.423	1.057
22	5.46	387.5	774.5	0.5780	147.826	0.414	0.423	0.932
23	5.43	387.0	772.0	0.5682	136.277	0.439	0.426	0.953
24	5.45	418.5	834.5	0.6455	152.397	0.447	0.411	1.020
25	5.42	377.5	756.0	0.5162	148.795	0.469	0.389	1.112
26	5.43	394.0	772.5	0.5770	162.708	0.370	0.380	1.017
27	5.41	393.0	767.5	0.6063	168.035	0.307	0.392	1.207

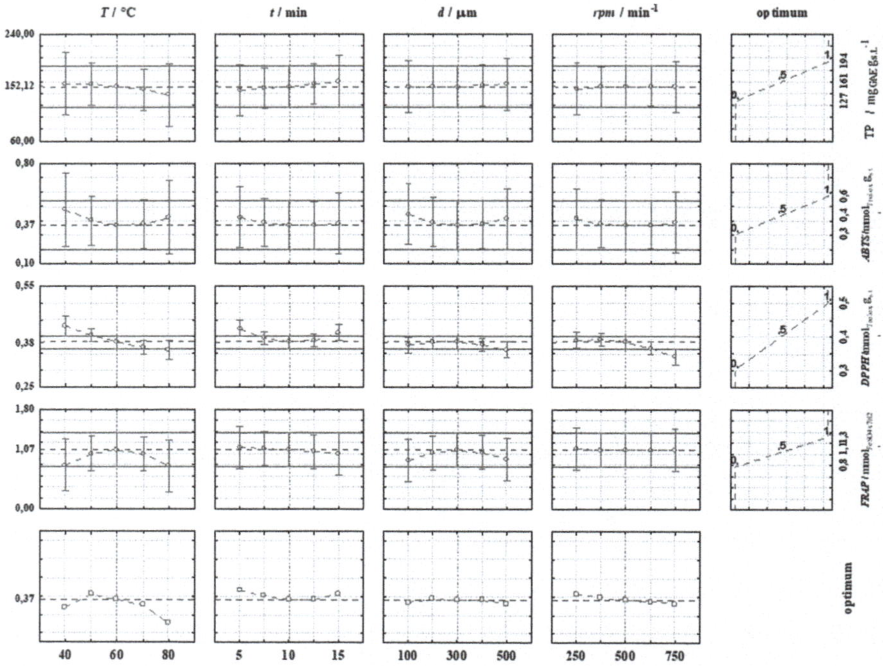

Fig. 1 RSM results for optimization of bioactives extraction from olive leaves

5 Application of Linear Optimization (LP)

When linear optimization is used, the model that leads to the optimal result has a structure where the goal function (Eq. 3) is subjected to constrains which define the expected range of observed parameters as follows:

$$min \quad \text{or} \quad max \quad F_G = c^T \cdot x \tag{3}$$

Subject to

$$Y_1 \cdot x \leq z_1 \tag{4}$$
$$Y_2 \cdot x \geq z_2 \tag{5}$$
$$Y_3 \cdot x = z_3 \tag{6}$$

The goal function, F_G aims at minimizing or maximizing an observed parameter. This function is defined as the product of a transposed vector c (c^T) and the vector x presenting the treated samples in different experiments ($x_1 - x_{27}$). In most cases, the goal function is used to minimize the costs (Stich et al. 2017; Liu et al. 2017) or maximize the profit (Wang et al. 2017). In this study, the goal function was used to maximize the content of polyphenols based on the ranges of parameters that are presented with equalities and inequalities (Eqs. 4–6). The matrix Y presents the content of other parameters in treated samples during different experiments.

The search for maximal content of total polyphenols based on experimental data showed slightly different results than the one based on the simple linear optimization where the goal function (F_G) was maximization of TP subjected to the experiment conditions (different t, T, rpm and d). In this optimization process the relations of any equality (Eq. 6) are not used.

The aim is to maximize the content of total polyphenols:

$$\max \Sigma c(TP)_i \cdot x_i \tag{7}$$

where i presents the number of the experiment, $i = 1, 2,..., 27$.

The goal function is subjected with the four experimental conditions ($j = 4$) in the range from the minimal to the maximal value of the observed experimental condition:

$$\min j \leq j_k \cdot x_i \leq \max j \tag{8}$$

where k is the first, second or third value of the observed parameter j ($k = 1, 2, 3$) form Table 1. A simple optimization structure is presented in Fig. 2, as a LINDO program.

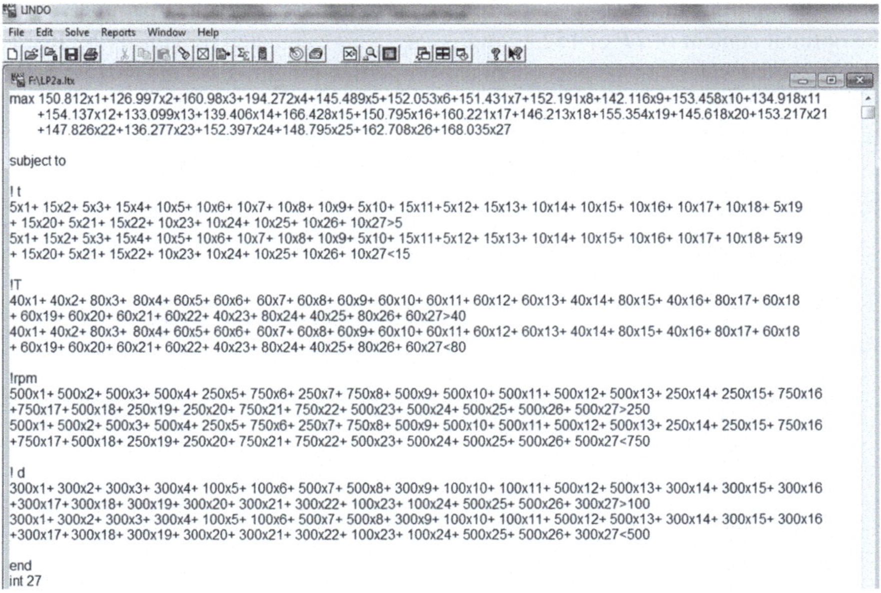

Fig. 2 Linear model applied in defining experimental conditions to maximize the content of total polyphenols

The maximized content of *TP* of 194.27 $mg_{GAE}\ g_{d.m.}^{-1}$ could be reached if the experiment is carried out at 40 °C, for 10 min at 500 rpm and with particle diameter of 300 μm.

The disadvantage of applied linear optimization is the necessity to repeat the optimization procedure if the aim is to find the maximum of the antioxidant activities (ABTS, DPPH, FRAP) as well as the change of c^T (Eq. 3).

However, if an equality (Eq. 6) is introduced as the previously calculated optimal *TP* content, the linear model will allow to calculate the expected antioxidant activity data under experimental conditions that would allow the maximum *TP* content (194.27 $mg_{GAE}\ g_{d.m.}^{-1}$).

Again, the goal function is aimed to maximize the content of total polyphenols (Eq. 7: max $\Sigma\ \{c(TP)_i \cdot x_i\}$), subjected to the experimental values of parameters from Table 2, presented as constrains limited with maximal and minimal values of the observed parameter:

$$\min l \le l_i \cdot x_i \le \max l \tag{9}$$

where *i* presents the number of the experiment $i = 1, 2, \ldots, 27$, while the 8 observed parameters are observed with the parameter *l*. The constraints are now the experimental results (Table 2) and introduced equality is shown in Fig. 3.

Expected physicochemical parameters for the optimal content of *TP* (194.27 $mg_{GAE}\ g_{d.m.}^{-1}$) are following: $pH = 5.44$; $TDS = 330.5$ mg L^{-1}; $G = 861.49$ μS cm^{-1}; *d. m.* = 0.55; $ABTS = 0.442$ $mmol_{Trolox}\ g_{d.m.}^{-1}$; DPPH = 0.502 $mmol_{Trolox}\ g_{d.m.}^{-1}$ and FRAP = 1.141 $mmol_{FeSO4x7H2O}\ g_{d.m.}^{-1}$.

Linear optimization was also used for expected physicochemical parameters for the optimal content of *TP* of 152.12 $mg_{GAE}\ g_{d.m.}^{-1}$, as calculated previously by RSM. The expected values of physicochemical properties for this case are: $pH = 5.36$;

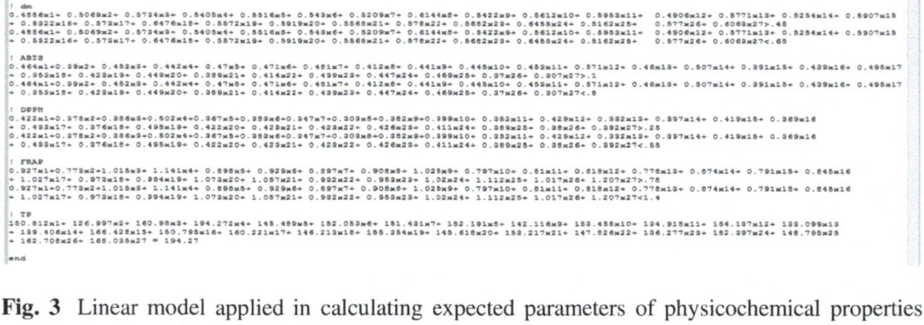

Fig. 3 Linear model applied in calculating expected parameters of physicochemical properties when the content of total polyphenols optimal (maximized value)

TDS = 389.5 mg L^{-1}; *G* = 782.22 μS cm^{-1}; *d.m.* = 0.52; *ABTS* = 0.4 $mmol_{Trolox}$ $g_{d.m.}^{-1}$; *DPPH* = 0.42 $mmol_{Trolox}$ $g_{d.m.}^{-1}$ and *FRAP* = 0.91 $mmol_{FeSO4x7H2O}$ $g_{d.m.}^{-1}$. In direct comparison of results obtained by LP and RSM, it was observed that the differences in antioxidant capacities for the same optimal TP content are in the range from 8% to 15%. Linear optimization is in the agricultural sector mostly used in maximizing energy usability (Zhang et al. 2016). But development of the industry which exploits natural raw materials and "waste" in the food industry, has inspired application of linear optimization by Lesellier et al. (2015) in developments approaches in supercritical fluid chromatography applied to the analysis of cosmetics. The disadvantage of linear approach is that the optimal solution is one exact number or a set of conditions expressed as exact number and for another combination i.e. new solution, the optimization process must be repeated after some settings in the program have been changed (Orešković et al. 2015). Christen and his co-workers presented in the study of fuzzy cognitive mapping help in agricultural policy design and communication (2015) showed how optimization in the agricultural sector does not have to be solely and rigid. Fuzzy logic (FL) deals with reasoning that is approximate instead of fixed and exact. In contrast with Boolean logic theory, where binary sets have two-valued logic: true or false, fuzzy logic variables may have a truth value that ranges in degree between 0 and 1. Fuzzy logic has been extended to handle the concept of partial truth, where the truth value may range between completely true and completely false (Agah 2014). This implies that, when linguistic variables are used, these degrees may be managed by specific functions (Kupka and Rusnok 2017). The fuzzy approach in the optimization introduces an innovation—instead of having one goal function—all observed parameters are observed with equal weight, which is often described as Pareto optimization (Gajdoš Kljusurić et al. 2012, 2014).

Although FL is primarily used in technical sciences, its application has extended to almost all scientific disciplines: e.g. fuzzy approach in menu planning was presented by Wirsam and Hahn (1999) when they modelled membership functions of daily recommendations of energy. Such "optimization" gives the same weight to all observed variables, at the same time respecting the recommendations for the observed variables and allowing the optimization in the Pareto sense. This presents the multiobjective optimization based on reasoning that is approximate instead of true or false as it is within Boolean logic theory. While variables in mathematics usually take numerical values, in fuzzy logic applications, the non-numeric linguistic variables are often used to facilitate the expression of rules and facts (Christen et al. 2015). The basic tendency is to achieve maximal value (value 1) of the membership function μ, for each observed parameter (Gajdoš Kljusurić et al. 2012).

In the Boolean logic theory, as crisp values, a range of allowed or expected values, x_a, of an observed parameter $\boldsymbol{a}$ would be explained as followed:

$$\boldsymbol{x}_{\boldsymbol{a},min} \leq \boldsymbol{x}_{\boldsymbol{a}} \leq \boldsymbol{x}_{\boldsymbol{a},max} \tag{10}$$

This rule is used in the example of applied LP, where an observed parameter was defined in two constrains (Eqs. 4 and 5). Projection of the same example through application of fuzzy logic, an expected concentration of an observed parameter is defined by a characteristic membership function $\mu(x_a)$:

$$\mu(x_a)\begin{cases} 1, \text{for}\, \boldsymbol{x_{a,min}} \le \boldsymbol{x_a} \le \boldsymbol{x_{a,max}} \\ 0, \text{unacceptable} \end{cases}. \tag{11}$$

Linguistic variable related to the membership function would be, for instance, "low concentration", "high concentration" and "very high concentration" of total polyphenols. As presented in the sample, fuzzy set is used to show the inherent imprecision and fuzziness of quantities as well as to model the gradual boundaries of the optimal expected values associated with the extraction method.

Fuzzy logic optimization has to follow two basic steps:

1. Fuzzification
2. Defuzzification.

The modelling process where crisp values are related with the grades of membership for linguistic terms of fuzzy sets is called fuzzification (Wirsam and Hahn 1999).

In the fuzzy set modelling of an observed parameter is crucial to construct the function following the basic properties of a fuzzy set (Gajdoš Kljusurić et al. 2012; Agah 2014) where the membership function must be defined by its core, height and support (Fig. 4).

The subset of a universal set X with the property $\mu_A(x) = 1$ is the core of fuzzy set A:

$$core\ (A) = \{(x \in X \mid \mu_A(x = 1)\} \tag{12}$$

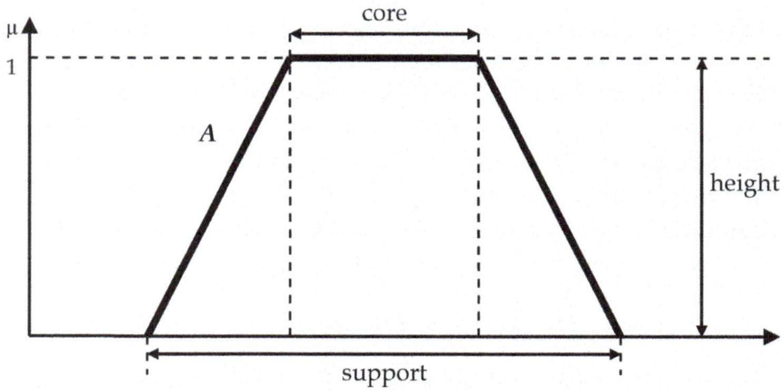

Fig. 4 Properties of a fuzzy set *A*

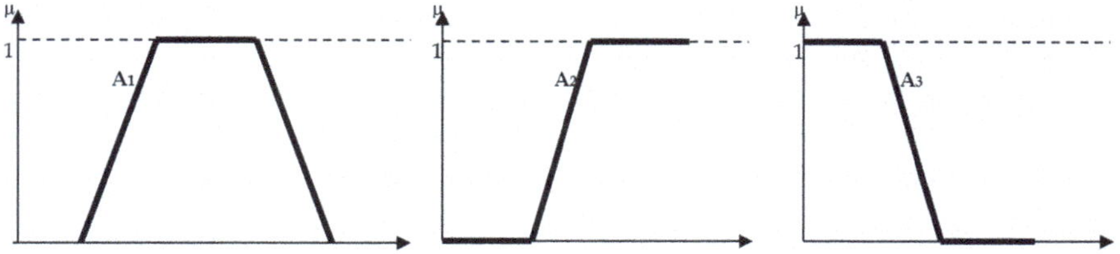

Fig. 5 Shape of a fuzzy set when the content of the observed parameters is preferred to be in an expected range $\mu(A_1)$, to be high $\mu(A_2)$ or, to be as lower as possible $\mu(A_3)$

The subset of the universal set *X* with nonzero membership grades is the support of the fuzzy set *A*:

$$supp\ (A) = \{(x \in X \mid \mu_A(x > 1)\} \tag{13}$$

The supremum (maximum) of the membership grades of *A* is the height of a fuzzy set, defined as:

$$hgt(A) = max_{x \in X} \mu_A(X) \tag{14}$$

Fuzzy set A is normal if the height is equal to 1 and defined in the interval [0, 1] and the core and support of a fuzzy set are ordinary subsets of X.

Three shapes of the fuzzy set are commonly used: (i) the so called "bell-shape" (Fig. 5, A_1), (ii) S (Fig. 5, A_2) and (iii) Z shape (Fig. 5, A_3).

A fuzzy set has to be explained with bools values while a linguistic term can be presented differently. For example, the sentence *Eat "more" dark chocolate* because it is rich in polyphenols contains a linguistic variable ("more") that could be presented as 10 or 100 g. This process of producing a quantifiable result in fuzzy logic is called defuzzification and is necessary because humans are more familiar with crisp values.

Wirsam and Hahn (1999) suggested the application of modified harmonic mean that defuzzifies the optimization and this crisp value is called Prerow value (*PV*):

$$PV = \mu(x_i)_{min} \cdot \left[\frac{(n-1)}{\sum_{i \neq i_{min}} \left(\frac{1}{\mu(x_i)} \right)} \right] \tag{15}$$

where $\mu(x_i)$ are the fuzzy sets for *i* observed parameters. The Prerow value is the measure of closeness to the recommended or expected value. Preferred PV values are greater than 0.7, and the optimal result would result in a PV value > 0.9. The optimization process by fuzzy logic was conducted using the program WR Mathematica v. 8. (Wolfram Research, USA).

After import of experimental data, all parameters undergo the fuzzification viz a bell-shaped membership function was modelled for each parameter.

Possible combinations of experiment conditions are presented in the algorithm (Wolfram Research Inc., 2017) as follows:

```
t = 3; T = 3; rpm = 3; d = 3;
exper = {t1, t2, t3, T1, T2, T3, rpm1, rpm2, rpm3, d1, d2, d3};
nkomb = t * T * rpm * d;
nvar = t + T + rpm + d;
var = Table[0, {i, nkomb}];
index = 1;
Do[{x = Table[0, {L, 1, nvar}], x[[i]] = 1,
  Do[{x[[j + t]] = 1,
    Do[{x[[k + t + T]] = 1, Do[{x[[m + t + T + rpm]] = 1, var[[index]] = x, index = index + 1, x[[m + t + T + rpm]] = 0},
      {m, 1, d}], x[[k + t + T]] = 0}, {k, 1, T}], x[[j + t]] = 0}, {j, 1, rpm}]}, {i, 1, t}]
```

The combinations of temperature, time, rpm and diameter that resulted with a crisp value (PV value) within the range of 0.7–1 were considered as acceptable experiment conditions which were expected to lead to the maximal extraction of total polyphenols.

It can be seen from the algorithm line that there can be 12 different variables ($3 \times t$, $3 \times T$, $3 \times rpm$ and $3 \times d$) which lead to 81 possible combinations ($3t \times 3T \times 3rpm \times 3d$). But defuzzification will show how many of them are acceptable ($PV > 0.7$) based on their bioactive profile (Table 3).

Sorted results, from those which are less appreciated to those that are highly recommended, have shown that all combinations are acceptable ($PV > 0.7$) but some of experiment conditions will lead to somewhat better extraction of the biologically active component, *TP* (Table 3).

The nonlinear approach where fuzzy logic was used, resulted in an optimal *TP* content of 155.4 $mg_{GAE}\ g^{-1}_{d.m.}$ if the aqueous extract of olive leaves (diameter 300 mm, d_2) is heated to 60 °C (T_2) for 5 min (t_1) and mixed at 250 rpm (rpm_1).

The developed algorithm also allowed the analysis of the expected contents of all other parameters after the applied nonlinear approach (Table 4).

All the fuzzy set membership values ($\mu(x_i)$ in Table 4) were used as inputs in Eq. (15) and the *PV* value equals the one for the last combination shown in Table 3.

Table 3 Combinations sorted according PV value (3 worst and 4 best)

Extraction condition combination		Prerow value (PV)
1	t_3, T_3, rpm_3, d_3	0.845
2	t_2, T_3, rpm_3, d_3	0.846
3	t_1, T_3, rpm_3, d_3	0.847
⋮	⋮	⋮
78	t_3, T_2, rpm_1, d_2	0.863
79	t_1, T_3, rpm_1, d_1	0.863
80	t_1, T_1, rpm_1, d_3	0.864
81	t_1, T_2, rpm_1, d_2	0.865

Table 4 Membership values for each parameter and the PV value for the best combination of experiment conditions (t_1, T_2, rpm_1, d_2)

Parameter (x_i)	$\mu(x_i)$	Value
pH	0.91	5.4
TDS	0.92	385.5 mg L^{-1}
G	0.93	771 µS cm^{-1}
d.m.	0.94	0.56%
TP	1.00	155.4 $mg_{GAE}\ g_{d.m.}^{-1}$
ABTS	0.96	0.42 $mmol_{Trolox}\ g_{d.m.}^{-1}$
DPPH	0.98	0.5 $mmol_{Trolox}\ g_{d.m.}^{-1}$
FRAP	0.93	0.99 $mmol_{FeSO4x7H2O}\ g_{d.m.}^{-1}$

$$\mathrm{PV} = \mu(x_i)_{min} \cdot \left[\frac{(n-1)}{\sum_{i \neq i_{min}} \left(\frac{1}{\mu(x_i)} \right)} \right]$$

$$= 0.91 \cdot \left[\frac{(8-1)}{\frac{1}{0.92} + \frac{1}{0.93} + \frac{1}{0.94} + 1 + \frac{1}{0.96} + \frac{1}{0.98} + \frac{1}{0.93}} \right] = 0.865$$

The linear approaches (RSM and LP) have shown that such models are simple to use with no special data preparation. However, the application of LP will result in just one possible solution for the appropriate experiment conditions for extraction of bioactive compounds from olive leaves. For any other combination of extraction conditions, solving of the LP should be repeated with the exclusion of the previous solution. This often stands out as a disadvantage of LP application. On the other hand, RSM decreases possible number of combinations to 1/3 (27 out of 81 possible combinations) and, as shown by the use of LP, the same input data will not lead to the same optimal solution.

Result of the fuzzy optimization process was a set of combinations of experiment conditions (temperature, time, rpm and diameter) that could reach the acceptable content of one or more target(ed) parameter(s) in 65% of the cases. Although the

Table 5 Optimal experiment conditions (*t, T, rpm, d*) for maximizing the extraction of total polyphenols, based on the linear (RSM or LP) and non-linear approach (fuzzy logic, FL)

	Experiment condition			
Optimization	*t* (min)	*T* (°C)	*rpm* (min^{-1})	*d* (μm)
RSM	10	60	500	300
LP	10	40	500	300
FL	5	60	250	300

whole set of parameter combinations are in the set as a final result, the disadvantage of this method is the complexity of the fuzzification (modelling of membership functions). The main advantage of application of fuzzy logic in the optimization is a set of combinations of experiment conditions from which the one that leads to the targeted concentration of a bioactive compound (polyphenols and/or antioxidant activity) can be chosen.

Is there possibility that another computing approach could be more effective, can be answered only then if the same problem is solved with another tool. To be sure, the optimal experiment conditions of all optimization processes in the decision-making should be compared. Those results are presented in Table 5. The optimal solution will give an answer how long (*t*, time/min) should the aquatic extract of a certain particle size of olive leaves (d/μm) be cooked and per which temperature (*T*/°C) and Revolutions per minute (*rpm*/min^{-1}).

Three different approaches lead to three different optimal conditions for optimal extraction of TP. The only parameter that has been optimal in all three optimization processes was the particle size of the olive leaves, 300 μm. But as presented in Table 3 where even the "worst" combination of experiment condition resulted with an acceptable PV value (>0.85), there is no doubt that optimal solutions where the RSM or LP were used are also very acceptable because they would also result in a high share of extracted total phenols and other observed parameters.

6 Conclusions

This chapter showed that, simple or complex, i.e. linear or nonlinear approach(es) in the search for optimal experiment conditions in extraction of bioactive compounds from olive leaves, will lead to an optimal solution, but the engineer will decide which approach is suitable for further application. Linear optimization and application of fuzzy logic resulted with the best possible offer per set limitations. However, the application of fuzziness allowed the extension of the set of acceptable experiment combinations to achieve the best extraction of a bioactive component.

References

Agah, A. (2014). *Medical applications of artificial intelligence*. New York: CRC Press/Taylor and Francis.

Anderson-Cook, C. M., Borror, C. M., & Mongomery, D. C. (2009). Response surface design evaluation and comparison. *Journal of Statistical Planning and Inference, 39*, 629–641.

AOAC. (1995). *Official methods of analysis of the association of official agricultural chemists*. Washington, DC: Association of Analytical Chemists.

Benzie, I. F., & Strain, J. J. (1996). The ferric reducing ability of plasma (FRAP) as a measure of "antioxidant power": The FRAP assay. *Analytical Biochemistry, 239*, 70–76.

Brand-Williams, W., Cuvelier, M. E., & Berset, C. (1995). Use of free radical method to evaluate antioxidant activity. *Food Science and Technology, 28*, 25–30.

Christen, B., Kjeldsen, C., Dalgaard, T., & Martin-Ortega, J. (2015). Can fuzzy cognitive mapping help in agricultural policy design and communication? *Land Use Policy, 45*, 64–75.

Dehe, B., & Bamford, D. (2015). Development, test and comparison of two multiple criteria decision analysis (MCDA) models: A case of healthcare infrastructure location. *Expert Systems with Applications, 42*(19), 6717–6727.

Derrien, M., Badr, A., Gosselin, A., Desjardins, Y., & Angers, P. (2017). Optimization of a green process for the extraction of lutein and chlorophyll from spinach by-products using response surface methodology (RSM). *Food Science and Technology, 79*, 170–177.

Gajdoš Kljusurić, J., Rumora, I., & Kurtanjek, Ž. (2012). Application of fuzzy logic in diet therapy – Advantages of application. In E. P. Dadios (Ed.), *Fuzzy logic – Emerging technologies and applications*. Rijeka: InTech.

Gajdoš Kljusurić, J., Orešković, P., Tušek, K., & Kurtanjek, Ž. (2014, June). *Linear programming versus fuzzy logic optimization in nutrition: Case of gestational diabetes*. OPT-i an international conference on engineering and applied sciences optimization, Kos, 03-07. paper 3631-CD.

Jurinjak Tušek, A., Benković, M., Belščak Cvitanović, A., Valinger, D., Jurina, T., & Gajdoš Kljusurić, J. (2016). Kinetics and thermodynamics of the solid-liquid extraction process of total polyphenols, antioxidants and extraction yield from Asteraceae plants. *Industrial Crops and Products, 91*, 205–214.

Kaushik, R., Pradeeo, N., Vamshi, V., Geetha, M., & Usha, A. (2010). Nutrient composition of cultivated stevia leaves and the influence of polyphenols and plant pigments on sensory and antioxidant properties of leaf extracts. *Journal of Food Science and Technology, 47*, 27–33.

Klatt, M. D., Sieck, C., Gascon, G., Malarkey, W., & Huerta, T. (2016). A healthcare utilization cost comparison between employees receiving a worksite mindfulness or a diet/exercise lifestyle intervention to matched controls 5 years post intervention. *Complementary Therapies in Medicine, 27*, 139–144.

Kljusurić, G. J., Orešković, P., Tušek K., & Kurtanjek Ž. (2014, June). *Linear programming versus fuzzy logic optimization in nutrition: Case of gestational diabetes*. OPT-i an international conference on engineering and applied sciences optimization, Kos, 03-07. paper 3631.

Kupka, J., & Rusnok, P. (2017). Regression analysis based on linguistic associations and perception-based logical deduction. *Expert Systems with Applications, 67*, 107–114.

Lesellier, E., Mith, D., & Dubrulle, I. (2015). Method developments approaches in supercritical fluid chromatography applied to the analysis of cosmetics. *Journal of Chromatography A, 1423*, 158–168.

Liu, Y., He, L., & Shen, J. (2017). Optimization-based provincial hybrid renewable and non-renewable energy planning – A case study of Shanxi, China. *Energy, 128*(1), 839–856.

Mogale, D. G., Kumar, S. K., García Márquez, F. P., & Tiwari, M. K. (2017). Bulk wheat transportation and storage problem of public distribution system. *Computers & Industrial Engineering, 104*, 80–97.

Orešković, P., Gajdoš Kljusurić, J., & Šatalić, Z. (2015). Computer-generated vegan menus: The importance of FCDB choice. *Journal of Food Composition and Analysis, 37*, 112–118.

Pinelo, M., Rubilar, M., Sineiro, J., & Nunez, M. (2005). A thermal treatment to increase the antioxidant capacity of natural phenols: Catechin, resveratrol, and grape extract cases. *European Food Research and Technology, 221*(3–4), 284–290.

Re, R., Pellegrini, N., Proteggente, A., Pannala, A., Yang, M., & Rice-Evans, C. (1999). Antioxidant activity applying an improved ABTS radical cation decolorisation assay. *Free Radical Biology and Medicine, 26*, 1231–1237.

StatSoft Inc. (2008). *STATISTICA (data analysis software system), version 8.0.* www.statsoft.com

Stich, J., Ramachandran, S., Hamacher, T., & Stimming, U. (2017). Techno-economic estimation of the power generation potential from biomass residues in Southeast Asia. *Energy, 135*(15), 930–942.

Wang, Y., Ma, X., Liu, M., Gong, K., Liu, Y., Xu, M., & Wang, Y. (2017). Cooperation and profit allocation in two-echelon logistics joint distribution network optimization. *Applied Soft Computing, 56*, 143–157.

Wirsam, B., & Hahn, A. (1999). Fuzzy methods in nutrition planning and education and clinical nutrition. In *Soft computing in human-related sciences*. Florida: CRC Press. ISBN 0849316359.

Wolfram Research Inc. (2017). *Mathematica, version 11.2*. IL: Champaign.

Zhang, T., Zhu, T., Gao, N., & Wu, Z. (2016). Optimization design of capacity and operation strategy for building level distributed energy system. *Procedia Engineering, 146*, 231–238.

Methodology and Criteria for the Allocation of Reused Water in Agriculture

María M. Borrego-Marín, L. Riesgo, and J. Berbel

Abstract This chapter provides a methodology to analyse the allocation of reused water. The tool has been developed for the Guadalquivir River Basin Authority, allowing decision makers to rank the actions on the reutilization of urban water for agriculture. The decision support is based on four groups of attributes: (1) resource supply, (2) environmental impact, (3) technical and economic feasibility and (4) social and institutional impact. A multicriteria decision method is proposed to aggregate all selected indicators. The results allow the River Basin Authority to classify different water requests of reused water, according not only to their technical knowledge, but also to the experience of different experts and stakeholders in water management.

Keywords Wastewater reuse · Agriculture · Water management · Multicriteria decision analysis · River Basin

1 Introduction

The combination of different factors, such as population growth, increased urbanization, water extraction for agricultural use, droughts, and deterioration of water quality implies a greater pressure on water resources worldwide (European Environment Agency 2016b). The level of water extraction is reaching the natural limits, and consequently, a drastic change in the conventional concept of use is required (United Nations 2017). As a result, water reuse is a strategy that has been

M. M. Borrego-Marín (✉) · L. Riesgo
Department of Economics, Quantitative Methods and Economic History, Pablo de Olavide University, Seville, Spain
e-mail: mmbormar@upo.es

J. Berbel
Department of Economics, Sociology and Agricultural Policies, University of Cordoba, Cordoba, Spain

J. Berbel et al. (eds.), *Multicriteria Analysis in Agriculture*, Multiple Criteria Decision Making, https://doi.org/10.1007/978-3-319-76929-5_8

gaining acceptance in many parts of the world, and connects directly with the new paradigm of the Circular Economy (European Commission 2015).

The Spanish legal framework for water reuse dates back to the Royal Decree 1620/2007, which includes authorized and prohibited uses, as well as the quality conditions required for each use (Ministry of Presidence 2007).

This chapter focuses on the technical bases and the criteria required for the allocation of reused water in agriculture by the Guadalquivir River Basin Authority, considering that agriculture is the main user of water resource in the river basin. The Water Authority may guarantee the compatibility of the proposed use with the hydrological planning (specifically, with the fulfillment of the ecological flows) and, eventually, the allocation of the resources requested from the reserve established for this purpose in the Hydrological Plans.

The use of a Multi-Criteria Decision Making (MCDM) technique to support decision making in the field of water management has numerous examples. Hajkowicz and Collins (2007) showed that MCDM in water resource management is widespread and growing. They reviewed 113 studies published since 1973. It was found that the annual publication rate has been steadily growing since the late 1980s. The majority of applications are related to the fields of water policy, supply planning and the evaluation of major infrastructure.

Recently Alvarado et al. (2016) select the best wells for water provision to urban networks. Borrego and Riesgo (2016) analyse and compare the sustainability of the integral water cycle in the Spanish River basins using two multicriteria decision-making methods, the Analytical Hierarchy Process (AHP) and the technique for order preference by similarity to ideal solution (TOPSIS). Sanguanduan and Nititvattananon (2011) apply MCDM methods for urban water reuse decision. Regarding wastewater reuse, Hadipour et al. (2016) apply a MCDM model (based on AHP) in order to find the best alternative for using wastewater in Iran as a case study. Kalavrouziotis et al. (2011) apply a MCDA model to the management of the Wastewater Treatment Plant (WWTP) of a Greek city with the aim of finding an optimum solution of the wastewater and bio sludge disposal.

2 Background and Case Study Description

The Guadalquivir River is the longest river in southern Spain with a length of around 650 km. Its basin covers an area of 57,527 km^2 and a population of 4,107,598 inhabitants. The basin has a Mediterranean climate with a heterogeneous precipitation distribution, annual average temperature is 16.8 °C, and the annual precipitation averages at 573 mm, with a range between 260 mm and 983 mm (standard deviation of 161 mm). The average renewable resources in the basin amount to 7043 (arithmetic mean) or 5078 hm^3/year (median), ranging from a minimum of 372 hm^3/year to a maximum of 15,180 hm^3/year. Agriculture is the main water user in the basin according to the Guadalquivir River Basin Management Plan (Fig. 1).

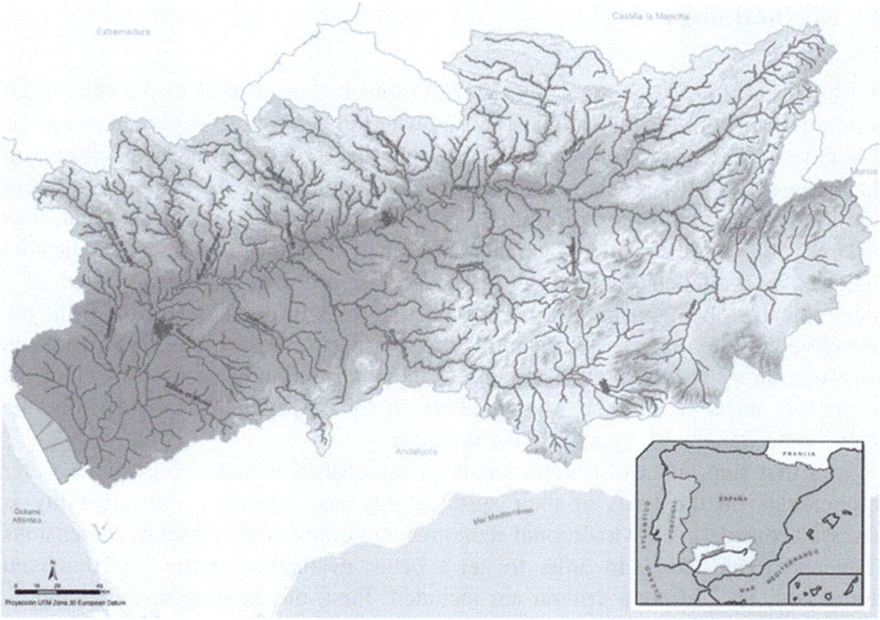

Fig. 1 Guadalquivir River Basin

Guadalquivir River Basin is probably the largest basin with severe scarcity problems in Europe (Berbel et al. 2012). Reused water in the Guadalquivir River Basin cannot be considered "new" water as the volume is already accounted in the water balance due to the integrated nature of the river basin where all return flows (either from agriculture, industry or urban) are an input for the next user downstream and support the environmental flow.

The growing demand for water resources has been driven by profitable agribusiness (Berbel et al. 2012). Within the Guadalquivir basin, more than 50% of Spanish olive oil is produced, and almost 50% of the olive groves are irrigated with a prevalence of high-tech intensive cultivation. In addition, some high-value crops such as early season fruits and strawberries increase the demand for irrigation water, while urban development and industrial demand simultaneously intensify the pressures, a situation which led to the 'administrative closure' of the basin when the old policy of supply (storage) increase could not meet the gap between demand and supply of resources. Berbel et al. (2013) discuss changes over time in the basin's water supply and extraction. It means that all basin resources are already compromised, and therefore there is not possible to allocate additional resources.

As a result, the Guadalquivir Basin Management Plan (2015–2021), approved by Royal Decree 1/2016, of 8 January (Ministry of Agriculture, Food and Environment 2016), establishes a reserve of up to 20 hm^3/year of reused water. The River Basin Authority shall establish the process to be followed to allocate these resources. Therefore, River Basin Authority needs a useful tool to rank and to hierarchy the actions regarding urban water reuse in agriculture.

3 Methodology

A Multicriteria Decision Analysis (MCDA) tool has been used to prioritize each action or project in the allocation of reused water in agriculture in the Guadalquivir River Basin Authority. The selection of an appropriate method has been chosen on the basis of the different criteria and indicators considered to evaluate different possibilities of reuse water allocation. As it will be presented below in detail, four different criteria are considered to reallocate water, as well as a number of indicators that can be classified within each criterion. The problem can be specified as a hierarchical problem, from a general objective and different criteria to fulfil the objective till a number of indicators to assess each criterion (see Fig. 2). Such structure allows specifying the problem as the contribution of each of the elements to the full objective, and the Analytical Hierarchy Process (AHP) seems to be the appropriate framework to analyse the problem.

In a first step, a set of criteria has been established in order to evaluate water applications on the basis of their sustainability and adequacy. Sustainability is assessed considering the traditional economic, environmental and social dimensions (Tan and Egan 2017). In order to get a better evaluation of the environmental dimension, two different criteria are included. First, the existing resource supply and secondly the environmental impact of the allocation of reused water. Then, all the criteria to evaluate the sustainability and suitability of water requests are the following:

(a) Resource supply: availability and quality of effluent resources.
(b) Environmental impact: environmental sustainability and compatibility with the environmental objectives of the water bodies.
(c) Technical and economic feasibility of the action, both in terms of initial investment and subsequent maintenance and operation.
(d) Territorial equity and social criteria.

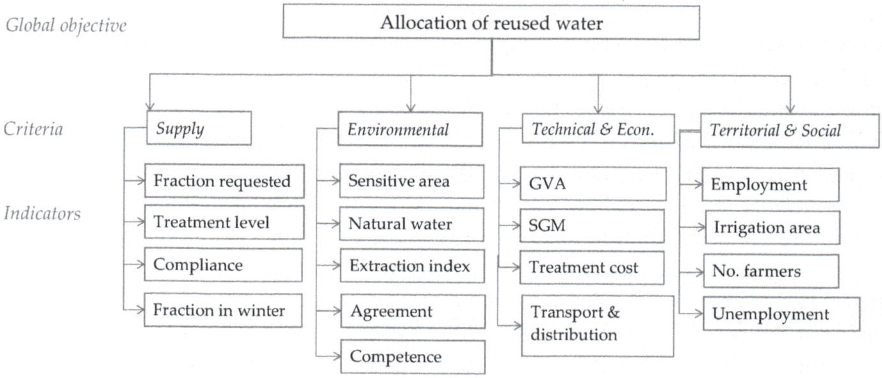

Fig. 2 AHP structure

To evaluate each criterion a number of indicators has been selected based on both the literature review and the expertise of a panel of experts. Table 1 shows the indicators used to evaluate the criteria.

The *supply dimension* is measured through four different indicators:

(a) Fraction of the total requested effluent: this indicator measures the requested and available volume of water in order to identify potential disequilibria between water demand and supply. In order to comply with the precautionary principle, a limitation has been set and the requested volume cannot exceed 80% of existing water. Considering this limitation, the following table (Table 2) shows the scores to each potential request.
(b) Current treatment level: according to EDARNet (Database of Waste-Water Treatment Plants), a high treatment level is highly recommended in order to avoid additional costs later on. Regeneration costs only include filtering and

Table 1 Criteria and indicators to assess water applications

Criteria	Indicators
Supply	Fraction of total requested effluent
	Current treatment level
	Compliance (wastewater treatment plant, WWTP)
	Fraction of the total volume captured in winter
Environmental impact	Discharge to a sensitive area
	A natural surface water body to which effluents are discharged
	Extraction index of the surface water body from which discharge is drawn
	Agreement with the owner of the discharge
	Sufficiency of proposed regulation
Technical and economic feasibility	Gross value added (GVA)
	Standard gross margin (SGM)
	Treatment cost
	Transport and distribution cost
Territorial and social	Employment
	Irrigation area existing in the location
	Total number of farmers in irrigation communities
	Unemployment in the area

Table 2 Scores[a] depending on the fraction of the total requested effluent

Fraction requested under 40% of the total requested effluent	5
Fraction requested between 40% and 50% of the total requested effluent	4
Fraction requested between 50% and 60% of the total requested effluent	3
Fraction requested between 60% and 70% of the total requested effluent	2
Fraction requested between 70% and 80% of the total requested effluent	1

[a]All the scores have been established by the Guadalquivir River Basin Authority

Table 3 Scores depending on the current treatment level

Secondary treatment + Strict	5
Secondary treatment	2
Lower-level treatments	1

Table 4 Scores depending on the compliance

Compliant wastewater	5
Non-compliant wastewater	1

Table 5 Scores depending on water abstracted in winter

Water abstraction of more than 90% in winter	5
80%–90% of water abstraction in winter	4
70%–80% of water abstraction in winter	3
60%–70% of water abstraction in winter	2
Water abstraction of less than 60% in winter	1

Table 6 Scores depending on potential discharge to a sensitive area

Discharge to a sensitive area	2
Discharge to a non-sensitive area	1

disinfection issues. Each request will be evaluated according to the following scores (Table 3) for this indicator.

(c) Compliance: the lack of compliance of the wastewater implies malfunctioning and therefore may increase the costs. According to this indicator, the following table (Table 4) shows the potential scores to each request.

(d) Fraction of the total volume abstracted in winter (15 September–15 April): water abstraction in winter may reduce the probability of affecting other water uses. Each request will be evaluated according to the following scores (Table 5) for this indicator:

The *environmental impact dimension* is measured through five different indicators:

(a) Discharge to a sensitive area: it is considered beneficial the nutrients remove caused by water reuse, in special when the requested use is discharged to a sensitive area (Table 6).

(b) Natural surface water body to which effluents are discharged (Table 7): water abstraction to be reused in water bodies where the alteration is low is negatively considered. The indicator to measure this is the percentage of the Potential Useful Habitat (PUH) established in the Hydrological Plan as an objective to determine ecological flows.

(c) Extraction index of the surface body where water is abstracted: the aim of this indicator is to show the impact of the surface water body. Therefore, abstractions in water bodies with high frequency of abstractions are negatively evaluated, as can be seen below (Table 8).

(d) Agreement with the owner of the wastewater discharge (Table 9): the existence of an agreement with the owner of the wastewater use right to collaborate in the

Table 7 Scores depending on the natural surface water body

High hydrological alteration (PUH 30%)	5
Medium hydrological alteration (PUH 50%)	3
Low hydrological alteration (PUH 80%)	1

Table 8 Scores depending on the extraction index of the surface water body

Extraction index lower than 0.50	5
Extraction index between 0.50 and 0.60	4
Extraction index between 0.60 and 0.70	3
Extraction index between 0.70 and 0.80	2
Extraction index higher than 0.80	1
Undetermined	2

Table 9 Scores depending on the agreement with the user

Agreement	5
Chance of agreement	3
Negotiation procedure	1

Table 10 Scores depending on the competence of the regulation

Volume requested/estimated volume needed $\geq$0.80	5
Volume requested/estimated volume needed between 0.67 and 0.80	4
Volume requested/estimated volume needed between 0.50 and 0.67	3
Volume requested/estimated volume needed between 0.33 and 0.50	2
Volume requested/estimated volume needed between 0.25 and 0.33	1

maintenance of the wastewater treatment plant or WWTP when the water reuse is requested is considered as positive.

(e) Competence of the regulation: water reuse request must be justified through a project, including the characteristics of the infrastructure for storage and distribution of recycled water from the point of delivery of regenerated water to the location where water is going to be used. Such infrastructure is going to be evaluated on the basis of their capacity to reduce the impact of recycled water on water bodies (Table 10).

The *technical and economic feasibility* includes four different indicators:

(a) Gross Added Value, GAV (€): this indicator is calculated as the multiplication of the unit GAV of each crop (GAV/m^3) times the water use of such crop. The GAV allow us to calculate water contribution to local and regional GDP (Table 11).
(b) Standard Gross Margin, SGM (€): this indicator shows the profitability of the farm (Table 12).

Table 11 Scores depending on the GAV

GAV >5,000,000 €	5
GAV between 2,000,000 and 5,000,000 €	4
GAV between 600,000 and 2,000,000 €	3
GAV between 200,000 and 600,000 €	2
GAV <200,000 €	1

Table 12 Scores depending on the SGM

SGM >3,500,000 €	5
SGM between 1,200,000 and 3,500,000 €	4
SGM between 400,000 and 1,200,000 €	3
SGM between 120,000 and 400,000 €	2
SGM <120,000 €	1

Table 13 Scores depending on the treatment cost

Treatment cost <0.09 €	3
Treatment cost between 0.09 and 0.12 €	2
Treatment cost >0.12 €	1

Table 14 Scores depending on the transportation and distribution costs

Transportation and distribution costs <0.05 €	4
Transportation and distribution costs between 0.05 and 0.10 €	3
Transportation and distribution costs between 0.10 and 0.20 €	2
Transportation and distribution costs >0.20 €	1

(c) Treatment cost (€/m^3): the recycling water costs (amortization of investment as a function of useful life, use and maintenance of infrastructures). All required treatments are included on the basis of the water use (Table 13).
(d) Transportation and distribution costs: these costs include all the costs from water abstraction (WWTP discharge) to the irrigation area, such as water pump, discharge pipelines to the pool, energy pumping costs and operating costs from the pool to the farm. Amortization of investments and maintenance are also considered (Table 14).

The *Territorial and social dimension* is measured through four different indicators:

(a) Annual labour. To assess such indicator the crops grown in the farm are taken into account (Table 15).
(b) Pre-existing irrigated area as a percentage of the village area. Those villages with small irrigated area are better evaluated since new projects are promoted in such areas (Table 16).
(c) Total number of water users in the irrigated using reused water. The higher number of water users involved the higher the valuation (Table 17).
(d) Average unemployment in 2015 (% of registered population): requests from areas with high unemployment rates are better evaluated, due to the capacity of agriculture to create jobs (Table 18).

Table 15 Scores depending on labour

Labour >130,000 man.days	5
Labour between 50,000 and 130,000 man.days	4
Labour between 20,000 and 50,000 man.days	3
Labour between 10,000 and 20,000 man.days	2
Labour <10,000 man.days	1

Table 16 Scores depending on existing irrigated area

Irrigated area <1% village area	5
Irrigated area between 1 and 7% village area	4
Irrigated area between 7 and 15% village area	3
Irrigated area between 15 and 34% village area	2
Irrigated area >34% village area	1

Table 17 Scores depending on total number of water users

More than 100 water users	5
Between 51 and 100 water users	4
Between 26 and 50 water users	3
Between 6 and 25 water users	2
Less than 6 water users	1

Table 18 Scores depending on average employment in 2015

Unemployment >12% registered population	5
Unemployment between 10 and 12% registered population	4
Unemployment between 8 and 10% registered population	3
Unemployment between 5 and 8% registered population	2
Unemployment <5% registered population	1

In a second step, considering the indicators mentioned above, the Analytical Hierarchy Process (AHP) is used to get the importance of each criteria and each indicator to assess each action or project received in the Guadalquivir River Basin Authority for the use of reused water in agriculture.

AHP allows to aggregate the performance of each attribute in each criterion, and afterwards to get a score based on the aggregation of each one. This aggregation of indicators and criteria considers the importance given by stakeholders.

The AHP method was created by Saaty (1980) as a structured but flexible technique for making decisions in a multi-criteria context. This method is based on dealing with complex decision problems using a hierarchical structure. Figure 2 shows the three-level structure considered for our case study.

Within this hierarchical structure, the relative importance or weightings (w_k) of criteria or sub-criteria hanging on each node are obtained from pair-wise comparisons between them. In order to perform these pair-wise comparisons, a 1–9 scale is used (Table 19), as proposed by Aguarón and Moreno-Jiménez (2000).

Scores of these comparisons are used to build the Saaty matrices ($A = a_{jk}$), which are employed to determine the vector of priorities or weights ($w_1, \ldots w_k, \ldots w_n$). Although different procedures to estimate these weights have been proposed, for

Table 19 Table of relative scores

Value of ajk	Scale meaning
1	j and k are equally important
3	j is slightly more important than k
5	j is more important than k
7	j is strongly more important than k
9	j is absolutely more important than k
2, 4, 6, 8	Middle values of the above
Reciprocal	$a_{jk} = 1/a_{kj}$

this case we select the simplest one: the geometric mean method, as proposed by Easly et al. (2000).

The AHP decision technique was originally designed for individual decision-makers, but was promptly extended for group decisions, such as our case study. Thus, in order to determine the weights attached to each criterion we have to consider the judgements of a group of people (p), each with his/her own pair-wise comparison matrix ($A_p = a_{jkp}$) and its related weights (w_{kp}). This individual information is suitably treated in order to obtain a synthesis of aggregated weights (w_k).

For this purpose, Forman and Peniwati (1998), and Whitmarsh and Palmieri (2009), suggest that group decision making should be done by aggregating individual priorities using the geometric mean:

$$w_k = \sqrt[m]{\prod_{p=1}^{p=m} w_{kp}} \tag{1}$$

For indicators weighting, we employed a panel of 12 experts in water management, sustainability (university lecturers, members of agricultural research centres, civil servants in charge of water policy implementation, environmental associations and farmers).

Before aggregating priority scores, the consistency of respondents' pairwise choices is tested by means of the consistency ratio (CR) based on the eigenvalue method proposed by Saaty and Vargas (2000). In this paper, we consider only CR lower than 0.1, as suggested by Bozoki and Rapcsak (2008).

4 Results

The application of the AHP allows us to get a classification of different actions according to their score and the performance of each project in every particular criterion. It also illustrates the importance of each indicator in contributing to it.

Table 20 shows the results of the application of the AHP method. First, we can see the weights for the sustainability dimensions according to the preferences of the group of experts. The environmental dimension is playing the most important role in the whole sustainability (44%), followed by the territorial and social criteria (25%)

and supply (18%). The dimension of technical and economic feasibility is the least important for sustainability of each project (14%) according to the panel of experts.

Using the weights showed above a global ranking of the different water requests can be got, and consequently, reused water will be allocated to the proposal with the highest score.

First, we need to evaluate each of the indicators according to what is included in the water request project. For each indicator the scores established by the Guadalquivir River Basin Authority (see Tables 2–18) are used (i.e., according to each water proposal, the indicator on water fraction requested would be evaluated from 1 to 5 (see Table 2), the indicator on treatment level would be valued from 1 to 5 (see Table 3), etc.).

Once every indicator is assessed, the next step consists of including the weights of each indicator and dimension on the basis of the weights established by the panel of experts (see Table 20), according to the following process:

$$score_p = \sum_m \sum_j d_m \cdot i_j \cdot score\_indicator_{m,j} \quad (2)$$

where $score_p$ is the score of each water request, $score\_indicator_{m,j}$ is the result of evaluating each indicator by the Guadalquivir River Basin Authority (Tables 2–18), i_j is the weight given by the panel of experts to each indicator (Table 20), and d_m is the weight of each dimension (Table 20).

Following this process, a ranking of projects will be obtained, and the water proposal selected for the allocation of reused water would be that positioned on the first place in the ranking (Table 21).

Table 20 Normalised weights for dimensions/criteria and indicators

Dimensions		Indicators	
Supply (d_a)	0.18	Fraction requested (i_{a1})	0.10
		Treatment level (i_{a2})	0.56
		Compliance (i_{a3})	0.27
		Fraction in winter (i_{a4})	0.07
Environmental (d_b)	0.44	Sensitive area (i_{b5})	0.36
		Natural water (i_{b6})	0.29
		Extraction index (i_{b7})	0.20
		Agreement (i_{b8})	0.05
		Competence (i_{b9})	0.10
Technical and economic (d_c)	0.14	Gross value added (i_{c10})	0.20
		Standard gross margin (i_{c11})	0.24
		Treatment cost (i_{c12})	0.30
		Transportation and distribution (i_{c13})	0.25
Territorial and social (d_d)	0.25	Employment (i_{d14})	0.40
		Irrigation area (i_{d15})	0.13
		Number of farmers (i_{d16})	0.15
		Unemployment (i_{d17})	0.33

Table 21 Global ranking of projects

Applications	$Score_p$	Ranking
Project 3	Highest score	1
Project 7	Second highest score	2
⋮		⋮
Project n	Lowest score	.

5 Concluding Remarks

Water reuse can be better integrated in water planning management (European Environment Agency 2016a). It should be more systematically considered by Member States in implementation of EU water legislation and the water scarce regions worldwide. It can be considered a measure to address water scarcity and achieve good status under the Water Framework Directive, and in the investment in the treatment of effluent as required by the Urban Waste Water Treatment Directive. For this purpose, guidelines on Integrating Water Reuse into Water Planning and Management in the context of the WFD were developed by the Commission, Member States and stakeholders, within the Common Implementation Strategy of the WFD. These guidelines (European Environment Agency 2016a) are based on existing practice in the EU and third countries. They contain recommendations on how to better integrate water reuse in water planning and management within the EU policy framework and taking into account underlying environmental and socio-economic benefits.

This chapter proposes a set of indicators that may be considered by the River Basin Authority to analyze different proposals in order to allow increased use of reused water in agriculture. The use of alternative water resources, technologies and governance models is a top priority in the Partnership for Research and Innovation in the Mediterranean Area stated by PRIMA Strategic Research and Innovation Agenda (CSA 4PRIMA), so that our proposal aims to enforce the governance and management instrument to reach these objectives.

Therefore, this study offers a useful tool for the River Basin Authority to rank different proposals regarding urban water reuse for agriculture and to facilitate water management in the river basin. This tool includes not only the preferences of the River Basin Authority based on technical criteria, but also the preferences of a group of experts in water management. Considering such preferences, the Guadalquivir River Basin Authority would be able to classify the water proposals according to both technical criteria and preferences of the society related to water management. Consequently, a higher acceptability of the water action is expected by those agents involved in the process.

However, this method may not be seen as a general framework for any allocation of reuse water, since most of the indicators thresholds (e.g., scores depending on the fraction of the total requested effluent, on the current treatment level, compliance, etc.) have been set by the competent authority, i.e. the Guadalquivir River Basin Authority. Any application of this methodology to another River Basin would require an updating of such thresholds to the current situation of such basin.

Acknowledgments The research behind this study has been financed by Guadalquivir River Basin Authority under the Project "Definition of allocation criteria for the reserve of up to 20 hm^3/year of reused water. CU(PH)-5489" and the authors have received financial support from MINECO-Grant AGL2014-53417-R. This paper does not represent any official position or endorsement of the funding organization.

References

Aguarón, J., & Moreno-Jiménez, J. M. (2000). Stability intervals in the analytic hierarchy process. *European Journal of Operational Research, 125*(114), 133.

Alvarado, A., Esteller, M. V., Quentin, E., & Expósito, J. L. (2016). Multi-criteria decision analysis and GIS approach for prioritization of drinking water utilities protection based on their vulnerability to contamination. *Water Resources Management, 30*(4), 1549–1566.

Berbel, J., Kolberg, S., & Martin-Ortega, J. (2012). Assessment of the draft hydrological basin plan of the Guadalquivir river basin (Spain). *International Journal of Water Resources Development, 28*(1), 43–55.

Berbel, J., Pedraza, V., & Giannoccaro, G. (2013). The trajectory towards basin closure of a European river: Guadalquivir. *International Journal of River Basin Management, 11*(1), 111–119.

Borrego-Marín, M. M., & Riesgo, L. (2016). Measuring the sustainability of water plans in inter-regional Spanish river basins. *Water, 8*(8), 342.

Bozoki, S., & Rapcsak, T. (2008). Saaty's and Koczkodaj's inconsistencies of pairwise comparison matrices. *Journal of Global Optimization, 42*(2), 157.

Coordination and Support Action "CSA 4PRIMA". (2017). Retrieved from http://4prima.org/sites/default/files/publication/PRIMA%20SRIA.pdf

Easley, R. W., Madden, C. S., & Dunn, M. G. (2000). Conducting marketing science: The role of replication in the research process. *Journal of Business Research, 48*(1), 83:92.

European Commission. (2015). *Closing the loop - an EU action plan for the circular economy.* COM/2015/0614 final. European Commission, communication from the commission to the European Parliament, The Council, The European Economic and Social Committee and the Committee of the Regions.

European Environment Agency. (2016a). *Guidelines on integrating water reuse into water planning and management in the context of the WFD.* Retrieved from https://www.eea.europa.eu/policy-documents/guidelines-on-integrating-water-reuse

European Environment Agency. (2016b). *Use of freshwater resources.* Retrieved from http://www.eea.europa.eu/data-and-maps/indicators/use-of-freshwaterresources-2/assessment-1

Forman, E., & Peniwati, K. (1998). Aggregating individual judgments and priorities with the analytic hierarchy process. *European Journal of Operational Research, 108*, 165:169.

Guadalquivir Basin Management Plan. (2015–2021). Retrieved from http://www.chguadalquivir.es/opencms/portalchg/planHidrologicoDemarcacion/demarcacionGuadalquivir/segundoCicloPlanificacion/

Hadipour, A., Rajaee, T., Hadipour, V., & Seidirad, S. (2016). Multi-criteria decision-making model for wastewater reuse application: A case study from Iran. *Desalination and Water Treatment, 57*(30), 13857–13864.

Hajkowicz, S., & Collins, K. (2007). A review of multiple criteria analysis for water resource planning and management. *Water Resources Management, 21*(9), 1553–1566.

Kalavrouziotis, I. K., Filintas, A. T., Koukoulakis, P. H., & Hatzopoulos, J. N. (2011). Application of multicriteria analysis in the management and planning of treated municipal wastewater and sludge reuse in agriculture and land development: The case of sparti's wastewater treatment plant, Greece. *Fresenius Environmental Bulletin, 20*(2), 287–295.

Ministry of Agriculture, Food and Environment. (2016). *BOE 2016 Royal Decree 1/2016, 439, 2972-4301*. Retrieved from https://www.boe.es/boe/dias/2016/01/19/pdfs/BOE-A-2016-439.pdf

Ministry of Presidence. (2007). *BOE 2007 Royal Decree 1620/2007, 294, 50639-50661*. Retrieved from http://www.boe.es/boe/dias/2007/12/08/pdfs/A50639-50661.pdf

Saaty, T. L. (1980). *The analytic hierarchy process*. New York: McGraw-Hill.

Saaty, T. L., & Vargas, L. G. M. (2000). *Methods, concepts and applications of the analytic hierarchy process*. Boston: Kluwer.

Sanguanduan, N., & Nititvattananon, V. (2011). Strategic decision making for urban water reuse application: A case from Thailand. *Desalination, 268*(1–3), 141–149.

Tan, L. K., & Egan, M. (2017). The public accountability value of a triple bottom line approach to performance reporting in the water sector. *Australian Accounting Review*. https://doi.org/10.1111/auar.12173.

United Nations. (2017). *World water development report*. Retrieved from http://www.unwater.org/publications/world-water-development-report-2017/

Whitmarsh, D., & Palmieri, M. G. (2009). Social acceptability of marine aquaculture: The use of survey based methods for eliciting public and stakeholder preferences. *Marine Policy, 33*, 452.

Simulating Farmers' Decision-Making with a Cobb-Douglass MAUF: An Application for an *Ex-Ante* Policy Analysis of Water Pricing

Nazaret M. Montilla-López, José A. Gómez-Limón, and Carlos Gutiérrez-Martín

Abstract Classical economic theory relies on the assumption that farmers' behavior can be modeled by maximizing profits or any utility function with profits as a single attribute. However, farmers' decision-making processes are actually driven by various typically conflicting criteria, in addition to the expected profit. Therefore, it must be assumed that producers' behavior is guided by the maximization of a multi-attribute utility function (MAUF) in which all relevant attributes considered for decision-making are condensed. The objective of this paper is to provide more in-depth knowledge about simulating farmers' behavior by using non-linear MAUFs, developing a new non-interactive method to elicit Cobb-Douglas MAUFs based on farmers' actual behavior that overcomes some shortcomings of traditional additive MAUFs. Moreover, this approach is compared with two others that are widely used: the profit maximization and additive MAUF approaches. This procedure is implemented for illustrative purposes to analyze the feasible impacts of water pricing in an irrigated district in southern Spain. The results obtained show that simulations using the Cobb-Douglas utility function are more reliable than the alternatives already used in the literature. In this regard, two pieces of evidence justify this assessment: the calibration is more precise, and the resulting water-demand curve is smoother than in the other two alternative simulation approaches considered.

Keywords Farmers' behavior · Mathematical programming · Multi-attribute utility theory · Non-linear MAUF · Simulation analysis · Water policy

N. M. Montilla-López · J. A. Gómez-Limón · C. Gutiérrez-Martín (✉)
Universidad de Córdoba. Water, Environmental and Agricultural Resources Economics (WEARE) research group, Córdoba, Spain
e-mail: carlos.gutierrez@uco.es

J. Berbel et al. (eds.), *Multicriteria Analysis in Agriculture*, Multiple Criteria Decision Making, https://doi.org/10.1007/978-3-319-76929-5_9

1 Introduction

Classical economic theory relies on the assumption that farmers' behavior can be modeled by maximizing profit or any utility function with profit as a single attribute, as assumed by *expected utility theory* (EUT). In fact, EUT has become one of the most popular approaches to simulating farmers' decision-making, being implemented using several mathematical programming tools (Chavas et al. 2010). However, there is large amount of evidence not only supporting the consideration of expected profit for farmers' decision-making processes but also agreeing that these processes are driven by various—typically conflicting—criteria related to their economic, social, cultural, and natural environmental criteria (for recent empirical studies confirming this idea see Berkhout et al. 2011; Mandryk et al. 2014). Hence, it can be assumed that producers' decision-making is guided by the maximization of a multi-attribute utility function (MAUF), in which all of the relevant attributes considered are condensed. This is the main idea underlying *multi-attribute utility theory* (MAUT), an approach largely developed after the publication of the seminal study by Keeney and Raiffa (1976) to overcome the limitations of single-attribute (profit-related) utility functions. This alternative approach has also been widely implemented in simulating farmers' behavior, as shown by Sumpsi et al. (1997), Amador et al. (1998), Gómez-Limón and Berbel (2000) and Gómez-Limón et al. (2004), among others.

Most empirical approaches to implementing MAUT to simulate farmers' decision-making have relied on additive MAUFs ($U = w_1 \cdot f_1 + w_2 \cdot f_2 + \ldots + w_m \cdot f_m$, where $f_1, f_2, \ldots f_m$ are the different attributes considered and $w_1, w_2, \ldots w_m$ are the weights given by farmers to each attribute) since these linear specifications of the utility function are easier to elicit and to interpret. These MAUFs have typically been estimated using a non-interactive procedure based on weighted goal programming (WGP), as shown by Sumpsi et al. (1997), Amador et al. (1998) or Gómez-Limón et al. (2004). This approach makes it possible to obtain the weights (w_a) of every single attribute with respect the total utility by solving an equation system in which a linear combination of the optimum of each attribute equals the observed attribute levels (for further details, see Sect. 2.2). However, this additive specification has some shortcomings from an economic perspective, with the most relevant being the assumptions regarding the constant marginal rate of substitution among attributes due to the consideration of linear indifference curves and the total compensation among attributes. This makes additive MAUFs inaccurate when simulating actual decision-making (Hardaker et al. 2007). For this reason, in this paper, we propose the elicitation of a Cobb-Douglas utility function ($U = f_1^{\alpha_1} \cdot f_2^{\alpha_2} \cdot \ldots \cdot f_m^{\alpha_m}$, where $\alpha_1, \alpha_2, \ldots \alpha_m$ are parameters related with the relevance given by farmers to each attribute) as a sounder approach, as already suggested by Gutiérrez-Martín and Gómez-Gómez (2011) and Gómez-Limón et al. (2016). This choice is justified because this function shape is more coherent with economic theory since it meets conditions of Inada (1963) that guarantee that there is a global optimum when the efficient frontier is convex, and it is consistent with the postulate of decreasing marginal utility for every attribute.

Gutiérrez-Martín and Gómez-Gómez (2011) and Gómez-Limón et al. (2016) have recently provided two different non-interactive approaches to eliciting the alpha coefficients (α_a) for Cobb-Douglas MAUFs. In general, the methods developed in both papers are based on the elicitation of the alpha coefficients by equating the marginal rate of transformation between attributes on the efficient frontier and the marginal rate of substitution between attributes on the indifference curves. However, both approaches involve a relatively complex operational burden that makes their implementation in a real-world setting difficult.

The general objective of this paper is to provide more in-depth knowledge about simulating farmers' behavior by using non-linear MAUFs. To that end, this objective is twofold. First, a new and simpler method is developed to elicit Cobb-Douglas MAUFs based on farmers' actual behavior. Second, this method is implemented for illustrative purposes to simulate farmers' behavior in case a water pricing policy were in place to show that this method is easier to implement than actual simulation exercises. In addition, the results obtained in this manner are compared with those resulting from simulation models that use profit and additive MAUF maximization, confirming the advantages of the Cobb-Douglas MAUF approach for simulation purposes.

To reach the objectives described above, this paper is organized as follows. After this introductory section, the following section introduces the new method developed to elicit Cobb-Douglas MAUFs as a sounder approach compared to existing alternatives to simulating farmers' behavior such as profit or additive MAUF maximization. Section 3 is focused on a real case study considered for the methodological implementation. First, the rationale of irrigation water pricing is explained; second, the irrigation farm type considered for modeling purposes is described. The following section describes the model construction, showing the decision variables, the attributes to be included within the MAUFs, the model constraints and the water price scenarios considered. Section 5 presents the results, both those from the MAUF calibration procedures and those from the simulations of water prices implemented. Finally, in view of the results obtained, Sect. 6 concludes, reviewing the advantages of the Cobb-Douglas simulation approach and the procedure developed to elicit these MAUFs.

2 Simulating Farmers' Decision-Making: Alternative Approaches

Farmers' decision-making aims to choose productive alternatives (i.e., crop and agricultural technique mixes) that maximize the farmers' utility. This utility can come from a single attribute (profit, according to classic economic theory) or from various attributes (constituted by more complex utility functions, as assumed by

MAUT). Regardless of the assumption considered regarding farmers' behavior (mono- or multi-attribute-guided), farmers' utility functions are considered to be a structural feature of these producers; that is, these utility functions do not change when circumstances do. For this reason, these functions can be used to simulate future scenarios by maximizing the corresponding utility function while changing the scenario specific parameters.

Moreover, to simulate farmers' behavior using mathematical programming (MP), the entire set of decision variables (i.e., all feasible crop and agricultural technique mixes) and all physical, technical, market or legal constraints that narrow the set of feasible solutions must be taken into account. In this manner, MP models for simulating farmers' decision-making are built considering an objective function (a mono- or multi-attributed utility function that, in turn, depends on the decision variables) to be maximized and the set of constraints limiting farmers' choices. In this section, we introduce three different approaches to modeling farmers' decision-making in this manner, with the last approach being a new contribution to simulating farmers' behavior.

The methodological approaches used in this paper are explained following an evolutionary rationale. First, an MP model relying on a classic mono-attribute objective function with profit as the only relevant attribute for decision-making is introduced.

Considering the multi-criterial nature of farmers' decision-making, the second approach presented is an MP model that maximizes an additive MAUF, elicited following the WGP method proposed by Sumpsi et al. (1997). This approach implies linear specifications for MAUFs that involves some shortcomings that warrant discussion (Hardaker et al. 2007). Considering additive MAUFs implies linear indifference curves (also called iso-utility curves or iso-preference curves), a condition involving a constant marginal rate of substitution among attributes that leads to oversimplified simulations of farmers' behavior. Furthermore, additive MAUFs allow total compensability among attributes; this is, lower values of every particular attribute can be compensated for by higher values of any other attribute, even if the former reach unacceptably low levels for farmers. This implication also makes additive MAUFs inaccurate when simulating farmers' actual decision-making.

Both limitations of additive MAUFs can be overcome with other utility specifications. In this sense, André and Riesgo (2007) have shown how the application of multiplicative utility functions can be more successful than additive utility functions in reproducing farmers' behavior. For this reason, Gutiérrez-Martín and Gómez-Gómez (2011) and Gómez-Limón et al. (2016) have proposed the use of Cobb-Douglas utility functions as a general and flexible multiplicative form for MAUFs that allow more real indifference curves and partial compensation between attributes. Moreover, this function is coherent with neoclassic economic theory since it guarantees that there is a global optimum when the efficient frontier is convex, and this formulation is consistent with the postulate of decreasing marginal utility for every attribute. Thus, the third approach proposed for simulating farmers' behavior is an MP model that maximizes a Cobb-Douglas MAUF. As noted above, to that end, a new and simpler procedure for eliciting Cobb-Douglas MAUFs is explained.

2.1 Profit Maximization

The more classical modeling approach to simulating farmers' behavior is to construct an MP model that considers profit (π) as the unique attribute to be maximized under a set of constraints. Thus, in this case, the model is as follows:

$$Max\ \ U(\boldsymbol{X}) = \pi(\boldsymbol{X}) \tag{1a}$$

$$\begin{aligned} \text{s.t.}\quad \boldsymbol{AX} &\leq \boldsymbol{B} \\ \boldsymbol{X} &\geq 0 \end{aligned} \tag{1b}$$

where $\boldsymbol{X}$ ($n \times 1$) is the vector of decision variables (the area devoted to each crop-technique mix), $U(\boldsymbol{X})$ is the objective function (total utility) to be maximized and $\pi(\boldsymbol{X})$ is the profit function. The model constraints are built based on matrix $\boldsymbol{A}$ ($p \times n$) of technical coefficients of the allocable resource constraints and vector $\boldsymbol{B}$ ($p \times 1$) of the available resource levels.

2.2 Additive MAUF Maximization: WGP Approach

The empirical implementation of MAUT for modeling purposes make it necessary to consider some assumptions. The most relevant assumption is that all of the attributes (f_a) contained within the MAUF (U) must be utility-independent.[1] This allows the entire utility $U = U(f_1,\ldots f_a,\ldots f_m)$ to became a separable function: $U = g[u_1(f_1), u_2(f_2),\ldots u_m(f_m)]$. Moreover, if both total and partial utility functions take values in the range of 0–1, then the MAUF takes either the additive form ($U = \Sigma\, w_a\, u_a(f_a)$) or the multiplicative form ($U = [\Pi\ (K \cdot w_a \cdot u_a(f_a)+1)-1]/K$), where $0 \leq w_a \leq 1$ and $K = f(w_a)$. If attributes are mutually utility-independent and $\Sigma w_a = 1$, then $K = 0$, and the utility function is additive. By contrast, if $\Sigma\, w_a \neq 1$, then $K \neq 0$, and the mathematical form is multiplicative (Fishburn 1982).

Considering that farmers' decision-making is guided by m attributes f_a and the abovementioned requirements for an additive MAUF are fulfilled, the simulation MP model can be specified as follows:

$$Max\ \ U(\boldsymbol{X}) = \sum_{a=1}^{m} w_a \cdot u_a(f_a(\boldsymbol{X})) = \sum_{a=1}^{m} w_a \cdot nf_a(\boldsymbol{X}) \tag{2a}$$

[1] According to Keeney and Raiffa (1976), attribute i is defined as the utility independent of attribute j when the conditional preferences for lotteries on attribute i given the attribute j do not depend on the particular level of attribute j (p. 226).

$$\text{s.t.} \quad \begin{aligned} &\sum_{a=1}^{m} w_a = 1 \\ &\boldsymbol{AX} \leq \boldsymbol{B} \\ &\boldsymbol{X} \geq 0 \end{aligned} \tag{2b}$$

This objective function depends on a set of m single or partial utility functions ($u_a(f_a(\boldsymbol{X}))$ that consider all relevant attributes for producers' decision-making, and w_a denotes the weight of each attribute, expressing its relative importance.

For operational and comparability purposes, we consider that all relevant attributes are related to objectives to be maximized (i.e., more-is-better attributes). This assumption does not imply any loss of generality. A less-is-better attribute (objective to be minimized) can be transformed into a more-is-better attribute simply by multiplying it by -1. If the attribute is to precisely reach a certain target (goal), then it can be written as an objective minimizing the distance (or maximizing the opposite of the distance) from the attained value to the target value, so that it can be formulated as a less-is-better (or more-is-better) objective. Therefore, the formulation proposed, which considers all attributes as objectives to be maximized, allows us to address any problem involving any of the relevant types of attributes (objectives or goal types) considered in the farmer's MAUF.

Moreover, it is assumed that each single-attribute or partial utility function ($u_a(f_a(\boldsymbol{X}))$ is equal to the corresponding attribute $f_a(\boldsymbol{X})$. This assumption implies linear utility-indifferent curves (constant partial marginal utility), a somewhat strong assumption that can be regarded as a close enough approximation if the attributes vary within a constrained range (Hardaker et al. 2007). Huirne and Hardaker (1998) show how the slope of the single-attribute utility function has little impact on the ranking of alternatives. Similarly, Amador et al. (1998) analyze how linear and quasi-concave functions yield almost the same results. Consequently, we assume this simplification in the elicitation of the MAUFs. Finally, for operational purposes, the attribute functions are properly normalized to be bounded between 0 and 1 ($nf_a(\boldsymbol{X})$).

Sumpsi et al. (1997) describe a widely used non-interactive process for eliciting the values of the calibrating parameters w_a in additive MAUFs. Following these authors, the crop mix selection ($\boldsymbol{X}$) can be viewed as a multi-objective programming (MOP) decision-making problem. Because the preferences of decision-makers should belong to the efficient frontier, a first approximation can be assessed through the pay-off matrix, which is obtained by maximizing each of the objectives (in our case, partial utility functions, $nf_a(\boldsymbol{X})$) separately, subject to the constraints set (expressions 2b). To obtain the relative weight of each attribute (w_a), a system of equations to make the sum of the weighted elements of the pay-off matrix for each attribute be equal to their observed value (nf_a^{obs}) is built as follows:

$$\begin{bmatrix} nf_{11}=nf_1^* & nf_{12} & \cdots & nf_{1m} \\ nf_{21} & nf_{22}=nf_2^* & \cdots & nf_{2m} \\ \vdots & \vdots & \ddots & \vdots \\ nf_{m1} & nf_{m2} & \cdots & nf_{mm}=nf_m^* \end{bmatrix} \cdot \begin{bmatrix} w_1 \\ w_2 \\ \vdots \\ w_m \end{bmatrix} = \begin{bmatrix} nf_1^{obs} \\ nf_2^{obs} \\ \vdots \\ nf_m^{obs} \end{bmatrix} \tag{3a}$$

$$\sum_{a=1}^{m} w_a = 1 \tag{3b}$$

where nf_a^* is the normalized ideal value for attribute a, $nf_{aa'}$ are the normalized values of the elements in the pay-off matrix of attribute a when attribute a' is optimized, and nf_a^{obs} are the normalized observed values of each attribute.

Typically, however, there is not an exact solution to the above system; thus, it is necessary to solve the problem by minimizing the sum of the deviational variables that determine the set of w_a that provided the closest solution:

$$Min \sum (n_a + p_a) \tag{4a}$$

$$\begin{aligned} \text{s.t.} \quad & w_1 \cdot nf_{11} + w_2 \cdot nf_{12} + \ldots + w_m \cdot nf_{1m} + n_1 - p_1 = nf_1^{obs} \\ & w_1 \cdot nf_{21} + w_2 \cdot nf_{22} + \ldots + w_m \cdot nf_{2m} + n_2 - p_2 = nf_2^{obs} \\ & \ldots \\ & w_1 \cdot nf_{m1} + w_2 \cdot nf_{m2} + \ldots + w_m \cdot nf_{mm} + n_m - p_m = nf_m^{obs} \\ & w_1 + w_2 + \ldots + w_m = 1 \end{aligned} \tag{4b}$$

where n_a and p_a are the negative and positive deviations from the observed values for each attribute, respectively.

2.3 Cobb-Douglas MAUF Maximization: WGP Approach

Considering a Cobb-Douglas MAUF, the MP model proposed for simulating farmers' behavior takes the following form:

$$Max \ U(\boldsymbol{X}) = \prod_{a=1}^{m} \left[u_a(f_a(\boldsymbol{X}))^{\alpha_a} \right] = \prod_{a=1}^{m} \left[nf_a(\boldsymbol{X})^{\alpha_a} \right] \tag{5a}$$

$$\begin{aligned} \text{s.t.} \quad & \sum_{a=1}^{m} \alpha_a = 1 \\ & \boldsymbol{AX} \leq \boldsymbol{B} \\ & \boldsymbol{X} \geq 0 \end{aligned} \tag{5b}$$

where α_a denotes the calibration coefficients of each attribute, related to their relative importance.

As established above for the additive MAUF, in the case of the Cobb-Douglas specification, it is also assumed that all attributes are related to objectives to be maximized (i.e., more-is-better), and each single-attribute or partial utility function is equal to the corresponding attribute properly normalized to be bounded between

0 and 1. For purposes of comparability, the same normalization has been performed in all approaches, including profit maximization.

The new methodological approach proposed for eliciting the α_a coefficients in Cobb-Douglas MAUFs is also based on the WGP approach explained above. To that end, first, the Cobb-Douglas function is transformed into an additive expression. In mathematical terms, one of the advantages of the Cobb-Douglas function used as an objective function (expression 5a) is the possibility of being transformed into an additive function using logarithms without losing any of its features:

$$log[U(\boldsymbol{X})] = V(\boldsymbol{X}) = \sum_{a=1}^{m} \alpha_a \cdot log[nf_a(\boldsymbol{X})] \tag{6}$$

Following a procedure similar to that developed by Sumpsi et al. (1997), this transformation makes it possible to estimate the most appropriate alpha parameters by solving the following m+1 system of equations, in which the weighted sum of the elements of the pay-off matrix are equal to the observed values of the attributes, all of them properly normalized and transformed by the natural logarithms:

$$\begin{bmatrix} log(nf_{11}) = \log(nf_1^*) & log(nf_{12}) & \cdots & log(nf_{1m}) \\ log(nf_{21}) & log(nf_{22}) = log(nf_2^*) & \cdots & log(nf_{2m}) \\ \vdots & \vdots & \ddots & \vdots \\ log(nf_{m1}) & log(nf_{m2}) & \cdots & log(nf_{mm}) = log(nf_m^*) \end{bmatrix} \cdot \begin{bmatrix} \alpha_1 \\ \alpha_2 \\ \vdots \\ \alpha_m \end{bmatrix} = \begin{bmatrix} log(nf_1^{obs}) \\ log(nf_2^{obs}) \\ \vdots \\ log(nf_m^{obs}) \end{bmatrix} \tag{7a}$$

$$\sum_{a=1}^{m} \alpha_a = 1 \tag{7b}$$

As in the additive WGP approach, the previous system may not have an exact solution (this is typically the case). Therefore, it is necessary to solve the problem by minimizing the sum of deviational variables that determine the set of alpha parameters that lead to the closest solution:

$$Min \sum_{a=1}^{m} (n_a + p_a) \tag{8a}$$

$$
\begin{aligned}
\text{s.t.}\quad & \alpha_1 \cdot log(nf_{11}) + \alpha_2 \cdot log(nf_{12}) + \ldots + \alpha_m \cdot log(nf_{1m}) + n_1 - p_1 = log\left(nf_1^{obs}\right) \\
& \alpha_1 \cdot log(nf_{21}) + \alpha_2 \cdot log(nf_{22}) + \ldots + \alpha_m \cdot log(nf_{2m}) + n_2 - p_2 = log\left(nf_2^{obs}\right) \\
& \qquad\qquad\qquad\qquad \cdots \\
& \alpha_1 \cdot log(nf_{m1}) + \alpha_2 \cdot log(nf_{m2}) + \ldots + \alpha_m \cdot log(nf_{mm}) + n_m - p_m = log\left(nf_m^{obs}\right) \\
& \sum_{a=1}^{m} \alpha_a = 1
\end{aligned}
\tag{8b}
$$

where n_a and p_a are the absolute negative and positive deviations, respectively.

Once the alpha parameters are estimated by running model (8), the shape of the Cobb-Douglas MAUF to be used for modeling purposes is as follows:

$$
U(\boldsymbol{X}) = nf_1(\boldsymbol{X})^{\alpha_1} \cdot nf_2(\boldsymbol{X})^{\alpha_2} \cdot \ldots \cdot nf_m(\boldsymbol{X})^{\alpha_m} \tag{9}
$$

3 Case Study

3.1 Irrigation Water Pricing

At present, water resources are increasingly scarcer in Spain because of rising demand and declining availability due to climate change. Moreover, traditional supply-side water policy instruments, such as the construction of dams and other water infrastructure to increase water supply, cannot be further developed since new increases in the water supply are technically infeasible or economically unaffordable, a situation known as 'basin closure' (Molle et al. 2010). When basin development reaches the closure stage, any new water demand must be satisfied by reducing other existing water uses. Under these circumstances, demand-side water policy instruments such as water pricing or water markets are considered the most suitable solutions for allowing a more efficient reallocation of water resources (Lago et al. 2015).

Closed basins are found not only in Spain but also in other member states of the European Union (EU) and other countries worldwide. This situation has caused EU institutions to decide to develop a common policy for water management. The approval of the Water Framework Directive (WFD; Directive 2000/60/CE of the European Parliament and of the Council) it is considered the main achievement in this field (Kallis and Butler 2001). The WFD (article 9) proposes water pricing as the main policy instrument for addressing the demand for water within the EU (European Commission 2001).

The rationale on which water pricing is based is purely economic. In this sense, farmers in irrigated areas, according to economic theory, will respond to the introduction of (or an increase in) water prices by reducing their consumption, in accordance with a negatively sloped demand curve. In this manner, the water savings obtained would be re-distributed among other uses such as productive or environmental purposes (ecological flows in rivers, etc.), according to societal preferences. Such a reallocation of water resources would improve the efficiency of their use (Johansson et al. 2002; Tsur et al. 2004). The impact of water pricing on farmers' behavior has been widely studied in the literature. In this sense, it is worth noting the book edited by Dinar et al. (2015), who show the experiences of water pricing in several countries around the world (Australia, Brazil, Canada, Chile, China, Colombia, France, India, Israel, Mexico, The Netherlands, New Zealand and South Africa). Also relevant is the book edited by in Berbel and Gutiérrez-Martín (2004), where interested readers can find a detailed analysis of the impacts of water pricing on irrigated agriculture in the European Union, considering case studies in Spain, Italy, Greece and Portugal. Other interesting works addressing water pricing in European countries are the ones developed by Bontemps and Couture (2002) in France, focused on the estimation of water demand when water pricing is implemented; Manos et al. (2006) in Greece and Fragoso and Marques (2015) in Portugal, where the impacts of water pricing under alternative scenarios of European agricultural policy are analyzed; and Bartolini et al. (2007), Galioto et al. (2013) and Pérez-Blanco et al. (2016) in Italy, regarding the design of tariff strategies aiming at cost recovery. Likewise, recent empirical studies by Pérez-Blanco et al. (2015), Kahil et al. (2016) and Montilla-López et al. (2017) about irrigation water pricing in Spain are also worth to be cited.

Most empirical studies that analyze how farmers would react in case water prices were introduced (or increased) have relied on MP models to simulate the feasible behavior of these producers when the parameter in the model representing the volumetric water price is increased. In this paper, a similar *ex-ante* policy analysis of water pricing is implemented for a real case study. However, as noted above, this analysis is performed by using three different modeling approaches to illustrate the outperformance of the new method proposed based on a Cobb-Douglas MAUF.

3.2 Modeling Scope

The empirical application proposed as the case study is developed in the *Canal de la Margen Izquierda del Bembézar* irrigation district, located in the Guadalquivir River Basin in southern Spain (see Fig. 1). This irrigated area covers a total of 4009 ha divided into 163 farms with an average farm size of 66.2 ha.

Within the same agricultural system (e.g., an irrigation district), it can be easily assumed that all farms fulfill the following features: i) technological homogeneity (the same possibilities of production, the same types of resources, the same technological level and the same management capacity); ii) pecunious proportionality (proportional profit expectations for each activity); and iii) institutional proportionality (the availability of resources to the individual farm proportional to the average availability). Assuming a profit-maximizing behavior, if the abovementioned requirements are met, then all farms can be modeled together within the same MP model without aggregation-biased results since all of them are assumed to have a

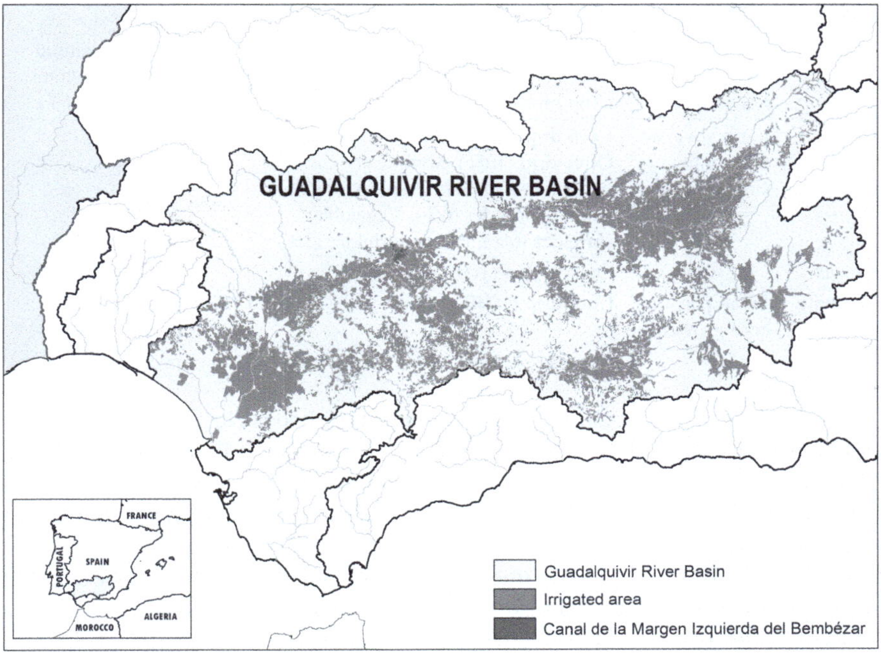

Fig. 1 Location of the selected irrigated area

similar productive behavior (Hazell and Norton 1986). However, real-world observations show that even within the same agricultural system, there exists heterogeneity between farms regarding the crop mixes and agricultural practices (in our case, irrigation techniques), mainly because of the differences in farmers' utility functions since profit maximization is only seldom the unique objective guiding farmers' decision-making (Pennings and Leuthold 2000; Berkhout et al. 2011; Karali et al. 2013). In fact, if an MAUT perspective is being considered, an additional homogeneity requirement is needed to avoid aggregation bias, i.e., homogeneity related to the MAUF shape (i.e., the values of the parameters defining additive or Cobb-Douglas MAUFs).

The MAUF shape is primarily based on the psychological characteristics of the decision-makers, which differ significantly from farmer to farmer. According to this perspective, the differences in decision-making (crop mix) among farmers in the same production area must be primarily due to the differences in their utility functions (in which the relative importance given to different criteria are condensed), rather than other differences related to the profits of economic activities or disparities in resource requirements or endowments. Thus, to avoid aggregation bias resulting from lumping together farmers with significantly different MAUFs, a classification of farmers into homogeneous groups with similar decision-making behavior (utility functions) is required.

Table 1 Farm types in *Canal de la Margen Izquierda del Bembézar*

	Label	Crop mix	Farm size (hectares)	Agricultural income/total income (%)
Cluster 1	Large diversified professional farmers	Corn-drip (32%), Orange-drip (25%), Olive-drip (12%), Sunflower-sprinkler (8%), Wheat-sprinkler (6%), Cotton-sprinkler (6%), Vegetables-sprinkler (6%), Potato-sprinkler (3%)	79.7	83.6
Cluster 2	Citrus growers	Orange-drip (100%)	47.0	65.5
Cluster 3	Small part-time corn growers	Corn-drip (100%)	13.1	42.9

For this type of classification, the most efficient method is cluster analysis (Berbel and Rodríguez-Ocaña 1998; Gómez-Limón and Riesgo 2004; Berkhout et al. 2011), taking farmers' real decision-making vectors (the actual crop mix) as the classification criterion. Thus, following this idea, in this paper, we use clustering techniques to assign individual farms to homogenous groups regarding their crop mixes (i.e., the MAUF shape). Within the possibilities that this technique contains, we have selected Ward's procedure as a criterion for aggregation and the Euclidean distance as a measure of the distance between farms of the irrigation district selected as the case study. Following this procedure, a dendrogram has been generated, clearing showing three different clusters; their average profiles (considering both crop mixes as classification variables and other structural variables such as farm size, farmer age, etc.) have been used to define the corresponding 'farm types', as shown in Table 1.

The homogeneous farms included in each cluster and represented by their own farm type can be properly modeled without aggregation biases. Thus, these farm types are considered decision units to be modeled in the individual MP models. Regardless, considering the illustrative purpose of this paper, from this point on, only the case of Cluster 1 (large diversified professional farmers) is considered for model building.

4 Model Building

4.1 Variables, Attributes and Objective Functions

The decision variables for the farmer are the area devoted to each alternative productive activity (***X***). These activities are denoted as $X_{i,j}$, where *i* means the crop and *j* the irrigation technique used. The combination of crops and the irrigation techniques considered as the decision variables for the case study analyzed includes the current irrigated activities shown in Table 1 and rain-fed alternatives (wheat, sunflower and olive groves). Thus, the model constructed will be able to simulate the impacts of the various pricing scenarios (the water-demand function) as the result of the famers' short-term production adjustments, simulating both the substitution of water-intensive crops by others and the cessation of irrigation and the introduction of rain-fed crops with no need for water.

For the elicitation of the utility functions, three attributes have been used as the most relevant attributes to model farmers' decision-making, considering the existing evidence (e.g., Gómez-Limón and Riesgo 2004; Pérez-Blanco and Gutiérrez-Martín 2017). These attributes are i) the profit in the short run, ii) the risk inherent to this profit, and iii) the managerial complexity associated with the crop mix. Attributes are defined as a mathematical function of the decision variables and become objectives when the direction of improvement of each attribute is set. That is, the objective related to each attribute will be profit maximization, risk minimization and managerial complexity minimization. Profit is defined by the expected total gross margin (the average value of the 2007–2013 times series) ($f_1(\boldsymbol{X})=GM(\boldsymbol{X})$). Risk is measured as the variance of the gross margin in the same period ($f_2(\boldsymbol{X})=VAR(\boldsymbol{X})$). Finally, total labor ($f_3(\boldsymbol{X})=TL(\boldsymbol{X})$) has been selected as a proxy for managerial complexity.

The expected gross margin ($GM(\boldsymbol{X})$) has been calculated as the sum of total income (the average crop price—p_i—multiplied by the average yield—$y_{i,\ j}$—plus coupled subsidies—s_i) minus the variable costs ($vc_{i,\ j}$) and the water cost from the water pricing, which is the product of the water quantity used ($wq_{i,\ j}$) and the water price (wp):

$$GM(\boldsymbol{X}) = \sum_i \sum_j \left\{ \left(p_i \cdot y_{i,j} + s_i - vc_{i,j} - wq_{i,j} \cdot wp\right) \cdot X_{i,j} \right\} \tag{10}$$

The variance of the gross margin in the time series considered ($VAR(\boldsymbol{X})$) is defined by Eq. (11), where $\boldsymbol{X}^t$ is the transposed vector $\boldsymbol{X}$ and $[\boldsymbol{cov}]$ is the variance-covariance matrix of the gross margins of productive activities per hectare during the 2007–2013 period.

$$VAR(\boldsymbol{X}) = \boldsymbol{X}^t \cdot [\boldsymbol{cov}] \cdot \boldsymbol{X} \tag{11}$$

Total labor is calculated as shown in Eq. (11), that is, as the sum of labor requirements per crop and the irrigation technique ($tl_{i,j}$) in the entire farm area.

$$TL(\boldsymbol{X}) = \sum_i \sum_j tl_{i,j} \cdot X_{i,j} \tag{12}$$

As noted above, the objectives related to the several attributes considered (partial utility functions) must be normalized for operational purposes to transform them into more-is-better and dimensionless functions (nf_a), whose values vary within the interval [0,1]. To that end, we propose transforming the original attribute functions into rates of success with respect to the ideal value of each attribute as follows:

$$nf_{GM}(\boldsymbol{X}) = \frac{GM(\boldsymbol{X})}{GM^*}; \quad nf_{VAR}(\boldsymbol{X}) = \frac{VAR^*}{VAR(\boldsymbol{X})}; \quad nf_{TL}(\boldsymbol{X}) = \frac{TL^*}{TL(\boldsymbol{X})} \tag{13}$$

where GM^*, VAR^* and TL^* are the optimal or ideal values for the gross margin, variance and total labor (their maximum and minimum, respectively). Note that whereas for the more-is-better attribute ($GM(\boldsymbol{X})$), the ideal value (the largest possible value) is in the denominator, for the less-is-better attributes ($VAR(\boldsymbol{X})$ and $TL(\boldsymbol{X})$), the ideal values (the smallest possible values) are in the numerator. Thus, it can be checked that operating in this manner, all normalized attributes (nf_a) are related to more-is-better objectives and that their values range between 0 and 1.

The values of the attributes, properly normalized, represent the partial utilities, which are combined in each utility function (objective functions in the MP models) according to each methodological approach described in Sect. 2. In this regard, Eqs. (14–16) represent the objective function in the case of profit maximization (Eq. 14), the additive MAUF (Eq. 15) and the Cobb-Douglas MAUF (Eq. 16):

$$Max\ U(\boldsymbol{X}) = nf_{GM}(\boldsymbol{X}) \tag{14}$$

$$Max\ U(\boldsymbol{X}) = w_{GM} \cdot nf_{GM}(\boldsymbol{X}) + w_{VAR} \cdot nf_{VAR}(\boldsymbol{X}) + w_{TL} \cdot nf_{TL}(\boldsymbol{X}) \tag{15}$$

$$Max\ U(\boldsymbol{X}) = nf_{GM}(\boldsymbol{X})^{\alpha_{GM}} \cdot nf_{VAR}(\boldsymbol{X})^{\alpha_{VAR}} \cdot nf_{TL}(\boldsymbol{X})^{\alpha_{TL}} \tag{16}$$

Note that the normalization of $GM(\boldsymbol{X})$ in Eq. (14) is not necessary but has been performed for the sake of homogeneity.

4.2 Model Constraints

Farmers' decision-making is subject to constraints that limit the feasible set of choices at hand. These constraints respond not only to the fact that resources are limited but also to other restrictions such as crop rotations, agricultural policy quotas, marketing channel limits, etc. The constraints limit the space of the solutions of the model to those that are attainable by the farmers (feasible solution set), explaining a large share of its behavior. Equations (17a–17d) show all restrictions taken into account by the farmers analyzed:

$$\sum_i \sum_j X_{i,j} \leq fa \tag{17a}$$

$$\sum_i \sum_j X_{i,j} \cdot wq_{i,j} \leq wa \cdot fa \tag{17b}$$

$$\boldsymbol{AX} \leq \boldsymbol{B} \tag{17c}$$

$$X_{i,j} \geq 0; \qquad \forall i,j \tag{17d}$$

Constraints (17a) and (17b) are related to land and water availability, respectively. The first limits the total area covered by the different alternatives to the farm size (fa). The water constraint establishes that irrigation water requirements cannot exceed water availability, with the former being the sum of water requirements per alternative and the latter the water allotment provided by the water agency considering farm size (*fa*) and water rights granted per hectare (*wa*). Moreover, Eq. (17c) denotes the rest of the constraints defining the feasible solution set, which constitute technical (agronomic and irrigation technology), policy (cotton quota) and market requirements:

a) *Agronomic constraints*. These include the rotational and frequency constraints actually followed by farmers as good agricultural practices.
b) *Permanent crops*. In the short run, it is not possible to increase or decrease the area devoted to permanent crops (in our case study, citrus and olive groves) because they are fixed assets that are only changeable in the long run. For this reason, permanent crops are not allowed to change. However, irrigated olive groves are allowed to change into rain-fed groves since this woody crop can be grown with and without irrigation.[2]
c) *Irrigation technique*. It is assumed that the specific equipment for each irrigation system (surface, sprinkler and drip) remains the same in the short term (new investments in irrigation technology are not considered). As a result, the maximum area irrigated by each of these systems is fixed. This fact is modeled by preventing the area covered by each irrigation technique from increasing by more than 5% compared to the observed values.
d) *Cotton quota*. The area devoted to cotton is limited to the maximum area observed in the period considered due to an agricultural policy constraint.
e) *Market constraints*. There are crops such as garlic and onions that are subject to limited marketing channels because they cannot be stored for extended periods (perishable products). The implication is that it is unlikely that farmers will significantly increase the area cultivated with such crops due to the inability of the market to absorb great variations in production in the short run. Thus, to model this constraint, an upper limit of the area cultivated with these crops was included on the basis of the maximum historical cultivation during the previous 7 years.

Finally, decision variables ($X_{i,j}$) are fixed as non-negative, as denoted by Eq. (17d).

4.3 Simulating Water Pricing

Simulations can be performed because the utility functions are considered a structural feature of farmers that does not change over the course of any simulation. Thus, the models built as explained in the two previous sections have been used to simulate

[2]This is not possible with orange groves since this permanent crop can be grown only under irrigation.

farmers' responses (in terms of the crop mix and water use) to an increment in the price of water (the parametrization of a volumetric water price affecting farming costs).[3]

The parametrization of the water price (*wp*) allows the water-demand curve to be built; that is, the model results will show the water quantity that farmers are willing to use at every simulated water price. Economic theory assumes that any increment in the water price will lead to a reduction in water use in accordance with a negatively sloped water curve. However, this change in the water quantity used will depend on the elasticity of the demand curve, which is related to its slope.

To perform the methodological comparison proposed by the estimation of the water-demand curves, we have parametrized the water price from 0.00 €/m^3 to 0.30 €/m^3 to simulate farmers' decision-making in the short run (only changing the crop mix) using the three approaches described in Sect. 2.

Finally, it must be noted that simulating the impact of water pricing using MP models will also make it possible to estimate a series of indicators of interest for policy decision-makers, covering economic (e.g., the aggregated gross margin), social (e.g., the aggregated agricultural labor demand) and environmental (e.g., the aggregated agrochemical use) issues (see for example the works by Gómez-Limón and Riesgo 2004; Gallego-Ayala et al. 2011). In fact, to support policy design and implementation, this type of *ex-ante* policy evaluation is very useful. However, considering the methodological main purpose of this paper, this policy analysis falls beyond its scope and is thus not reported in this chapter.

5 Results and Discussion

5.1 MAUF Calibration and Validation of Simulation Models

Running the calibration procedures as explained in model (4) for the additive MAUF and model (8) for the Cobb-Douglas MAUF, both sets of calibration parameters (weights—w_a—and alphas—α_a—, respectively) are obtained. This allows the elicitation of the objective functions in MP simulation models (2) and (5) built for the farm type considered. Thus, the formulation of expressions (2a) and (5a) became as follows:

$$U(\boldsymbol{X}) = 0.89 \cdot nGM(\boldsymbol{X}) + 0.06 \cdot nVAR(\boldsymbol{X}) + 0.04 \cdot nTL(\boldsymbol{X}) \tag{18}$$

$$U(\boldsymbol{X}) = nGM(\boldsymbol{X})^{0.89} \cdot nVAR(\boldsymbol{X})^{0.11} \tag{19}$$

In the case of the additive MAUF, all proposed attributes are included in the calibrated utility function. However, in the case of the Cobb-Douglas MAUF, only the expected gross margin and variance are taken into account, showing that the

[3]The current water cost is already included in the variable costs ($vc_{i,j}$).

Table 2 Model validation: attributes values

	Profit maximization	Additive MAUF	Cobb-Douglas MAUF	Observed
GM (€/ha)	2215.85	2214.69	2210.53	2103.19
Risk (€²/ha)	6598.47	6510.35	6316.97	6079.75
Labor (h/ha)	87.40	87.16	86.32	83.75
Mean squared error (MSE)	6.3%	5.6%	4.1%	

contribution of total labor to the total utility is negligible when this utility function is considered. Although weights and alpha parameters are not totally comparable (the alpha parameters are not exactly weights but, rather, a proxy), both approaches show the much greater relevance of the expected gross margin over the rest of the attributes in the decision-making process. As a consequence of this fact, for this case study, the solutions when maximizing both MAUFs will not be too far away from the solution to the first simulation approach (MP model (1)), in which only the expected gross margin is maximized.

To validate the simulation models built for the farm type considered, we proceed to compare the actual situation (observed levels) with the simulated results for the current scenario (Qureshi et al. 1999). These simulations for the current scenario are obtained by maximizing every objective function subject to the constraints considered, as shown in models (1), (2) and (5) for profit maximization, the additive MAUF and the Cobb-Douglas MAUF, respectively. To validate these models, the simulated results obtained in the space of attributes ($GM(\boldsymbol{X})$, $VAR(\boldsymbol{X})$, and $TL(\boldsymbol{X})$) and in the space of decision variables ($\boldsymbol{X}$) are compared.

The results of the comparison between the simulated results for the attributes under the current scenario with those related to the actual crop pattern are shown in Table 2. The last row of this table shows the values achieved by the mean squared error (MSE) for each simulation approach. This statistical indicator measures the average of the squares of the errors or deviations between the estimator and what is estimated between the observed and the simulated vectors of the attributes, following the formula below:

$$MSE = \sqrt{\frac{\sum_{a=1}^{m}\left(\frac{f_a^{obs}(\boldsymbol{X})-f_a(\boldsymbol{X})}{f_a^{obs}(\boldsymbol{X})}\right)^2}{m}} \tag{20}$$

This error aims to clarify which approach best approximates the observed attribute levels. In this regard, the MSE shows that the calibration with the Cobb-Douglas MAUF is the most accurate approach. In other words, the values of the attributes from the simulation using the Cobb-Douglas MAUF approach are closer to the actual values than those from the other approaches.

Additionally, the validation in the space of the decision variables attempts to analyze the capacity of the model to reproduce the farmers' actual crop mix. To that

Table 3 Model validation: decision variables (crop areas in hectares)

Crop mix	Profit maximization	Additive MAUF	Cobb-Douglas MAUF	Observed
Durum wheat-sprinkler	0.00	0.80	3.65	6.14
Corn-drip	27.14	27.14	27.14	25.85
Potato-sprinkler	13.25	12.45	9.60	2.69
Cotton-sprinkler	4.98	4.98	4.98	4.98
Sunflower-sprinkler	0.00	0.00	0.00	6.34
Garlic-sprinkler	1.79	1.79	1.79	1.70
Onion-sprinkler	3.27	3.27	3.27	2.73
Orange-drip	19.73	19.73	19.73	19.73
Olive-drip	9.49	9.49	9.49	9.49
PAD	31.3%	29.3%	22.2%	
FK index	84.3%	85.3%	88.9%	

end, we have calculated two indicators: i) the percentage absolute deviation (PAD) and ii) the Finger-Kreinin similarity index (FK, see Finger and Kreinin 1979), which are calculated as follows:

$$PAD\ index\ (\%) = \frac{\sum_i \sum_j \left| X_{i,j}^{obs} - X_{i,j} \right|}{fa} \tag{21}$$

$$FK\ similarity\ index = \sum_i \sum_j min\left(\frac{X_{i,j}}{fa}; \frac{X_{i,j}^{obs}}{fa} \right) \tag{22}$$

The PAD compares simulated and observed crop areas by adding all absolute deviations and expressing this summation in perceptual terms. Thus, this index can vary from 0% (perfect calibration fitting) to 200% (the worst possible calibration). Similarly, the FK similarity index compares the simulated and the observed shares of each crop mix, varying between 0% and 100%, with the latter being an exact match between the observed and the simulated crop mixes.

Table 3 shows the simulated crop mix for each approach and the observed levels of the different productive alternatives. Additionally, in the last two rows, the two similarity indexes calculated for every approach are presented.

According to these results, it is shown that the most accurate approach is the Cobb-Douglas MAUF since it reaches the lowest PAD (22.2%) and the highest FK index (88.9%). Thus, it is demonstrated that the Cobb-Douglas MAUF approach is once again the best among the approaches considered since for the reference scenario, it reproduces the farmers' behavior better than the other approaches. The additive MAUF approach is ranked second following these two indicators, also outperforming the profit maximization approach, which obtains the worst values in both the PAD and the FK index.

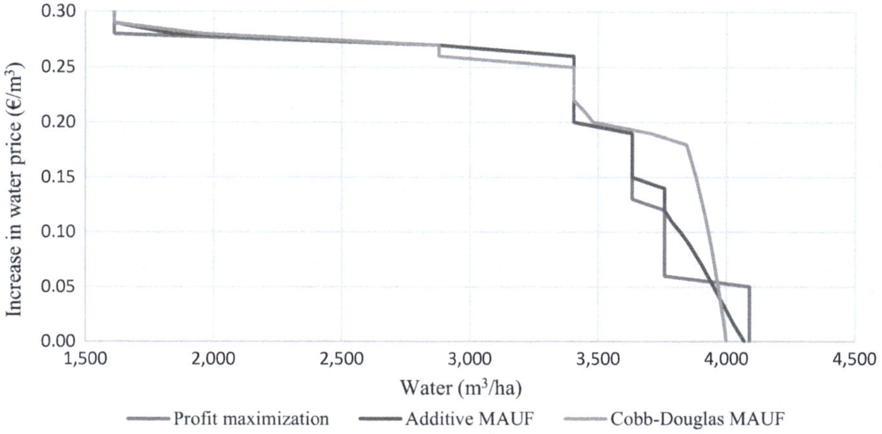

Fig. 2 Water demand considering profit maximization, the additive MAUF and the Cobb-Douglas MAUF approaches

Finally, it is worth noting that since the deviations in the objectives and in the decision variable spaces are sufficiently small in the case of the Cobb-Douglas MAUF approach, it can be affirmed that this modeling approach is a reasonably accurate enough approximation for simulating farmers' actual decision-making.

5.2 *Demand Curves*

After the calibrations, the resulting MP simulation models have been run to parametrize the water price from 0.00 €/m^3 to 0.30 €/m^3. From each iteration, the resulting water use has been recovered to construct the different demand curves. These curves show the changing behavior of farmers when an increase in water pricing is implemented, as shown in Fig. 2.

The first result worth noting is that the shapes of the three curves are somewhat similar, presenting a common inelastic segment (high slope) for low increases in the water price (less than 0.20 €/m^3); this is, relative high increases in the water price lead to relative low decreases in water use. This simulated behavior with a large initial inelastic segment can also be found in many previous empirical studies (e.g., Molle and Berkoff 2007; Wheeler et al. 2008; Montilla-López et al. 2017). From 0.20 €/m^3 on, the results obtained for all methods are almost the same.

These similarities in the three demand curves can be explained because of the great relevance of the attribute expected gross margin in this case study, which leads to very similar utility functions in all of the approaches tested. In case the farmers analyzed were more risk or managerial complexity adverse, the results would be greatly different.

Nevertheless, the demand curves show that the differences in the inelastic segments are worth taking into account, with the smoothness of these curves being the most relevant difference. As is widely known, simulations that use the profit maximization approach lead to an inertia in the vicinity of the reference situation and a 'jumpy' behavior that does not make this approach sufficiently reliable (Mérel and Howitt 2014). MAUF approaches, namely, the new method based on the Cobb-Douglass MAUF, provide much more credible simulation results, avoiding over-reactions to exogenous shocks, as the policy change proposed. In fact, the Cobb-Douglass MAUF approach shows the smoothest demand curve, which is known to be a good indicator of realism (Heckelei and Britz 2005).

6 Conclusions

The main contribution of this paper is the development of a new and simpler method to elicit Cobb-Douglas MAUFs. This method is a sounder approach than traditional additive MAUFs since this type of utility function assumes neither a constant marginal rate of substitution between attributes nor total compensation between attributes, thus being more coherent with economic theory.

This new methodological approach has been empirically implemented to simulate farmers' behavior in a real case study, and the results obtained have been compared with those derived from two other well-known approaches, profit maximization and the additive MAUF. This study reaches two main conclusions. First, the approach proposed to elicit Cobb-Douglas MAUFs can be easily implemented in real settings, and therefore, it can be a useful procedure for *ex-ante* simulations of policy instruments or any type of future scenario. Second, this new method proposed based on the maximization of the Cobb-Douglass MAUF can produce fruitful outcomes for policy analysis because it provides better simulation results than more traditional approaches. Two pieces of evidence justify this assessment. First, calibration is more precise using this approach than in the other approaches compared since the resulting MAUF better reproduces farmers' current behavior. Second, the resulting demand curve has a smoother and more credible shape than those obtained from previous approaches since farmers are expected to make marginal changes when facing marginal external shocks.

However, it is also worth pointing out that the method proposed is based on some rather restrictive assumptions that can be seen as potential shortcomings. The strongest ones are: i) the assumption regarding utility-independence, allowing the MAUF became a separable function, and ii) the assumption about the stability of the MAUF, i.e. the parameters of the utility function do not change when circumstances do (farmers behave the same way whatever occurs). Because of both potential limitations, further research is required to confirm that this new non-interactive method to elicit Cobb-Douglas MAUFs represents a reasonable enough approximation to simulate real farmers' behavior. In this regard, some others functional forms of the utility function could be elicited and tested, such as the constant elasticity of substitution function (CES, a more general form than the Cobb-Douglas function) and other ones not assuming utility-independence. Moreover, it would be worthwhile implementing experiments to test that MAUF parameters remain constant over time (by using multiple elicitation procedures with the same decision-makers in different time periods).

Acknowledgments The authors acknowledge the financial support from the Spanish Ministry of Economics and Competitiveness (MINECO) and the European Regional Development Fund (ERDF) through the research project MERCAGUA (AGL2013-48080-C2-1-R) and the associated predoctoral fellowship (BES-C-2014-0006).

References

Amador, F., Sumpsi, J. M., & Romero, C. (1998). A non-interactive methodology to assess farmers' utility functions: An application to large farms in Andalusia, Spain. *European Review of Agricultural Economics, 25*(1), 92–102.

André, F. J., & Riesgo, L. (2007). A non-interactive elicitation method for non-linear multiattribute utility functions: Theory and application to agricultural economics. *European Journal of Operational Research, 181*(2), 793–807.

Bartolini, F., Bazzani, G. M., Gallerani, V., Raggi, M., & Viaggi, D. (2007). The impact of water and agriculture policy scenarios on irrigated farming systems in Italy: An analysis based on farm level multi-attribute linear programming models. *Agricultural Systems, 93*(1–3), 90–114.

Berbel, J., & Gutiérrez-Martín, C. (2004). *Sustainability of European Agriculture under Water Framework Directive and Agenda 2000*. Brussels: European Commission.

Berbel, J., & Rodríguez-Ocaña, A. (1998). An MCDM approach to production analysis: An application to irrigated farms in southern Spain. *European Journal of Operational Research, 107*(1), 108–118.

Berkhout, E. D., Schipper, R. A., van Keulen, H., & Coulibaly, O. (2011). Heterogeneity in farmers' production decisions and its impact on soil nutrient use: Results and implications from northern Nigeria. *Agricultural Systems, 104*(1), 63–74.

Bontemps, C., & Couture, S. (2002). Irrigation water demand for the decision maker. *Environment and Development Economics, 7*(4), 643–657.

Chavas, J.-P., Chambers, R. G., & Pope, R. D. (2010). Production economics and farm management: A century of contributions. *American Journal of Agricultural Economics, 92*(2), 356–375.

Dinar, A., Pochat, V., & Albiac, J. (2015). *Water pricing experiences and innovations*. Cham: Springer.

European Commission (Ed.). (2001). *Pricing water: Economics, environment and society*. Luxembourg: Office for Official Publications of the European Communities.

Finger, J. M., & Kreinin, M. E. (1979). A measure of 'export similarity' and its possible uses. *Economic Journal, 89*(356), 905–912.

Fishburn, P. C. (1982). *The foundations of expected utility*. Dordrecht: Reidel Publishing Company.

Fragoso, R. M., & Marques, C. (2015). Alternative irrigation water pricing policies: An Econometric Mathematical Programming Model. *New Medit Mediterranean Journal of Economics, Agriculture and Environment, 14*(4), 42–49.

Galioto, F., Raggi, M., & Viaggi, D. (2013). Pricing policies in managing water resources in agriculture: An application of contract theory to unmetered water. *Water, 5*(4), 1502–1516.

Gallego-Ayala, J., Gómez-Limón, J. A., & Arriaza, M. (2011). Irrigation water pricing instruments: A sustainability assessment. *Spanish Journal of Agricultural Research, 9*(4), 981–999.

Gómez-Limón, J. A., & Berbel, J. (2000). Multicriteria analysis of derived water demand functions: A Spanish case study. *Agricultural Systems, 63*(1), 49–72.
Gómez-Limón, J. A., Gutiérrez-Martín, C., & Riesgo, L. (2016). Modeling at farm level: Positive multi-attribute utility programming. *Omega, 65*, 17–27.
Gómez-Limón, J. A., & Riesgo, L. (2004). Irrigation water pricing: Differential impacts on irrigated farms. *Agricultural Economics, 31*(1), 47–66.
Gómez-Limón, J. A., Riesgo, L., & Arriaza, M. (2004). Multi-criteria analysis of input use in agriculture. *Journal of Agricultural Economics, 55*(3), 541–564.
Gutiérrez-Martín, C., & Gómez-Gómez, C. M. (2011). Assessing irrigation efficiency improvements by using a preference revelation model. *Spanish Journal of Agricultural Research, 9*(4), 1009–1020.
Hardaker, J. B., Huirne, R. B. M., Anderson, J. R., & Lien, G. (2007). *Coping with risk in agriculture*. Wallingford: CABI Publishing.
Hazell, P. B. R., & Norton, R. D. (1986). *Mathematical programming for economic analysis in agriculture*. New York: MacMillan.
Heckelei, T., & Britz, W. (2005). Models based on Positive Mathematical Programming: State of the art and further extensions. In F. Arfini (Ed.), *Modelling agricultural policies: State of the art and new challenges* (pp. 48–73). Proceedings of the 89th European Seminar of the European Association of Agricultural Economics, University of Parma, Parma, Italy, February 2–5.
Huirne, R. B. M., & Hardaker, J. B. (1998). A multi-attribute utility model to optimise sow replacement decisions. *European Review of Agricultural Economics, 25*(4), 488–505.
Inada, K. (1963). On a two-sector model of economic growth: Comments and a generalization. *The Review of Economic Studies, 30*(2), 119–127.
Johansson, R. C., Tsur, Y., Roe, T. L., Doukkali, R., & Dinar, A. (2002). Pricing irrigation water: A review of theory and practice. *Water Policy, 4*(2), 173–199.
Kahil, M. T., Albiac, J., Dinar, A., Calvo, E., Esteban, E., Avella, L., et al. (2016). Improving the performance of water policies: Evidence from drought in Spain. *Water, 8*(2), 34.
Kallis, G., & Butler, D. (2001). The EU Water Framework Directive: Measures and implications. *Water Policy, 3*(2), 125–142.
Karali, E., Brunner, B., Doherty, R., Hersperger, A. M., & Rounsevell, M. D. A. (2013). The effect of farmer attitudes and objectives on the heterogeneity of farm attributes and management in Switzerland. *Human Ecology, 41*(6), 915–926.
Keeney, R. L., & Raiffa, H. (1976). *Decisions with multiple objectives: Preferences and value trade-offs*. New York: Wiley.
Lago, M., Mysiak, J., Gómez, C. M., Delacámara, G., & Maziotis, A. (2015). *Use of economic instruments in water policy: Insights from international*. Cham: Springer.
Mandryk, M., Reidsma, P., Kanellopoulos, A., Groot, J. C. J., & van Ittersum, M. K. (2014). The role of farmers' objectives in current farm practices and adaptation preferences: A case study in Flevoland, the Netherlands. *Regional Environmental Change, 14*(4), 1463–1478.
Manos, B., Bournaris, T., Kamruzzaman, M., Begum, M., Anjuman, A., & Papathanasiou, J. (2006). Regional impact of irrigation water pricing in Greece under alternative scenarios of European policy: A multicriteria analysis. *Regional Studies, 40*(9), 1055–1068.
Mérel, P., & Howitt, R. (2014). Theory and application of Positive Mathematical Programming in agriculture and the environment. *Annual Review of Resource Economics, 6*(1), 451–470.
Molle, F., & Berkoff, J. (2007). *Irrigation water pricing: The gap between theory and practice*. Wallingford: CABI Publishing.
Molle, F., Wester, P., & Hirsch, P. (2010). River basin closure: Processes, implications and responses. *Agricultural Water Management, 97*(4), 569–577.
Montilla-López, N. M., Gutiérrez-Martín, C., & Gómez-Limón, J. A. (2017). Impacto de la tarifación del agua de riego en el Bajo Guadalquivir. *ITEA. Información Técnica Económica Agraria, 113*(1), 90–111.

Pennings, J. M. E., & Leuthold, R. M. (2000). The role of farmers' behavioral attitudes and heterogeneity in futures contracts usage. *American Journal of Agricultural Economics, 82*(4), 908–919.

Pérez-Blanco, C. D., & Gutiérrez-Martín, C. (2017). Buy me a river: Use of multi-attribute non-linear utility functions to address overcompensation in agricultural water buyback. *Agricultural Water Management, 190*, 6–20.

Pérez-Blanco, C. D., Delacámara, G., & Gómez, C. M. (2015). Water charging and water saving in agriculture. Insights from a revealed preference model in a Mediterranean basin. *Environmental Modelling and Software, 69*, 90–100.

Pérez-Blanco, C. D., Standardi, G., Mysiak, J., Parrado, R., & Gutiérrez-Martín, C. (2016). Incremental water charging in agriculture. A case study of the Regione Emilia Romagna in Italy. *Environmental Modelling and Software, 78*, 202–215.

Qureshi, M. E., Harrison, S. R., & Wegener, M. K. (1999). Validation of multicriteria analysis models. *Agricultural Systems, 62*(2), 105–116.

Sumpsi, J. M., Amador, F., & Romero, C. (1997). On farmers' objectives: A multi-criteria approach. *European Journal of Operational Research, 96*(1), 64–71.

Tsur, Y., Dinar, A., Doukkali, R. M., & Roe, T. (2004). Irrigation water pricing: Policy implications based on international comparison. *Environment and Development Economics, 9*(6), 735–755.

Wheeler, S. A., Bjornlund, H., Shanahan, M., & Zuo, A. (2008). Price elasticity of water allocations demand in the Goulburn–Murray irrigation district. *Australian Journal of Agricultural and Resource Economics, 52*(1), 37–55.

Perceptions and Attitudes of Greek Farmers Towards Adopting Precision Agriculture: Case Study Region of Central Greece

Evagelia Koutridi, Olga Christopoulou, and Marie-Noëlle Duquenne

Abstract Modern scientific community after years of intensification of agricultural resources (soil, water, etc.) management, and with the actual risk of their depletion or degradation, is called to redefine standard agricultural practices with an environmentally friendly approach, focusing on their preservation, their enrichment and perpetuity of their yields. Also the nutritional stakes and environmental threats are high, brought about by the continuous growth of the world population which is expected to reach 10 billion in 2050 compared to 7.1 in 2013. In this context, the present study explores to what extent those directly involved, the crop producers, perceive the necessity of sustainable management of agricultural resources through the emerging practice of Precision Agriculture regarding the management of smaller parts of the fields according to the needs of each of them, while reducing inputs. This research aims to examine the concepts of crop producers regarding the prospects that arise through the adoption of Precision Agriculture in Greece, a country with problematic primary sector, with particular climatic conditions and varied microclimates while compete countries of low labour costs. The methodological approach is based on field research using questionnaires concerning a representative sample of crop producers in the Region of Central Greece. The choice of variables assessed as necessary for the adoption of the Precision Agriculture techniques by the producers, was based on empirical observations, as well as the use of literature sources. Then, an exploratory factor analysis is carried out on parameters that are considered necessary by producers to adopt new technologies and how they perceive the successive situation that will be shaped by the new digital revolution in agricultural practice. Finally, the possibility of restarting primary production is being discussed, now that, due to the economic recession, many young people more familiar with technology are returning to the province and undertake to cultivate the land in the absence of any other employment.

E. Koutridi (✉)
Lamia, Greece
e-mail: ekoutridi@uth.gr

O. Christopoulou · M.-N. Duquenne
Volos, Greece

J. Berbel et al. (eds.), *Multicriteria Analysis in Agriculture*, Multiple Criteria Decision Making, https://doi.org/10.1007/978-3-319-76929-5_10

Keywords Sustainable agricultural resources management · Crop production · New technologies and techniques · Precision agriculture · Factor analysis

1 Introduction

1.1 Sustainable Agricultural Practice: Precision Agriculture

The growing production of crop products over the past 50 years for food and clothing has led to severally aggravating environmental conditions such as water pollution, greenhouse gas emissions, soil erosion and chemical burden (Kurth et al. 2015).

Especially with regard to the agri-food sector, whilst arable land is degraded and diminished, recent surveys estimate the worldwide demand for cereals to rise by 75% between 2000 and 2050 (IAASTD 2009; FAO 2010).

As the requirements for changing land uses (urban fabric, industrial uses, road network, etc.) are increasing, it is possible to reduce the area occupied by agricultural land despite technological developments that may extend the limits of the territories that can be cultivated (Oliver 2013).

This possibility is particularly provided in developed countries, while the agricultural sector plays a huge role in the economies of developing countries, both because it contributes significantly to the national income but also because it mainly employs a large workforce.

Developed countries, in the first decade of the twenty-first century, focused in the direction of funding for the modernization of agricultural practices through Smart Farming Technologies (SFT), especially after the World Bank's report "Agriculture for Development" (World Bank 2007) and IAASTD's report "Agriculture at a Crossroads" (IAASTD 2009), triggered by three simultaneous global crises—food, climate and economic.

In the European Union (EU), with the implementation of the Common Agricultural Policy (CAP) and in particular with Agenda 2000, although food safety was ensured in a large part of it, environment was clearly degraded (Geiger et al. 2010). The mid-term review of 2003, incorporating Multiple Compliance, constricted everyday agricultural practice through norms, redirecting it towards the preservation of existing natural resources (Anthopoulou and Goussios 2007). With its revision in 2014, the weight was shifted from subsidizing quality products to environmentally friendly farming practices,—sustainable agriculture with the adoption of REG 1306/2013 (http://ec.europa.eu/).

Sustainable management of natural agricultural resources such as soil, water, seeds, coal, nitrogen, etc., by definition, deals with their continuous sustainable yield while keeping the environment operational and maintaining ecosystem's health. In other words, it strives to synthesize a denote positive sign equation for the environment, utilizing as mathematical operators natural resources, its high demand for productivity and its mostly feasible economy, so as to bring the well-being and

Image 1 Visualization of sustainable agriculture as a cross-section of four key assumptions. Source: National Research Council of the National Academies, USA, 2001

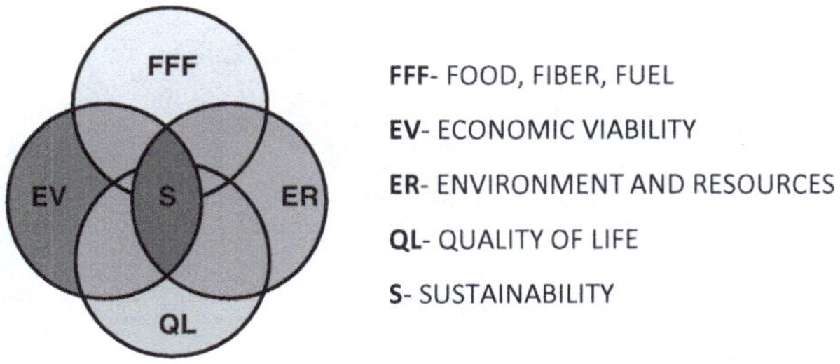

euphoria of the living organisms that consume and live in to it (Image 1) (Gerakis et al. 2008).

The case study, Precision Agriculture (PA), is a form of sustainable agriculture that is part of the integrated crop and livestock management systems, in which the territorial units—parcels—for cultivation are divided into management zones where interventions are carried out based on the specific characteristics of the available territorial resources in each spatial zone and the time required inputs from the cultivation. Or as Gebbens and Adamchuk in 2010 simply and descriptively stated: "apply the right treatment in the right place at the right time" (Gebbens and Adamchuk 2010).

For example, based on soil analysis and in conjunction with the available propagating material and water requirements, a crop is set up and inputs are decided by "smart systems". These systems derive information from sensors located in the parcel or remote sensing it and combine data into spatial databases (GIS), generating thematic maps. They develop strategies for dealing with crises (e.g. water stress), agricultural practices (e.g. pre-emergence fertilization) and finally produce results maps with the productivity of each zone and, therefore, the reduction of the cultivation production cost (Oliver 2013; Chen et al. 2014) (Image 2).

During the process of PA, initially, the variability of parcels is recorded, then management zones are created in them and then the inflows are applied, according to the needs of each zone. So, using PA techniques, it is understood that the disadvantage of intensive cultivation, which is the uniform application of inputs, eliminates. Inputs become limited and already—in the absence of the productive effect, i.e. the yield of the crop—significant economic and environmental benefits arise (Headley 2014).

According to Oliver, Bishop and Marchant (2013), "Precision agriculture (PA) involves the application of technologies and agronomic principles to manage spatial and temporal variation associated with all aspects of agricultural production in order to improve crop performance and environmental quality".

It is a method that uses the cutting edge of information technology, Geostatistics, to predict the outcome of each input, and is capable of making corrective actions in future time.

Precision Agriculture deals precisely with this term, in its mathematical meaning, from which it got its name. In order to understand precisely this term, hence the

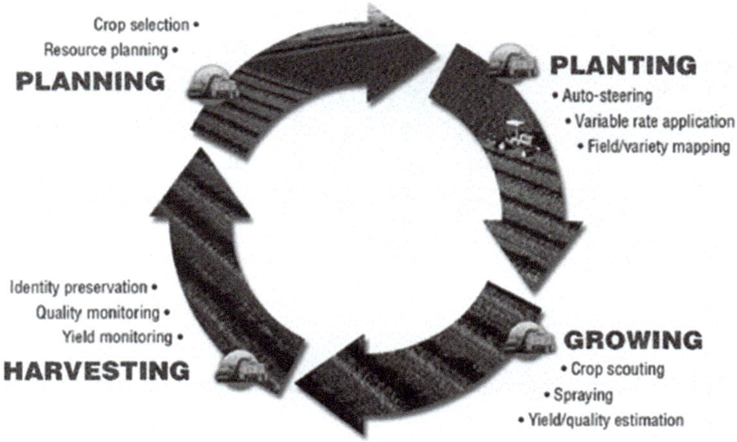

Image 2 Representation of an integrated crop management system using the Precision Agriculture technique. Source: www.ag.topconpositioning.com

purpose of this practice, the term "accuracy" should also be clarified. Accuracy is about how close you get to the right outcome. The accuracy is improved by properly calibrated machines and by constantly practicing a practice. Precision on the other hand is about how often you will achieve the same result using the same method. Precision is achieved with more reliable machines and minimum requirement calculations and corrections.

The US National Research Council, in a 1997 edition, defined Precision Agriculture as "strategic management using information technologies to analyze data from multiple sources and make decisions about agricultural production". It then defines its three main stages:

- the collection of scale data
- recording and analyzing data
- the decision to intervene in a specific spatial unit at the right time (National Research Council 1997).

The analysis and density of spatial information in the case of PA is also its great revolution, which means that it essentially involves changing the scale of manipulations. It is a methodology and a holistic approach that optimizes land use and preserves the environment (Auerhammer 2001; Katter et al. 2009).

The site-specific machinery used in Precision Agriculture are among the most specialized with high-speed processing and response sensors and always assisted by the methods of Geostatistical Science, and the use of data from Remote Sensing, Interpretation of Satellite imagery, Satellite Positioning Systems (GNSS Global Navigation Satellite Systems) and Geographic Information Systems (GIS). Thus, Precision Agriculture allows us to make decisions for small spatial segments—parcels based on their demands but also to continue to cultivate extensively with the use of large machinery (Fountas and Gemtos 2015).

1.2 Brief Introduction of the Research

The research presented focuses on investigating the hypothesis that those directly involved—crop producers in the Region of Central Greece, understand the necessity for sustainable management of agricultural resources through the emerging practice of Precision Agriculture which concerns not only a new cultivation technique but mostly a new holistic management philosophy of the spatial variation of crop fields—an innovational method.

For this reason, a field survey through a questionnaire was carried out on a random sample of 375 crop production producers active in the Region of Central Greece during the period April–June 2016. Individuals were able to complete the questionnaire form via a website or in a printed form distributed through certified submission agencies of Unitary Application for Aid Schemes (UAAS), stores for agricultural supplies and geotechnical consulting offices across the geographical area of the study.

In general, this research aims at examining producers' perceptions of usefulness regarding crop production in relation to the prospects arising from the adoption of PA. It also examines the features that a new integrated cultivation management system should have. They could be able to adopt it once they realize its usefulness on every day agronomical practices.

A first attempt for studying the phenomenon (Mourtzinis et al. 2007) demonstrated that only 9% in a sample of 130 producers all around Greece really knew about PA.

Another survey, aiming to investigate producers' perception in relation to managing their time and saving money from adopting Smart Farming Technologies (SFTs), was held in 2009. This comparative study on the attitude towards adopting sophisticated information systems by producers in four European Member States (Denmark, Finland, Germany and Greece) showed that in all countries producers were uncertain about the benefits that would result from the use of this technology (Lawson et al. 2011).

Michailidis et al. (2010) conducted another study in order to determine the reasons leading Greek producers to adopt or not PA. A sample of 2070 farmers, all over Greece, was called to clarify their familiarity and their interest to adopt SFT and PA. Of this initial sample 1195 producers were separated for further evaluation. They were those who owned these technologies as well as those who would adopt them in the near future. The greatest interest to adopt was found in farmers sited in Thessaly and Central Macedonia Region, as well as their familiarity to PA. Central Greece Region farmers expressed little familiarity and limited interest to adopt (Michailidis et al. 2010).

Kountios (2014) studied the educational needs of young farmers in Central Macedonia. These farmers were beneficiaries of the Rural Development Program of Greece (second term years 2000 to 2006—REG1268/1999 and third term years 2007 to 2013—REG.1698/2005) and the Agricultural Fund for rural Development (EU), in order to start a competitive agricultural business (Kountios 2014). In the same frame, examining the perceptions and educational needs of young farmers, at a most recent published study (Kountios et al. 2017) the PA knowledge rates were very high reaching 28% at the Regional Unit of Thessaloniki amongst young farmers. Young farmers were also asked about the sources affecting their decision to undertake PA methods. They replied that they were mostly affected by the opinions of other farmers, by successful examples and business consultants and less by public financial incentives.

So, a new research regarding the perceptions and attitudes of producers less interested to the subject was necessary to be attempted. Central Greece Region can be named a "late majority" area, towards adopting PA and SFTs according to Diffusion of Innovation Theory. Late majority people are skeptical of change, and will only adopt an innovation after it has been tried by the majority (Rogers 2003).

There was no previous research on the specific agronomic factors considered important by the observed population for the adoption of Smart Farming Technologies.

There wasn't also any other study aiming to draft policy proposals to incorporate new agricultural techniques and technologies into agricultural practice in the Central Greece Region.

The present aims to exam the perception of usefulness of PA and SFTs by using primary data from a field survey in a specific administrative area, and with specialized statistical analysis techniques tries to identify which factors can meet the demands of "late majority" crop producers in order for them to gradually adopt PA.

Finally, according to the collected information, a proposed policy strategy model, for adopting them, is attempted.

2 Study Area

2.1 Greek Crop Production

According to EUROSTAT (2014), the Greek territory covers an area of 131.621 km^2, 82.2% of which are rural areas. Its total population is about 11 million, 44.1% of which live in rural areas (http://ec.europa.eu/agriculture/rular-development2014-2020/country-files/el/factsheet-greece_en.pdf).

A key element of farm holdings is the small and much-sliced allotment that makes the agricultural production process difficult. 78% of the total utilized agricultural area (UAA), i.e. the total area of arable land, permanent pasture and meadows, permanent crops and vegetable gardens (http://ec.europa.eu/eurostat/ramon/nomen clatures), spans spatially within areas of natural constraints. 53.9% of the total UAA is in mountainous areas and only 19.8% is irrigated land. Although irrigated land is relatively small, 86% of water use in Greece is often used in agricultural activity with significant losses. The agricultural practice of organic farming is practiced at only 3.8% of the land (http://ec.europa.eu/agriculture/rular-development2014-2020/country-files/el/factsheet-greece_en.pdf).

The agricultural sector in Greece is characterized by small farms with 76.7% of them occupying an area of less than 50 acres. Of the 723,010 agricultural holdings, more than half occupy an area of less than 20 acres. The average of the agricultural allotment is just 6.8 ha, much less than the European (EU-26) average of 16.1 ha.

The population of those engaged in agriculture and livestock farming is quite aged with just 5.2% of Greek farmers in the age group of up to 35 years old. Young farmers in this age group (<35 years old) are headmen of agricultural holdings at a percentage of up to 12.6% of the total and only 3.5% of them have received vocational training in agricultural practice. The indicator of the standard of living of those active in the sector is 64.4% of the level of life of people employed in other sectors (http://www.europarl.europa.eu/studies).

The sector's contribution to Greece's Gross Added Value is 3.7%, with the average of the European Union of 26 Member States at 1.6%. The workforce permanently employed in the sector reaches 13.6% of the total employed population, whereas in the 26 Member States the average figure is only 4.7% (http://ec.europa.eu/agriculture/rular-development2014-2020/country-files/el/factsheet-greece_en.pdf).

The above evidence demonstrates the great economic importance of the sector in Greece. Although the general climatic conditions favor the differentiated production of crop and animal products, trade balance figures demonstrate that Greek economy is strongly dependent on imports, while the income of those active in the primary production sector, compared to the other sectors of the economy, becomes more and more unstable year-after-year, from year 2003 to the present.

2.2 *The Special Conditions of the Study Area*

The Region of Central Greece (NUTS2) is located in the center of mainland Greece (Map 1), it has a total area of 15,549.07 km^2, a population of 547,390 permanent residents (http://www.ststistics.gr/el/statistics/-/publication/SAM04/) and is administratively subordinated to the Decentralized Administration of Thessaly-Sterea Hellas. It is divided into five Regional Units (NUTS3): Viotia, Euboea, Evrytania, Fthiotida and Fokida.[1]

In the study area, the primary productive sector plays an important role both in the Region of Central Greece and in the five NUTS 2. The lands with agricultural use cover an area of 69,214 km^2, of which the cultivated plant species have an area of 3336 km^2 (Hellenic Statistical Authority (ELSTAT) 2014) and are traditionally covered by cropland areas with trees (mainly olives) in most of the Regional Units,

[1]According to the current EUROSTAT classification, NUTS 2 is referring to the basic regions for the application of regional policies.

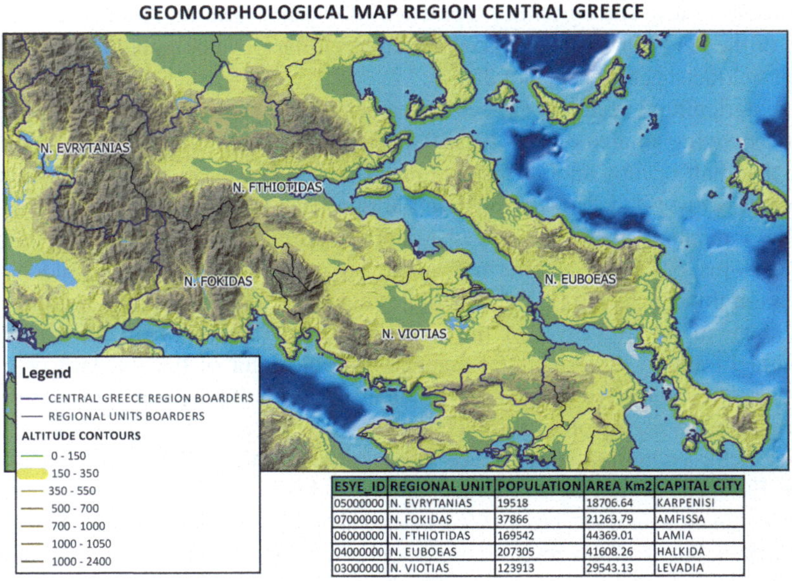

Map 1 Geomorphological map of study area with demographic data

Table 1 Accumulative data from all five Regional Units of the Region of Central Greece. Summarized from Municipality Level and pertaining crops and their area coverage in acres, as declared in the Uniform Applications for Aid Schemes (UAAS 2014)

Type of cultivation	Area in 1000 m^2	Percentage
Fallow land	290,022.3	11.83
Flower crops	244.2	0.01
Olive/vineyard/trees	748,569.4	30.54
Industrial crops	392,126.2	16.00
Animal feeds	243,310	9.93
Vegetables	63,407	2.59
Human consumption crops	712,561.6	29.07
Nurseries/greenhouses	542.9	0.02
Total	2,450,783.6	100.00

Source: www.opekepe.gr aggregated after own processing

and secondly with arable crops either for human consumption (cereals, potatoes, etc.) or for industrial process (cotton, energy crops, etc.). To a large extent, animal feeds are also cultivated in all Regional Units.

There are 69,660 agricultural holdings in the Region of Central Greece which on average exploit an area of 41.97 acres (Hellenic Statistical Authority (ELSTAT) 2014). The most recent data on the type of crops and land declared by producers in the Unitary Application for Aid Schemes (UAAS—EAE) 2014, from the statistical data maintained by the Greek Payment and Control Agency for Guidance and Guarantee Community Aid (OPEKEPE) and are posted on its website (www.opekepe.gr, 2016) are presented in Table 1.

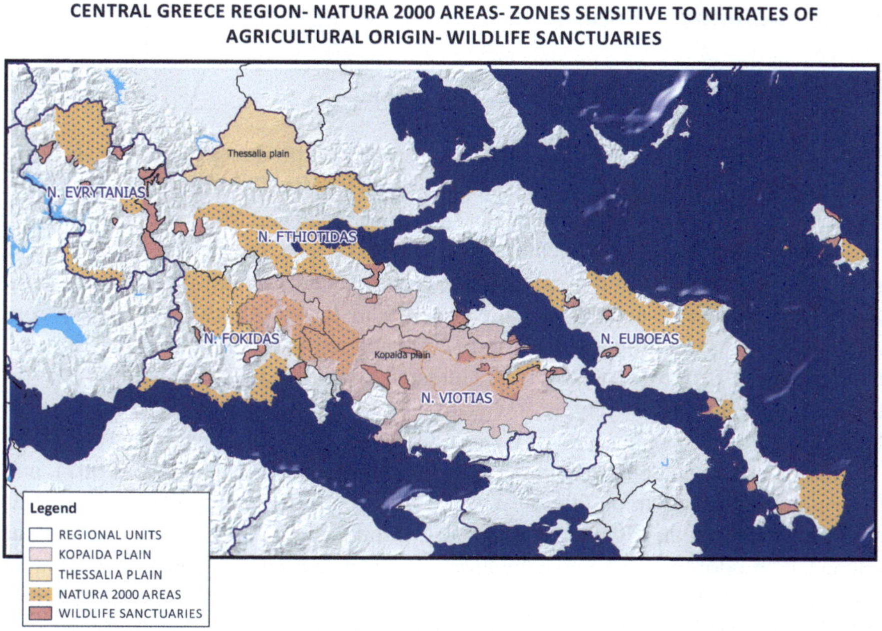

Map 2 Map of the region of Central Greece with NATURA 2000 areas, wildlife sanctuaries and sensitive zones to nitrates of agricultural origin

The study area was chosen because the Region of Central Greece, in addition to its large extent and its geographical location, is a typical example of the unevenness of the terrain relief, the diversity of the climate and the soils.

Also, within its geographical area, many areas of ecological interest are delineated, such as areas of the NATURA 2000 network and zones vulnerable to nitrates of agricultural origin (Map 2).

Finally, it was chosen due to the competitiveness of its products (olive oil, pistachios, figs, cotton, cereals, vegetables, livestock, milk, meat, etc.), as demonstrated on the latest data (2016) on the land regarding the crops declared by the producers in the UAAS 2015, from the statistical data maintained by OPEKEPE and are posted on www.iris.gov.gr.

It has a sufficient number of people employed in agriculture, several infrastructures, good access to the main road and a large number of products.

Thus, this Region possesses the elements that can create the conditions for the development of healthy agricultural enterprises with diversified production and good access of the products to the markets. However, development in an internationalized environment is not only achieved through means' availability but also through entrepreneurship and innovation.

3 Methodology: Sample

3.1 Materials and Methods

This study is based on a representative sample of producers engaged in crop production in Central Greece. The sample was 375 individuals/observations. Compared to the sample of Mourtzinis et al. (130 individuals) and Michailidis et al. (2070 individuals) that studied the phenomenon all over Greece; the sample of the present study was sufficient. The representativeness was tested through the comparison to the publicly available data base of the Greek Statistical Authority (ELSTAT) from the Census of Agriculture and Livestock 2009 and to the latest data provided by the Payment and Control Agency for Guidance and Guarantee Community Aid (OPEKEPE) (the plurality of year's 2015 UAAS—Uniform Applications for Aid Schemes to the Regional Units of Central Greece).

Taking into consideration the previous studies by Mourtzinis et al. (2007), and regarding their observation that only 9% of the producers were really aware of PA, and Michailidis et al. (2010) that refer to Central Greece Region as little familiar/interested in adopting; a re-investigation of the phenomenon of the perception of PA's and SFT's usefulness, is being attempted here. With 95% Confidence Interval and 3% sampling error, the respondents in the present study (375 crop producers) are considered sufficient to confirm the hypotheses.

Originally the profile of the participants concerning age, educational level, parallel employment, GPS possession and familiarization with new technologies (Smartphone, tablet, etc.), is presented through exploratory statistical analysis using frequency of responses as well as Cross Tabulation analysis (CTA). The CTA relates to creating contingency tables which exhibit the multivariate distribution of the frequency of responses after a Pearson Chi-Square test (Pearson 1900). Regarding the agricultural holding profile, the response frequencies of the individuals are presented concerning the type of the main culture (over 50%), the farm size in acres and the existence of mechanical equipment. Finally the responses of individuals are presented to questions relating to employment of other staff on their holdings and to their cooperation with geotechnical staff regarding either counseling on the production process, or compliance with the input-output files and a calendar listing of the operations under the European Commission REG 1306/2013 and the regulation of Multiple Compliance. These data are then examined according to the main characteristics of the respondents and holdings profiles (providing contingency tables and Pearson Chi-Square tests).

Finally, the data of two profiles were crossed using the same method (CTA and control of Pearson Chi-Square checks) in relation to:

i. their knowledge on the term Precision Agriculture,
ii. their understanding of the practice of Precision Agriculture
iii. and finally the extent that they perceive that, in the future, Precision Agriculture and Smart Farming Technologies will have to be adopted by farmers.

To detect factors that may support the decision by the target population for the adoption of Smart Farming Technologies (SFT) and Precision Agriculture (PA), an Exploratory Factor Analysis was performed on the basis of the producers' responses concerning their knowledge and understanding of the PA. That was extracted from their preferences of all the steps involved in this new technique (automatic navigation, recording technologies, variable dose applications and intelligent information systems of management—Computer Information Management Systems), whether they relate to offline or online data, as well as, the case that they get trained in the use of Precision Agriculture applications and the Smart Farming Technologies application means or get guided by the physical presence of specialized or non-specialized agronomist. The large number of items and the width of the sample led to the choice of the Principal Component Analysis method after first performing a Reliability Analysis test (Lei and Wu 2007), using the Reliability Indicator (internal demister type) Cronbach Alpha, to estimate the degree of cohesion or correlation of each variable/item with the scale and with the other variable and a Factor Analysis to confirm the relationship structure between the variables (Cronbach and Shavelson 2004; Hogan and Cannon 2007).

Principal Component Analysis (PCA) is an exploratory approach of observations (responses in the case study) on the phenomenon under examination and represents the linear combination that reveals the greater symmetry of the data variance. It concerns reduction of variables, expected based on the initial hypothesis to explain the factors that lead crop producers to adopt the techniques of the PA and the SFT. Concentration of information is achieved by creating new synthetic indicators/factors, called main components, and are essentially hyper-variables. The Principal Components result from the linear correlation of the initial one-dimensional variables strongly correlated with each other (Duquenne 2016).

The Factor Analysis (FA) as it has prevailed in international terminology is a statistical technique which allows for the investigation and interpretation of complex phenomena that depend on multiple components. This method is intended to reduce the overall information, through search and confirmation of relations between variables based on a recognized theoretical background while losing a minimum of initial information. In the present case study three Principal Components were extracted in order to provide empirical content to the policy proposals subsequently drafted (Rogerson 2001; Pison et al. 2003).

3.2 Phenomenon Dimensions: Hypotheses

The actual agricultural holdings in Greece mainly consist of medium-sized family farms, which in the context of the spatial households, in the sense[2] given by Anthopoulou and Goussios (2007), are fragmented into a smaller size of individual holdings—an average of 41.97 acres in the study area (own process—Hellenic Statistical Authority (ELSTAT) 2014).

It is clear that this structure affects, in addition to the size of the holding held individually by the active farmers of each family, the management of mechanical equipment, the pace of work, decisions on cultivation techniques and future investments, let alone investments in mechanical and technological equipment, a major requirement for the adoption of SFT and PA.

Also, in the context of family holdings, as observed by Goussios and Duquenne (2003), on spatial employment in the primary sector,

> the organization and management of the agricultural holding is adapted and integrated into a strategic family-based program structured within the framework of relations with bourgeois/ immigrants, relatives/shareholders and those who live in the village, by achieving to maintain this family holding embedded, within the social system of production of the village.

Here three groups of farmers are mentioned who, while appearing in statistical data, do not function in the same way. So, there are farmers who are locally based on the farm working at the holding full time and are the backbone of agricultural production, the near-distance exploiters who usually live in nearby towns and work in other sectors as well, and the long-distance exploiters living In large urban centers. The proximity between the place of residence and the site of the holding separates the third from the second category. All of them, however, are involved in decision-making on the exploitation and can influence any change in agricultural practice. This gradation is a phenomenon directly linked to the educational development of rural residents and urbanization, and as it becomes clear the size of agricultural holdings is misunderstood, since distance-farmers and family holdings are a Greek phenomenon (Goussios and Duquenne 2003).

Regarding age, persons mainly engaged in agricultural and livestock production in Greece belong to a large percentage to the ages of over 40 and have not been trained in agricultural practice, although the sector has a 13.6% share in a permanent workforce

[2]Ibid.

and 3.7 % of the country's Gross Added Value (http://ec.europa.eu/agriculture/rulardevelopment2014-2020/country-files/el/factsheet-greece_en.pdf).[3]

Demographically favored areas are those close to urban centers, seaside and mountain areas that use their natural resources also as tourist assets. Finally, the lowlands that are connected by road network and are served by transport networks (Anthopoulou and Goussios 2007; Chalkos 2013).

Adoption by Pannell et al. (2006) is the learning process in which information collected, then synthesized and finally evaluated for any brought results.

The farmer's adoption of the Precision Farming according to Katter et al. (2009) concerns:

> combined utilization of several site-specific technologies using Global Positioning Systems (GPS) such as auto guidance and variable rate applications (VRT) of inputs and/or yield mapping on farm. This definition does not imply that these practices have to be carried out by farm staff but can be offered by a third party as well.

In the same study, the methodology followed was based on Rogers' innovative theory hypothesis (2003) defining five categories of individuals based on adoption time: innovators, early adopters, early majority, late majority and laggards.

Due to the random sampling within the site that this survey was conducted and the fact that there were no responses from individuals that participated in the research, who have been in a learning or an adoption process, the method followed could not be supported from this theory and concerns exclusively the recording of existing trends and the detection of the following hypotheses with the aid of descriptive statistics aiming for the preparation of policy proposals for the adoption of SFTs.

The composition of the active population of crop producers within the study area has been maintained the same regarding the main age group active in the sector, but young producers have an increased level of education and adequate knowledge of new technologies, as it will emerge from the presentation of the participants' profiles.

The profile of the participants is presented in two stages. The first relates to their own data (gender, age, level of education, etc.) and the second relates to the data relevant to their farms (size of exploitation, type of main crop, etc.).

In the context of this study, three dimensions of the phenomenon examined will be detected and reported in corresponding hypotheses:

[3]The "spatial household": *"... expresses as a concept the relationships that a group of people, with family or other social ties, exploits and develops through the distribution of the agricultural land they own, either in the form of renting or direct exploitation, within the framework of the organization and operation of an agricultural business. At the same time it expresses: 1) the geographical distribution and spatial position of members who participate with land and/or labor and capital in the organization and operation of a farm holding, 2) the size of the land under control, and 3) the organization produced and managed by the coordinating agricultural family within the spatial household with the aim of regulating land and labor relations"* (Anthopoulou and Goussios 2007).

Table 2 Data used to calculate the individuals of the sample per Regional Unit

Calculation table per regional unit of the population sample			
Regional unit	Number of agricultural holdings (ELSTAT 2009)	Number of UAAS 2015 declared crop production (OPEKEPE 2015)	Questionnaires collected
Viotia	16,474	14,976	90
Euboea	23,828	12,877	70
Evrytania	2266	403	30
Fthiotida	20,518	22,472	126
Fokida	6574	3624	59
Total	69,660	54,352	375

Source: ELSTAT and OPEKEPE data (after own processing)

Hypothesis 1 Few of the farm producers operating in the Region of Central Greece know the term of PA.

Hypothesis 2 Even fewer farm producers perceive the practice of PA.

Hypothesis 3 There should be many actions in the future in order for producers to understand that PA and SFT will be a prerequisite for their prosperity in terms of sustainable management of agricultural resources.

The factors that can contribute to assessing the effectiveness of SFTs by the target population and are therefore considered to be able to push producers into adopting PA farming practices can be evaluated as to formulate a proposed policy model. The factors studied here are related to the purpose of the cultivation techniques (questions 15–26), the support and monitoring of SFT applications by geotechnical staff (questions 27–29), education in new technologies (question 30), the possibility of investing (Questions 31–34), and finally distance counseling (questions 35–38) (Table 14).

Beyond investigating the responses of individuals of the sample to verifying the hypotheses, the specific conditions that have been developed in this sector in the period after the data from the two surveys mentioned above are being sought, as they relate to the 2007–2009 period, before the financial crisis of the Greek public sector and the increase of unemployment rates that followed.

The classical methodology for an empirical applied research was applied. It included a review of the relevant literature, the creation of a questionnaire, the selection of the study area, the formation of a representative sample of population, the sampling, and finally the statistical analysis of the data.

The number of questionnaires that were to be collected per Regional Unit was cross-checked to the published data of the Hellenic Statistical Authority (ELSTAT) from the Agricultural and Livestock Census 2009 and the most recent data provided by OPEKEPE (number of Unitary Application for Aid Schemes, 2015) as presented in Table 2.

More specifically, the research aims to produce safe conclusions in order to verify or reject hypotheses originally made through:

- the particular characteristics of the active population of crop producers
- the perceptions of crop producers about the term and practice of PA
- their attitudes towards the adoption of Smart Farming Technologies
- the factors that can contribute to assessing the effectiveness of Smart Farming Technologies by the target population.

3.3 Questionnaire Structure

The questionnaire was set up in four parts to detect:

- the profile of farmers and the profile of agricultural holdings of crop production
- knowledge of the term and practice of Precision Agriculture (quantitative information) and its origin (qualitative information)
- the factors assessed as being sufficient, by the target population, for the adoption of Precision Agriculture and Smart Farming Technologies
- Observed population's opinion about the future use of PA and SFTs.

4 Results

4.1 The Profile of Producers and Holdings

The profiles of the participants are presented in two stages (Tables 3 and 7). The first are the data concerning themselves (gender, age, level of education, etc.) and the

Table 3 Summary presentation of the demographic profile of the individuals of the sample

Population characteristics	Categories/classes	Frequencies	Percentage
Gender	Male	253	67.5
	Female	122	32.5
Age	18–25	18	4.8
	26–35	81	21.6
	36–45	85	22.7
	46–55	100	26.7
	56–67	61	16.2
	>67	30	8.0
Age summary	N = 375; min = 18; max = >67; average = 46		
Education level	Without basic education	14	3.7
	Primary graduates	35	9.3
	Low secondary graduates	54	14.4
	Secondary graduates	153	40.8
	Sort circle tertiary graduates	74	19.7
	Tertiary graduates	45	12.0

Table 4 Education level in relation to the age profile of crop producers

Characteristic (education level)	χ^2	Degrees of freedom	p-Value
Age	209.879[a]	25	0.000***

Asterisks indicate the level of significance for p-value as: ***$p \leq 0.001$
[a]13 cells (36.1%) have expected count less than 5. The minimum expected count is 0.67

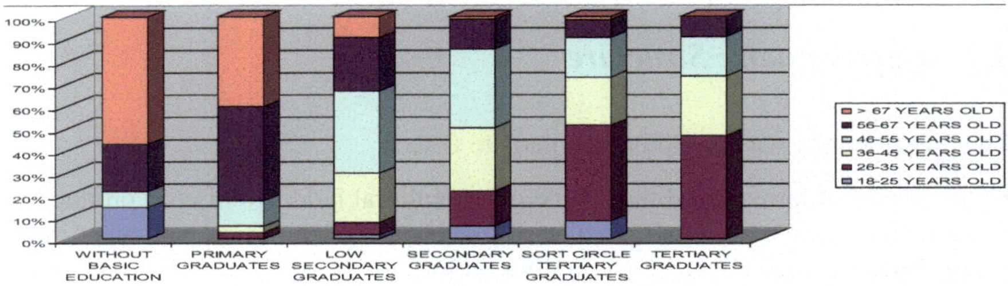

Fig. 1 Ranges of accumulated absolute response columns for the level of education versus the ages

Table 5 Hetero-employment in relation to the characteristics of crop producers regarding their level of education and age

Characteristic (hetero-employment)	χ^2	Degrees of freedom	p-Value
Education level	39.956[a]	5	0.000***
Age	14.964[b]	5	0.010**

Asterisks indicate the level of significance for p-value as: **$p \leq 0.01$, ***$p \leq 0.001$
[a]1 cell (8.3%) has expected count less than 5. The minimum expected count is 4.44
[b]0 cell (0.0%) has expected count less than 5. The minimum expected count is 5.71

second relates to the data on their agricultural holdings (size of holding, type of main crop, etc.).

The composition of the active population of crop producers within the study area has been maintained the same regarding the main age group operating in the sector, but young producers demonstrate a higher level of education and an adequate knowledge of new technologies.

The data on the level of education presented here (Table 4 and Fig. 1) give a significant differentiation of the young in age population regarding education. In the age groups 26–35 and 46–55, over 50% of the individuals have a Secondary Education (High School) certificate, while it is worth mentioning a percentage of 32.8% in this level of education of the age category of 56–67 years.

Graduates of Higher Education are also concentrated in the young ages 25–45 and with a clear lead of sort circle tertiary graduates. University graduates present a concentration in the age group of 26–35 years. This concentration of tertiary education graduates in the age groups 18–25 and 26–35 years old, is an element that reinforces the premise that many young people at the end of their studies in tertiary education return to the province and take over agricultural holdings, either seeking supplementary income, or changing career orientation, particularly amid the worst financial crisis (2009–2016) faced by the country for decades (Table 5).

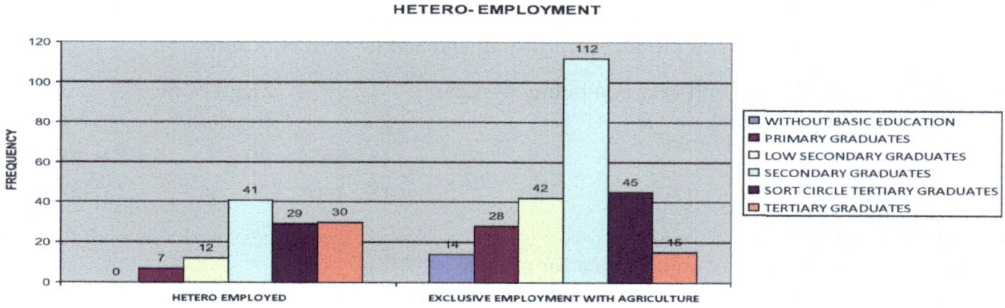

Fig. 2 Frequency response bar chart of hetero-employment and level education

Table 6 Age of respondents and possession of a Global Positioning System (GPS)

Characteristic (GPS possession)	χ^2	Degrees of freedom	p-Value
Age	21.138[a]	5	0.001***

Asterisks indicate the level of significance for p-value as: ***$p \leq 0.001$
[a]1 cell (8.3%) has expected count less than 5. The minimum expected count is 4.94

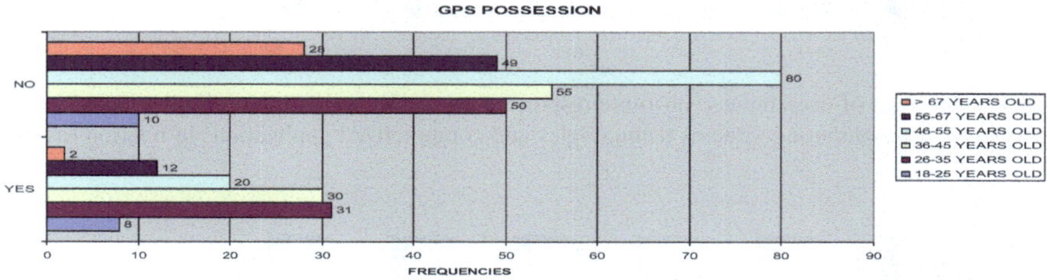

Fig. 3 Bar chart of absolute frequencies of responses in age groups and possession of Global Positioning System (GPS)

This hypothesis, regarding young people seeking supplementary income, is also reinforced by the evidence highlighted by the answers to the question on farmers' hetero-employment. In the question of whether they have a second occupation, 31.7% responded positively, and the cross-referencing data on their level of education, as shown here, shows a strong concentration (49.6%) of post-secondary education degree holders in relevance to their secondary employment (Fig. 2).

In relation to the question that is directly related to the understanding of the tools of the technology used for the application of Precision Agriculture, and concerns the possession of a GPS-Global Positional System, a necessary element for the mapping of information (spatial variability) and the application of the individual techniques (automatic navigation), positive responses support the hypothesis that the younger and the middle aged were already familiar with their use (Table 6 and Fig. 3).

Age and level of education are very much related to the familiarization of producers with new computing systems technologies and computerized applications

Table 7 Familiarity with new computing systems technologies and computerized applications in relation to the characteristics of producers regarding age and level of education

Characteristic (familiarity with new computing systems and computing applications)	χ^2	Degrees of freedom	p-Value
Age	187.665[a]	20	0.000***
Education level	189.440[b]	20	0.000***

Source: Own processing
Asterisks indicate the level of significance for p-value as: ***$p \leq 0.001$
[a]12 cells (28.6%) have expected count less than 5. The minimum expected count is 1.58
[b]13 cells (31.0%) have expected count less than 5. The minimum expected count is 1.23

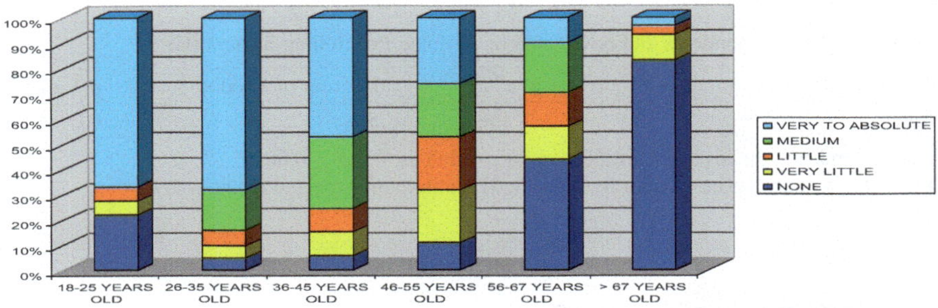

Fig. 4 Bar chart of accumulated absolute response frequencies columns for the degree of familiarity with new computing systems technologies and computerized applications in relation to ages

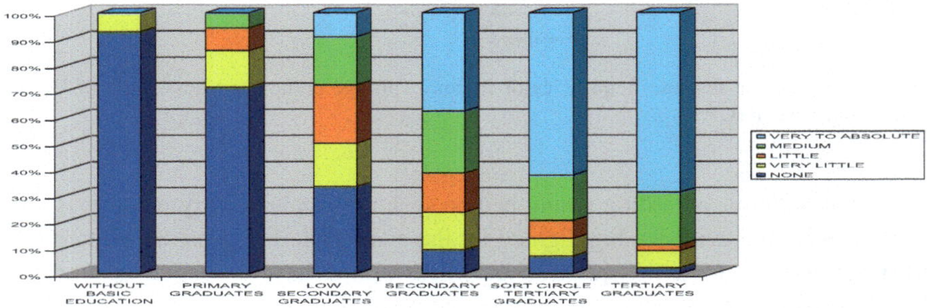

Fig. 5 Bar chart of accumulated absolute response frequencies columns for the degree of familiarity with new computing systems technologies and computerized applications in relation to the level of education

(Table 7). Thus, producers up to 35 years old presented the highest awareness of these technologies, at percentages ranging around 68%. Then, the age group from 36 to 45 years old presented a moderate to excellent awareness of them. As expected, individuals of the population who are not aware of them are those over the age of 56 (Fig. 4).

The level of education is strongly related to knowledge in new technologies, with the percentage increasing depending on the level of education completed by the

Table 8 Summary of the profile of the holdings of the sample individuals

Characteristics of agricultural holdings	Categories/classes	Frequencies	Percentage
Main crop type	Olive trees/vineyards/trees	152	40.5
	Arable crops for human consumption	52	13.9
	Arable crops for industrial process	75	20.0
	Crops for animal feed	66	17.6
	Vegetables	26	6.9
	Nurseries/greenhouses	4	1.1
	Flower crops	0	0.0
Farm size in 1000 m^2	0–10	42	11.2
	11–25	67	17.9
	26–50	78	20.8
	51–100	81	21.6
	101–150	35	9.3
	151–200	24	6.4
	>201	48	12.8
Agronomist consultant for production	No, I don't need	61	16.3
	No, I am in knowledge	48	12.8
	Auxiliary when needed	196	52.3
	Yes, constantly	66	17.6
	I don't know/I won't answer	4	1.0
Agronomist consultant for keeping files and registers	No, I don't need	101	26.9
	No, I keep on my own	105	28.0
	Auxiliary when needed	108	28.8
	Yes, constantly	41	10.9
	I don't know/I won't answer	20	5.3

population. Two of the three producers (62.2% and 68.9%) who have completed tertiary education were highly aware of these technologies (Fig. 5), while the secondary education graduates were familiar with this technology to a satisfactory up to an excellent degree at a percentage of over 50%.

As far as the agricultural holding profile is concerned, the response frequencies of the persons in terms of the size in acres and the type of the main crop (over 50%)—TEO/ Technical-Economic-Orientation (Table 8) is presented.

The study area is traditionally covered by large areas of crop production with trees (mainly olives) in most of the Regional Units, and secondly with large-scale crops, either for human consumption (cereals, potatoes, onions, etc.), or for industrial processing (cotton, energy plants, etc.). To a large extent of the land, animal feeds are also cultivated in all Regional Units.

Finally, the responses of individuals to questions concerning their cooperation with geotechnical staff are presented, either advisory on the production process or on

Table 9 Summary of population perceptions on precision agriculture and intelligent farming technologies

Characteristic	Categories/ classes	Frequencies	Percentage
Knowledge of the term precision agriculture	None	150	40.0
	Very little	56	14.9
	A little	50	13.3
	Medium	58	15.5
	Much to absolutely	61	16.3
Knowledge of the practice of precision agriculture	None	168	44.8
	Very little	53	14.1
	A little	55	14.7
	Medium	52	13.9
	Much to absolutely	47	12.5
Perception of future use of smart farming technologies and precision agriculture	None	21	5.6
	Very little	18	4.8
	A little	36	9.6
	Medium	72	19.2
	Much to absolutely	228	60.8

keeping records of inputs and outflows and the work calendar recording under Regulation 1306/2013 and the Information Decision Support Systems.

So, it appears that:

(A) on the question of whether they cooperate with an agronomist consultant for the production process, 52.3% said that their cooperation is on a subsidiary basis
(B) on the question of whether they cooperate with an agronomist consultant to keep records of inputs and outputs and records in the Work and Financial Data Logs as set out in the Codes of Good Agricultural Practice and the Reg. 1306/2013, 28.8% replied that their cooperation is also on a subsidiary basis.

However, the answer, which highlights in particular a subsection of the subject that is being considered—the adoption by the producers of computerized applications for their observance in the framework of the Systems of Information Management Systems, and concerns their observance by them, possesses the non-insignificant percentage of 28% of all individuals. This observation is encouraging for supporting the assumption that crop production farmers in the study area understand the need for keeping records and data for assessment for future use.

The first two hypotheses are presented initially through the frequencies of the answers to the questions that detect the perceived knowledge level of the population on the term and practice of the PA (Table 9). Crop producers were familiar to the term of Precision Agriculture at a percentage of 16.3%, but they were familiar with its practice at just 12.5%.

In the question regarding the perception of the crop producers on the necessity of adopting Precision Agriculture and Smart Farming Technologies in the future, they were extremely positively predisposed. They, therefore, believe highly to absolutely, that it will be necessary in the future to adopt the new technologies at a percentage of 60.8%.

Then, with the aid of CTA and Pearson Chi-Square test, the influence of the two profiles was examined, i.e. the objective characteristics of the population and their holdings, on the perception of the target population for future use of Precision Agriculture and Smart Farming Technologies.

Producer profile elements that faithfully reflect this attitude (Table 10) are gender, education level, GPS possession, and familiarity with new technologies on computing systems and computing applications. Their perception is not at all related to age or to whether they are hetero-employed in crop production (Figs. 6 and 7).

The perception of the future adoption of PA and SFTs is strongly related to the profile of their holdings, except for the employment of permanent staff on the holding, but in this category the answers are statistically significant (Figs. 8 and 9).

Women at a percentage of 32.8% and the age group of 46–55 years old are the strongest trend here, whereas, regarding the education level completed by sample individuals, tertiary education graduates and holders of Postgraduate Diploma are again superior (Fig. 6).

Only 9.7% of individuals in the possession of GPS speculates that Precision Agriculture and Smart Farming Technologies will not be necessary in the future, and even the ones who poorly know the new technologies regarding information computing systems and computerized applications believe, at a percentage of 34.1%, that their use will be needed in future (Fig. 7).

Table 10 Perception for the adoption of SFTs and PA in relation to the characteristics of the profiles of producers and holdings

Profile	Characteristics	χ^2	Degrees of freedom	p-Value
A' Crop producers	Gender	9.572[a]	4	0.048**
	Age	7.730[a]	16	0.956
	Education level	54.134[a]	16	0.000***
	Hetero-employment	4.299[a]	4	0.367
	GPS possession	17.987[a]	4	0.001***
	Familiarity with new computing systems and computing applications	66.709[a]	16	0.000***
B' Farm holdings	T.E.O./Technical-Economic Orientation	33.105[a]	16	0.007***
	Size of household	53.360[a]	20	0.000***
	Employment of staff on the holding	11.525[a]	4	0.021**
	Employment of seasonal staff on the holding	15.709[a]	4	0.003***
	Agronomist consultant for production	32.621[a]	4	0.000***
	Agronomist consultant for keeping files and registers	31.347[a]	4	0.000***

Asterisks indicate the level of significance for p-value as: $**p \leq 0.01$, $***p \leq 0.001$

[a]0-6 cells (0%–20%) have expected count less than 5 for each one of the characteristics

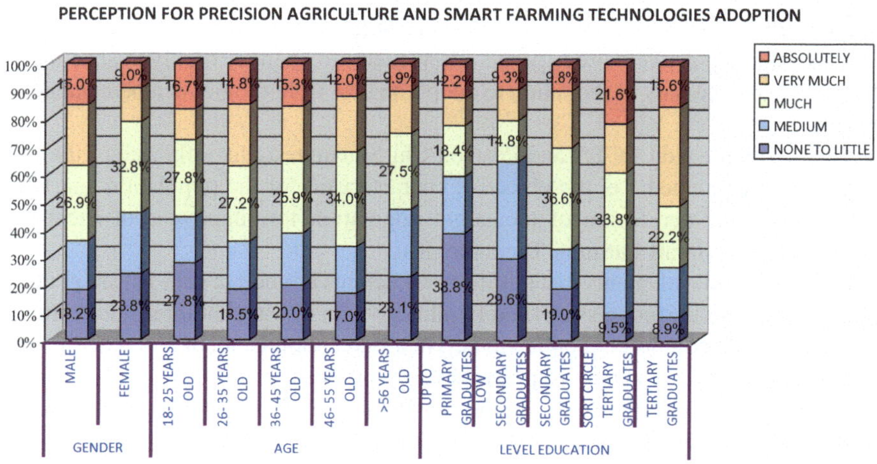

Fig. 6 Bar chart of accumulated absolute response frequencies columns for the perception for PA and SFTs adoption relating to the gender, age and level of education of sample individuals

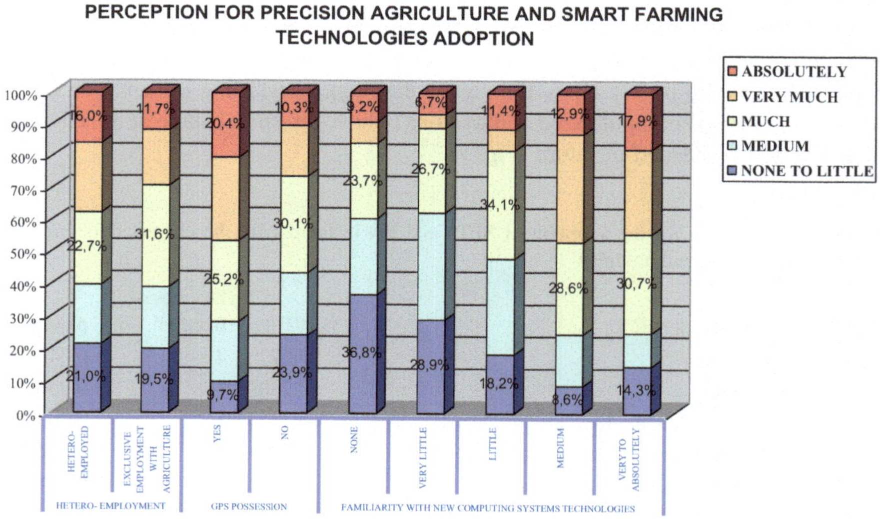

Fig. 7 Bar chart of accumulated absolute response frequencies columns for the perception for PA and SFTs adoption relating to hetero-employment, GPS possession and familiarity with new technologies regarding computing systems and computerized applications

In the case of the farm holdings profile, it is clear that new techniques are of no interest to producers who exploit a small area (42.9%) and grow olives, trees, vineyards (25.7%). Producers exploiting an area of 101–150 acres believe at a 40% percentage that Precision Agriculture and Smart Farming Technologies are

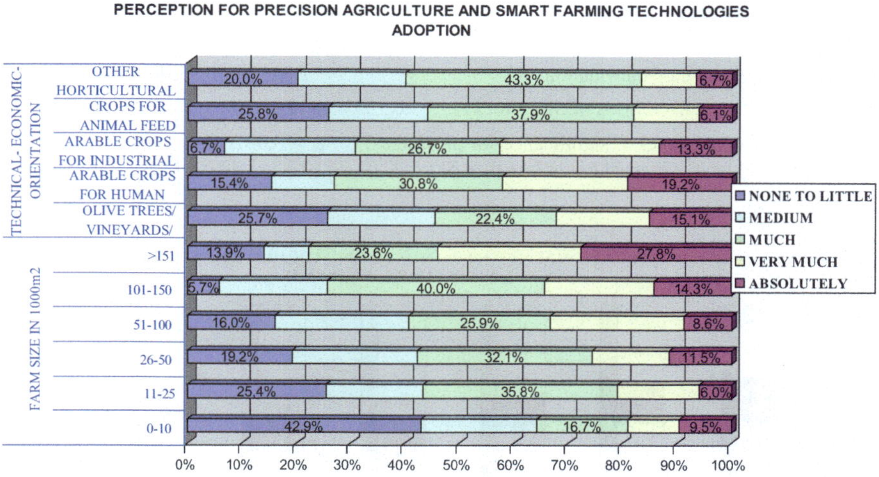

Fig. 8 Bar chart of accumulated absolute response frequencies columns for the perception for PA and SFTs adoption relating to TOP and farm size

most likely to be necessary in the future, and exploiters of over 151 acres do not consider them necessary at a percentage of 13.9% (Fig. 8).

The extremely positive view of producers for the future use of Precision Agriculture and Smart Farming Technologies is formed at high rates, when their farms are supported by an agronomic consultant, either for the production or for keeping records and data, and can reach individual rates of 36.9% (Fig. 9).

The perceived notion of individuals about the future usefulness of SFTs and PA as it emerged through the Contingency Table between the source of information and the perception of the future usefulness of PA and SFTs (Table 11), highlighted that agronomists and companies active in the field hold the lion's share (40.8%) of the absolutely positive perception of producers for them. The media also help in this

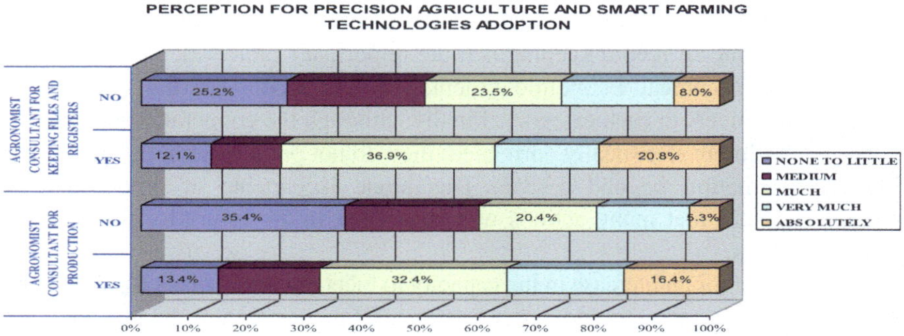

Fig. 9 Bar chart of accumulated absolute response frequencies columns for the perception for PA and SFTs adoption relating to cooperating with an agronomist

Table 11 Producers' perception of the future usefulness of SFTs and PA in relation to their source of information about PA

Characteristic (source of information)	χ^2	Degrees of freedom	p-Value
Perception of future usefulness of SFT and PA	40.153[a]	12	0.000***

Asterisks indicate the level of significance for p-value as: ***$p \leq 0.001$
[a]6 cells (20.0%) have expected count less than 5

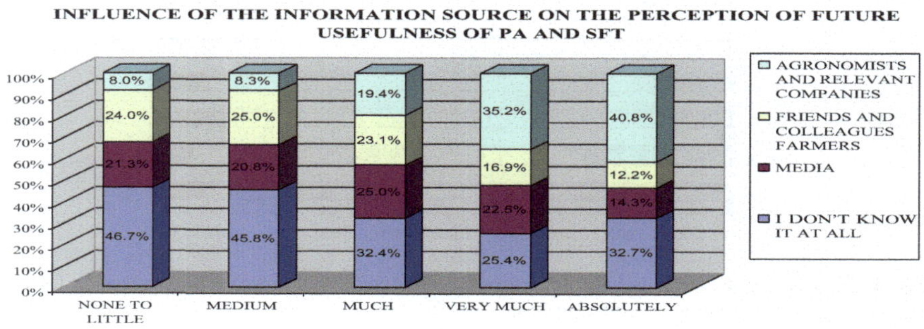

Fig. 10 Bar chart of accumulated absolute response frequencies columns for the influence of the information source on the perception of the future usefulness of SFTs and PA

direction, but far less as far as the intensity of the perception is concerned. So, although the category "high" intensity is at 25%, there is a strong core of the order of 21.3% that is not convinced of the usefulness of these technologies even if they were informed about them by the media. Discussions with friends and colleagues form a very modest perception among the individuals about the usefulness of SFTs and PA in the future (Fig. 10).

Control with a statistical analysis of Contingency Tables and Pearson Chi Square's tests of the individuals' profile data and the source of information demonstrated that statistically significant profile variable is gender, age, level of education and familiarity with new technologies. As far as the second profile of the farms is concerned, all its elements are important for the origin of the information they receive, but the cooperation with an agronomist either as a production consultant or as a consultant for record keeping is more important (Table 12).

Regarding Technical Economic Orientation, the media are the mainstay of informing producers in each category. Finally, although the cooperation with agronomists is absolutely statistically correlated, they do not participate in the dissemination of information beyond 13.3%. This alone is capable of reinforcing the hypothesis (No3) that many actions will be needed in the future for the adoption of SFTs and PA.

It is common for an increase in the capacity to meet the needs of each professional (in this case the farmers) to undergo a first stage of recognition of the new field that they can expand in. More simply, in order to develop a new model of agricultural practice, farmers of crop production would seek out more than one source of information on new and emerging techniques and technologies of primary

Table 12 The origin of PA awareness in relation to the characteristics of producers profiles (A) and farms (B)

Profile	Characteristics	χ^2	Degrees of freedom	p-Value
A' Crop producers	Gender	12.265[a]	4	0.015**
	Age	46.894[a]	16	0.000***
	Education level	61.692[a]	16	0.000***
	Hetero-employment	1.516[a]	4	0.824
	GPS possession	48.552[a]	4	0.000***
	Familiarity with new computing systems and computing applications	96.185[a]	16	0.000***
B' Farm holdings	T.E.O./Technical-Economic Orientation	28.184[a]	16	0.030**
	Size of household	40.558[a]	20	0.004**
	Employment of staff on the holding	24.678[a]	4	0.000***
	Employment of seasonal staff on the holding	36.868[a]	4	0.000***

Asterisks indicate the level of significance for p-value as: $^{**}p \leq 0.01$, $^{***}p \leq 0.001$
[a]0-6 cells (0%–20%) have expected count less than 5 for each one of the characteristics

Table 13 Multiple or not, information on PA in relation to producers profiles data (A) and holdings (B)

Profile	Characteristics	χ^2	Degrees of freedom	p-Value
A' Crop producers	Gender	14.640[a]	2	0.001***
	Age	47.446[a]	8	0.000***
	Education level	56.504[a]	8	0.000***
	Hetero-employment	0.193[a]	2	0.908
	GPS possession	39.568[a]	2	0.000***
	Familiarity with new computing systems and computing applications	91.943[a]	8	0.000***
B' Farm holdings	TEO/Technical-Economic Orientation	15.071[a]	8	0.058*
	Size of household	14.983[a]	10	0.133
	Employment of staff on the holding	17.497[a]	2	0.000***
	Employment of seasonal staff on the holding	26.430[a]	2	0.000***

Asterisks indicate the level of significance for p-value as: $^{*}p \leq 0.05$, $^{***}p \leq 0.001$
[a]0-6 cells (0%–20%) have expected count less than 5 for each one of the characteristics

production. From the statistical analysis of Double Input Tables and Pearson Chi Square's test regarding the producers profile data, it emerged that multiple information on PA is highly dependent on gender, age, level of education, GPS ownership and familiarity with new technologies. It also depends on the cooperation with an agronomist, whether he has the role of a production consultant or a scholar who monitors and keeps records of the holding (Table 13).

Naturally, the contribution of agronomic consultants to triggering producers' interest in obtaining information through more sources is not positive, since farmers working with agronomists obtain information from only one source, at percentages of 55–63%.

4.2 Factor Analysis

The factors, that can contribute to assessing the effectiveness of PA and SFTs by the target population and are therefore considered to be able to impel producers into adopting PA farming practices and SFTs, can be evaluated and shape a proposed policy model. The factors studied are related to the purpose of the function of cultivation techniques, the support and supervision of SFT applications by geotechnical staff, training in new technologies, investment potential, and finally remote counseling.

Sample individuals were asked to respond to the preferred intensity of use of the various techniques and technologies that make up PA and SFTs, their desire to be trained in them or to be guided by consultants, as well as their willingness to adopt and invest in the new technique and technology, in questions 15–38 of the questionnaire (Table 14).

In Order to detect the perception of the target population for Precision Agriculture and Smart Farming Technologies, using the method of Factor Analysis a model was developed—as presented here—, which meets up to their preferences for the adoption of all their stages.

The production of the composite markers presented here is based on Exploratory Factor Analysis Methods and therefore the selection of the final markers requires a systematic evaluation of the results based on the multiple criteria recommended in the international literature (Duquenne 2016).

When using the total of the original independent variables (Questions 15–38), the ratio between the number of observations and the number of variables is in the order of 15 (375: 24), which is an ideal condition for the application of the Exploratory Factor Analysis, as long as the observations/responses are $n > 200$.

There is excellent consistency between the variables, with the Kaiser-Meyer-Olkin Index (KMO) being 0.952. Also, the independent variables show not only a significant degree of variability, with the CV variability factor systematically greater than 38% (Table 15), but also high correlation with p-values <0.5 (5%) (Table 16). The composition of the 24 independent variables leads to a satisfactory model in the sense of KMO and the limited number of composite indicators (four), which reflect 72% of the total inertia, i.e. a loss of information of 28%.

However, the initial positive evaluation of the model should not conceal its weaknesses, especially in terms of correlations and degrees of participation of the variables (Table 17).

With the exception of the V37 (on-line counseling versus actual) variable, all the others participate in a satisfactory (H2 > 0.500) to a very good degree (H2 > 0.800) in the composite indicator configuration. In this case, it is recommended to remove

Table 14 Presentation of the independent variables of Factor Analysis

Variable	Variable description
Question No 15	Attitude towards the need to detect the possible cultivation by computerized application with spatial variation mapping method (Recording Technologies) [None/1 to Absolutely/7].
Question No 16	Attitude towards the need for adoption of auto steering technology for planting/sowing [None/1 to Absolutely/7].
Question No 17	Attitude towards the need for adoption of Variable Rate application—VRA for fertilizers inputs [None/1 to Absolutely/7].
Question No 18	Attitude towards the need for adoption of Variable Rate application—VRA for irrigation [None/1 to Absolutely/7].
Question No 19	Attitude towards the need for adoption of Variable Rate application—VRA for pesticide and herbicide inputs [None/1 to Absolutely/7].
Question No 20	Attitude towards the need for adoption of Computer Information Management Systems to register inputs [None/1 to Absolutely/7].
Question No 21	Attitude towards the need to adopt computerized application with spatial variation mapping method (Recording Technologies) during harvesting [None/1 to Absolutely/7].
Question No 22	Attitude towards the need for adoption of Computer Information Management Systems to register outputs [None/1 to Absolutely/7].
Question No 23	Attitude towards the need for adoption of Computer Information Management Systems to register financial data [None/1 to Absolutely/7].
Question No 24	Attitude towards the need for adoption of Computer Information Management Systems to register farming/cultivation practices [None/1 to Absolutely/7].
Question No 25	Attitude towards the need for adoption of Computer Information Management Systems to register Balance sheets, debts, sales and warehouses [None/1 to Absolutely/7].
Question No 26	Attitude towards the need for adoption of Computer Information Management Systems to make future decisions on cultivation practice [None/1 to Absolutely/7].
Question No 27	Attitude towards the need to apply the above-mentioned techniques through a non-assisted online platform [None/1 to Absolutely/7].
Question No 27.1	Auxiliary/sporadically in answer to the question 27 [Yes/1, No/2, I don't wish to answer/3]
Question No 28	Attitude towards the need to apply the above-mentioned techniques with the assistance of an agronomist [None/1 to Absolutely/7].
Question No 28.1	Auxiliary/sporadically in answer to the question 28 [Yes/1, No/2, I don't wish to answer/3].
Question No 29	Attitude towards the need to apply the above-mentioned techniques with the assistance of a specialized agronomist [None/1 to Absolutely/7].
Question No 29.1	Auxiliary/sporadically in answer to the question 29 [Yes/1, No/2, I don't know or don't wish to answer/3]
Question No 30	Attitude towards the need to apply the aforementioned techniques after training [None/1 to Absolutely/7].
Question No 31	Interest in investing in new technologies [None/1 to Absolutely/7].
Question No 32	Estimation to the potential payback of capital invested [None/1 to Absolutely/7].

(continued)

Table 14 (continued)

Variable	Variable description
Question No 33	Estimation of the amortization time of the invested capital (1–15 years).
Question No 34	Estimation of the amount of money that can be invested in relation to the annual profits of the existing holding (percentage).
Question No 35	Interested in private counseling/guidance through mobile phone or personal computer in new technologies [None/1 to Absolutely/7].
Question No 35.1	Auxiliary/sporadically in answer to the question 35 [Yes/1, No/2, I don't know or don't wish to answer/3].
Question No 36	Interested for public counseling/guidance (e.g. Ministry of Rural Development and Food) through mobile phone or personal computer in new technologies [None/1 to Absolutely/7].
Question No 36.1	Auxiliary/sporadically in answer to the question 36 [Yes/1, No/2, I don't know or don't wish to answer/3].
Question No 37	Assessment of whether private or public distance counselling would replace the work of an agronomist as a physical presence [None/1 to Absolutely/7].
Question No 38	Assessment of whether guidance should be provided by a specialized agronomist in new technologies and techniques [None/1 to Absolutely/7].

Table 15 Variability degrees of the independent model variables

Independent variables	Mean	Std. deviation	CV = (Std/mean)%
V15_planing_Recording_Technologies	4.14	1.809	44
V16_auto_steering	4.24	1.813	43
V17_fertiliser_VRA	4.59	1.777	39
V18_water_VRA	4.59	1.824	40
V19_pest/herb_cides_VRA	4.77	1.811	38
V20_inputs_smart_view	4.27	1.880	44
V21_harvest_crop_Recording_Technologies	4.22	1.840	44
V22_yield_smart_view	4.15	1.835	44
V23_outputs_smart_view	4.39	1.816	41
V24_work_calenter_smart_view	4.27	1.763	41
V25_balance_sheet_smart_view	4.29	1.827	43
V26_predict_fut_crop_smart_view	4.22	1.862	44
V27_internet_smart_view	3.77	1.831	49
V28_prefer_agriclt_advisor	4.03	1.641	41
V29_prefer_specialist_agriclt_advisor	4.25	1.669	39
V30_prefer_to_learn	3.93	1.880	48
V31_tendency_to_invest	3.34	1.596	48
V32_estimate_to_regain	3.41	1.655	49
V33_estimate_time_to_regain	3.58	1.785	50
V34_estimate_value_over_annual_income	2.93	2.441	83
V35_internet_private_counseling	2.91	1.719	59
V36_internet_public_counseling	3.70	1.847	50
V37_online_counseling_vrs_actual	3.11	1.651	53
V38_specialized_servises	4.34	1.694	39

Table 16 Evaluation table of the correlations between the initial variables (p-value)

Correlation matrix[a]

Independent variables	V15	V16	V17	V18	V19	V20	V21	V22	V23	V24	V25	V26	V27	V28	V29	V30	V31	V32	V33	V34	V35	V36	V37	V38
V15_planing_Recording_Technologies																								
V16_auto_steering	.000																							
V17_fertiliser_VRA	.000	.000																						
V18_water_VRA	.000	.000	.000																					
V19_pest/herb_cides_VRA	.000	.000	.000	.000																				
V20_inputs_smart_view	.000	.000	.000	.000	.000																			
V21_harvest_crop_Recording_Technologies	.000	.000	.000	.000	.000	.000																		
V22_yield_smart_view	.000	.000	.000	.000	.000	.000	.000																	
V23_outputs_smart_view	.000	.000	.000	.000	.000	.000	.000	.000																
V24_work_calenter_smart_view	.000	.000	.000	.000	.000	.000	.000	.000	.000															
V25_balance_sheet_smart_view	.000	.000	.000	.000	.000	.000	.000	.000	.000	.000														
V26_predict_fut_crop_smart_view	.000	.000	.000	.000	.000	.000	.000	.000	.000	.000	.000													
V27_internet_smart_view	.000	.000	.000	.000	.000	.000	.000	.000	.000	.000	.000	.000												
V28_prefer_agriclt_advisor	.000	.000	.000	.000	.000	.000	.000	.000	.000	.000	.000	.000	.000											
V29_prefer_specialist_agriclt_advisor	.000	.000	.000	.000	.000	.000	.000	.000	.000	.000	.000	.000	.000	.000										
V30_prefer_to_learn	.000	.000	.000	.000	.000	.000	.000	.000	.000	.000	.000	.000	.000	.000	.000									
V31_tendency_to_invest	.000	.000	.000	.000	.000	.000	.000	.000	.000	.000	.000	.000	.000	.000	.000	.000								
V32_estimate_to_regain	.000	.000	.000	.000	.000	.000	.000	.000	.000	.000	.000	.000	.000	.000	.000	.000	.000							
V33_estimate_time_to_regain	.012	.010	.082	.034	.172	.024	.035	.006	.025	.045	.016	.058	.004	.333	.470	.473	.000	.000						
V34_estimate_value_over_annual_income	.482	.389	.324	.497	.379	.498	.499	.419	.363	.380	.464	.456	.269	.304	.420	.291	.034	.276	.000					
V35_internet_private_counseling	.000	.000	.000	.000	.000	.000	.000	.000	.000	.000	.000	.000	.000	.000	.000	.000	.000	.000	.009	.235				
V36_internet_public_counseling	.000	.000	.000	.000	.000	.000	.000	.000	.000	.000	.000	.000	.000	.000	.000	.000	.000	.000	.453	.305	.000			
V37_online_counseling_vrs_actual	.000	.000	.000	.000	.000	.000	.000	.000	.000	.000	.000	.000	.000	.001	.000	.000	.000	.000	.126	.005	.000	.000		
V38_specialized_servises	.000	.000	.000	.000	.000	.000	.000	.000	.000	.000	.000	.000	.000	.000	.000	.000	.000	.000	.099	.002	.000	.000	.000	

[a]Determinant = 1.768E-012

Table 17 Indices of participation of the 24 independent variables

Participation indices		
Independent variables	Initial	Extraction
V15_planing_Recording_Technologies	1.000	0.764
V16_auto_steering	1.000	0.819
V17_fertiliser_VRA	1.000	0.844
V18_water_VRA	1.000	0.789
V19_pest/herb_cides_VRA	1.000	0.811
V20_inputs_smart_view	1.000	0.824
V21_harvest_crop_Recording_Technologies	1.000	0.743
V22_yield_smart_view	1.000	0.834
V23_outputs_smart_view	1.000	0.790
V24_work_calenter_smart_view	1.000	0.811
V25_balance_sheet_smart_view	1.000	0.772
V26_predict_fut_crop_smart_view	1.000	0.820
V27_internet_smart_view	1.000	0.655
V28_prefer_agriclt_advisor	1.000	0.756
V29_prefer_specialist_agriclt_advisor	1.000	0.859
V30_prefer_to_learn	1.000	0.510
V31_tendency_to_invest	1.000	0.655
V32_estimate_to_regain	1.000	0.638
V33_estimate_time_to_regain	1.000	0.700
V34_estimate_value_over_annual_income	1.000	0.557
V35_internet_private_counseling	1.000	0.513
V36_internet_public_counseling	1.000	0.573
V37_online_counseling_vrs_actual	1.000	0.474
V38_specialized_servises	1.000	0.707

Extraction method: Principal Component Analysis

the variable because, according to international literature, this variable is characterized as "irregular" in relation to the others.

Regarding the necessary correlations between variables—a prerequisite for applying the Factorial Analysis, there is an issue with two variables: V33 (estimated time to regain) and V34 (estimated value over annual income). These exhibit an absolute correlation between them (p-value = 0.000) and generally weak correlations with all others (Table 16). This result was expected since they are expressed not only on a different scale but they also refer to a different approach, i.e. they do not concern perception or interest but purely economic evaluation. The chances of producing a strong indicator of the economic dimension of the phenomenon under consideration would increase if the questionnaire included more economic variables (questions) and not just two.

The weaknesses mentioned above are confirmed when carefully examining the structure of the Main Components Table (Table 18).

The solution produced is not simple, i.e. each independent variable is not associated with a single component. Also, the fourth index participates in the overall

Table 18 Main component table of the four indices produced by the initial 24 variables

Rotated component matrix[a]	Component			
	1	2	3	4
V17_fertiliser_VRA	0.873			
V19_pest/herb_cides_VRA	0.858			
V16_auto_steering	0.854			
V18_water_VRA	0.831			
V20_inputs_smart_view	0.815			
V15_planing_Recording_Technologies	0.792			
V24_work_calenter_smart_view	0.774			
V22_yield_smart_view	0.771	0.406		
V23_outputs_smart_view	0.766	0.401		
V26_predict_fut_crop_smart_view	0.749	0.451		
V21_harvest_crop_Recording_Technologies	0.740			
V25_balance_sheet_smart_view	0.730	0.441		
V31_tendency_to_invest		0.703		
V27_internet_smart_view	0.459	0.661		
V32_estimate_to_regain		0.658		
V35_internet_private_counseling		0.629		
V37_online_counseling_vrs_actual		0.592		
V30_prefer_to_learn	0.413	0.578		
V36_internet_public_counseling	0.516	0.527		
V29_prefer_specialist_agriclt_advisor			0.868	
V28_prefer_agriclt_advisor			0.780	
V38_specialized_servises			0.719	
V33_estimate_time_to_regain				0.815
V34_estimate_value_over_annual_income				0.731

[a]Extraction method: Principal Component Analysis. Rotation method: Varimax with Kaiser normalization

Table 19 Test of α-Cronbach from the reliability analysis of the seven significant loads of the second index (without the V25 and V26 variables)

Reliability statistics		
Cronbach's alpha	Cronbach's alpha based on standardized items	N of items
0.854	0.854	7

variation by 5% and is therefore incomplete for the interpretation of the phenomenon. Finally, variable V36 (internet public counseling) is associated with almost the same load as the first two components.

The sequential check process confirms that both V37 and V36 variables must be removed from the analysis. Especially for V37, the test of α-Cronbach indicates that without this variable the alpha index increases while at the same time there is a difference regarding the average evaluation of the seven variables (scale mean if Item deleted). Below (Tables 19 and 20), the Tables of Reliability Analysis for the

Table 20 Total descriptive statistical survey of the seven significant loads of the second indicator (without the V25 and V26 variables)

Item-total statistics					
	Scale mean if item deleted	Scale variance if item deleted	Corrected item-total correlation	Squared multiple correlation	Cronbach's alpha if item deleted
V27_internet_smart_view	20.40	56.578	0.701	0.515	0.820
V30_prefer_to_learn	20.25	57.898	0.623	0.424	0.832
V31_tendency_to_invest	20.83	59.805	0.685	0.641	0.824
V32_estimate_to_regain	20.76	59.652	0.659	0.633	0.827
V35_internet_private_counseling	21.27	60.185	0.604	0.414	0.835
V36_internet_public_counseling	20.47	57.614	0.650	0.459	0.828
V37_online_counseling_vrs_actual	21.06	65.790	0.401	0.175	0.862

Table 21 Aggregate data of Exploratory Factor Analysis under the conditions (a) to (c)

	KMO	H^2	Number of components (total variation)	Remarks
(a)	0.952	V35 < 0.500	4 (73.6%)	4th component: 5.9%
(b)	0.951	V35 < 0.500	4 (72.5%)	4th component: 6.1%
(c)	0.952	V35 < 0.500	4 (74.6%)	4th component: 6.1% (here the components configuration improves)

Table 22 Aggregate data of Exploratory Factor Analysis under the condition (d)

	KMO	H^2	Number of components (total variation)	Remarks
(d)	0.950	$V_{1,2,\ldots,6} > 0.500$	4 (76%)	4th component: 6.4%

second Index with seven significant loads without the V25 and V26 variables, which contribute substantially to the first Index, are presented.

Tables 21 and 22 show the figures of the Kaiser-Meyer-Olkin Index (KMO), the least satisfactorily observed in a variable degree of Participation (H^2), the Generated Indicators with the percentage of their total variance in the model and observations for the participation of the fourth Indicator in the overall variance of the model, when performing the Exploratory Factor Analysis, are successively subtracted as appropriate:

(a) V37 – and V36 remains
(b) V36 – and V37 remains
(c) V37 and V36 (Table 21)

And finally (d) V37, V36 and V35 (Table 22), since as observed in Table 21, by eliminating V37 and V36, V35 consistently presented a degree of participation below satisfactory ($H^2 < 0.500$).

Table 23 Factor Analysis load table for the four components of the model (d) without the variables V37, V36 and V35

Rotated component matrix[a]

	Component			
	1	2	3	4
V17_fertiliser_VRA	0.863			
V16_auto_steering	0.852			
V19_pest/herb_cides_VRA	0.848			
V20_inputs_smart_view	0.830			
V18_water_VRA	0.820			
V24_work_calenter_smart_view	0.792			
V15_planing_Recording_Technologies	0.788			
V23_outputs_smart_view	0.787			
V22_yield_smart_view	0.784			
V26_predict_fut_crop_smart_view	0.766	0.414		
V21_harvest_crop_Recording_Technologies	0.757			
V25_balance_sheet_smart_view	0.751			
V31_tendency_to_invest		0.802		
V32_estimate_to_regain		0.779		
V27_internet_smart_view	0.499	0.611		
V30_prefer_to_learn	0.426	0.580		
V29_prefer_specialist_agriclt_advisor			0.875	
V28_prefer_agriclt_advisor			0.789	
V38_specialized_servises			0.730	
V34_estimate_value_over_annual_income				0.800
V33_estimate_time_to_regain				0.798

Extraction method: Principal Component Analysis. Rotation method: Varimax with Kaiser normalization

[a]Rotation converged in seven iterations

The model finally adopted is based on 21 initial items (Variables V15–V34 plus V38) presenting significant variability (CV $< 38\%$) and significant interaction (correlation) between them (p-value $= 0.000$***). Through the implementation of the Factor analysis, it was possible to extract 4 indicators expressed as a linear function of the initial variables (Table 23). The model is characterized by an excellent consistency between the variables, with the Kaiser-Meyer-Olkin Index (KMO) having a value of 0.953 and their participation is very satisfactory with a minimum communality index for the variable V30 $= 0.548$.

The first principal component explains 63.6% of the total variance (Fig. 11) while it is related to 12 initial variables which present a particularly high score of the α-Cronbach statistics (Table 24), confirming the clear internal consistency between these 12 initial variables. The second component (7.4% of the total variance) is mainly related to four initial variables presenting also a very good level of internal consistency (α higher than 0.800) while the third component contributing around 6% to the total variance depends mainly on three initial variables (α higher than 0.800). The fourth principal component does not present an acceptable α-Cronbach, since, as

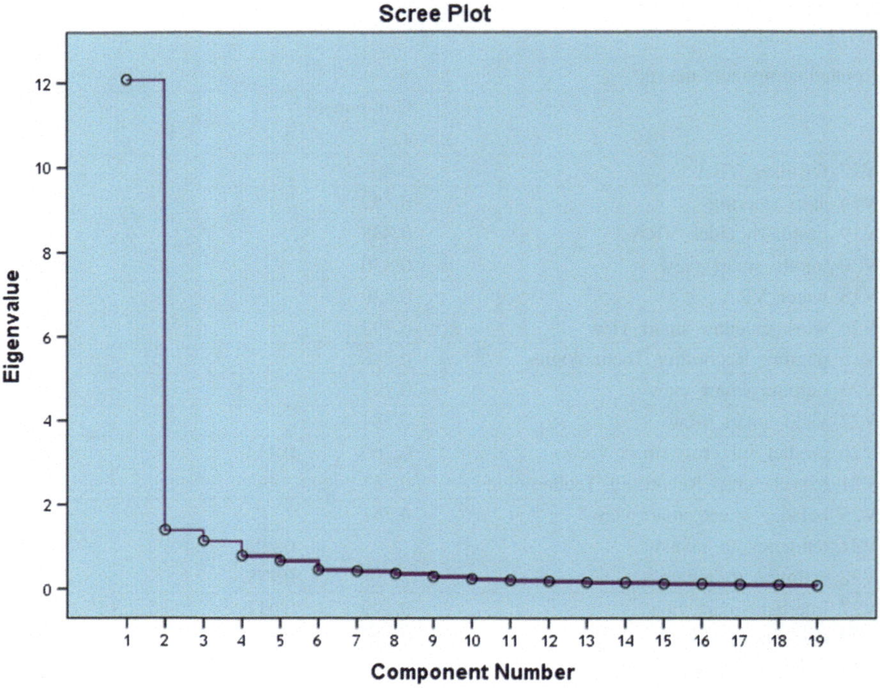

Fig. 11 Scree plot of the series (x-axis) and the eigenvalue value (y-axis) of the composite variables of the Factor Analysis

Table 24 Summary table for the reliability of the independent variables involved in each of the four indices deriving from the Factor Analysis of the independent variables V17–V34 and V38

Components	No of variables	α-Cronbach	Remarks
1st	12	0.976	Same average scale for the 12 variables. No need to remove variables.
2nd	4	0.842	Nearly the same average scale for the four variables. No need to remove variables.
3rd	3	0.848	Almost the same average scale for the three variables. Marginal problem with V38 ($\alpha = 0.861$).
4th	2	0.466	Not a reliable variable due to the different approach and scale ($\alpha < 0.700$)

already mentioned above; it has a different scale and concerns a different assessment-approach. Therefore, the Hyper-Variables/Indicators that are significant are in essence limited to three (Table 25).

Finally, the model based on the three first principal components, explains almost 77% of the phenomenon (Table 26), the loss of information is clearly limited (rate of up to 23%). The loads of the initial variables are over 0.4 and the majority of them

Table 25 Summary table of produced indicators/hyper-variables of Factor Analysis with the loads involved in each independent variable (Questions 15–32 and 38)

Rotated component matrix[a]	Component		
	1	2	3
V17_fertiliser_VRA	0.852		
V16_auto_steering	0.841		
v19_pestherb_cides_VRA	0.835		
V18_water_VRA	0.806		
v20_inputs_smart_view	0.801		
V15_planing_Recording_Technologies	0.763		
V24_work_calenter_smart_view	0.754	0.412	
V22_yield_smart_view	0.747	0.430	
V23_outputs_smart_view	0.747	0.420	
V21_harvest_crop_Recording_Technologies	0.717	0.416	
V26_predict_fut_crop_smart_view	0.716	0.484	
V25_balance_sheet_smart_view	0.708	0.460	
V31_tendency_to_invest		0.810	
V32_estimate_to_regain		0.787	
V27_internet_smart_view	0.423	0.683	
V30_prefer_to_learn		0.658	
V29_prefer_specialist_agriclt_advisor			0.880
V28_prefer_agriclt_advisor			0.797
V38_specialized_servises			0.751

Extraction method: Principal Component Analysis. Rotation method: Varimax with Kaiser normalization

[a]Rotation converged in six iterations

Table 26 Interpretation of the total variance through the analysis of main components

Total variance explained						
	Extraction sums of squared loadings			Rotation sums of squared loadings		
Component	Total	% of variance	Cumulative %	Total	% of variance	Cumulative %
1	12.082	63.588	63.588	7.859	41.363	41.363
2	1.396	7.346	70.934	3.866	20.345	61.707
3	1.140	5.998	76.932	2.893	15.224	76.932

are in the range 0.7–0.8 in the first and third indicator. The second indicator is equally strong—it explains cumulatively the phenomenon with the first one by 70%—but because it reflects dual information, entrepreneurship and innovation, terms compatible from a point of view, but also obscure by the general public, tends to concentrate loads on the spectrum 0.6–0.7.

Table 27 A summary presentation of the factors comprising Factor F1, regarding the required data as considered by the crop producers to be met by PA for the Cultivation Practice

Factor 1—Cultivation Practice Index	Factor loadings
V17_fertiliser_VRA	0.852
V16_auto_steering	0.841
V19_pestherb_cides_VRA	0.835
V18_water_VRA	0.806
V20_inputs_smart_view	0.801
V15_planing_Recording_Technologies	0.763
V24_work_calenter_smart_view	0.754
V22_yield_smart_view	0.747
V23_outputs_smart_view	0.747
V21_harvest_crop_Recording_Technologies	0.717
V26_predict_future_crop_smart_view	0.716
V25_balance_sheet_smart_view	0.708
V27_internet_smart_view	0.423

4.3 Interpretation of Hyper-values/Factors

From the responses/observations and the Exploratory Factor Analysis methods, four indicators initially derived, reflecting the whole spectrum of the independent variables tested. From these Hyper-Variables/Indicators, the fourth indicator, which constituted the attitude for investment in PA and SFTs, was excluded from the interpretation of the phenomenon, since, on one hand, due to the very different variables (in their nature) and the different measurement range, it did not show Consistency with the other independent variables, and, on the other hand, did not contribute more than 5% to the total variance (inertia).

The three resulting indicators eventually reflect 76.9% of the total variance, i.e. a loss of information of the order of 23% is recorded.

4.3.1 Factor 1: Cultivation Practice Index

The first factor [F (actor) 1], hereinafter referred to as the Cultivation Practice Index, and has been thus named since it aggregates all those independent variables that refer to the questions that focus on the main stages of the Cultivation Practice and Computer Information Management Systems and marginally the process of online counseling. Thus, it concerns the observations of questions 15–27. Table 27 and Fig. 12 show the Factor Loadings of factor F1.

Exercising the cultivation practice, with the contribution of the Computer Information Management Systems, as well as data on the application of techniques from web sites, is a fairly widespread process in the case of developing countries that are gradually orienting producers to familiarity with PA. It is also a key influx of entrepreneurship and innovation to increase the knowledge pool.

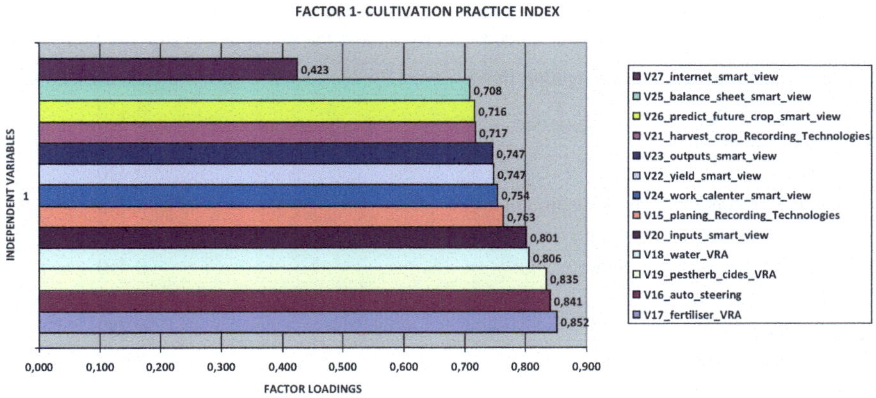

Fig. 12 Section Bar-chart for loads carried by the independent variables that make up the F1—Cultivation Practice Index

Crop producers evaluate elements relating to routine farming practices (fertilization, soil preparation, plant protection and irrigation) and the recording of inputs with a weight (charge) of more than 80% as much required (as it was expected). That is, they evaluate the technology of VRA, of Auto-Steering and input monitoring, more positively than the Recording Technologies and the registrations of other elements of their exploitation (>70%). The factor concerning assistance from a web site, related to the techniques and technologies of PA, is considered marginally (42.3%) important.

4.3.2 Factor 2: Entrepreneurship/Innovation Index

In order for new technologies and techniques to be adopted by the agricultural productive population, they should be evaluated through daily farming practice and then through their business performance in the context of an innovative investment.

The second index (F2) exported, concerned data regarding entrepreneurship and innovation. Trends for evolution and for the introduction of new innovative technical processes and organizing forms are weighted in this Index, along with the willingness for education and investing, and finally the provision/estimation for depreciation of the capital expended (Table 28).

The corresponding Factor 2 loadings as depicted in Fig. 13, which bear the evaluated gravity by the sample population, accrue high scores (>65%) in the predisposition for investment, evaluation for depreciation, assistance by a web site for implementation of the new techniques and the intention to train for their implementation, the main elements, namely, to increase profitability and to improve competitiveness. The existence of the V7 (internet smart view) variable at this

Table 28 Summary presentation of the factors comprising Index F2 regarding the data required, as evaluated by crop producers, to be complied with by the PA for Entrepreneurship and Innovation

Factor 2—Entrepreneurship/Innovation Index	Factor loadings
V24_work_calenter_smart_view	0.412
V22_yield_smart_view	0.430
V23_outputs_smart_view	0.420
V21_harvest_crop_Recording_Technologies	0.416
V26_predict_future_crop_smart_view	0.484
V25_balance_sheet_smart_view	0.460
V31_tendency_to_invest	0.810
V32_estimate_to_regain	0.787
V27_internet_smart_view	0.683
V30_prefer_to_learn	0.658

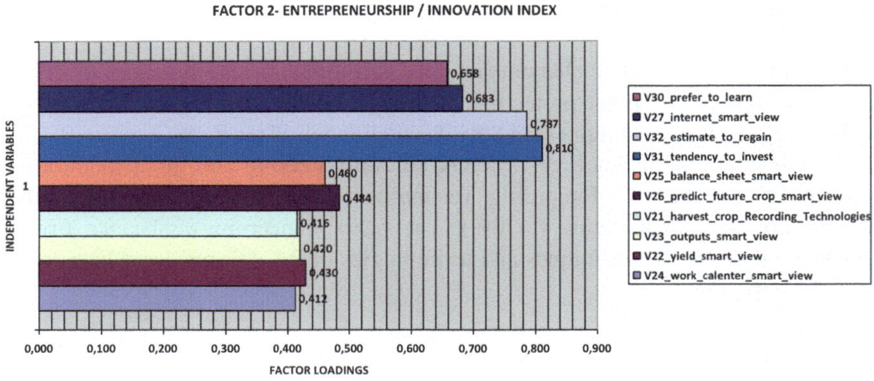

Fig. 13 Sections Bar-chart for loads carried by the independent variables that make up the F2—Entrepreneurship Index

indicator, and even braced by a score of 68.3%, reinforces the observation that PA and SFTs are treated as an input of innovation in agricultural holdings.

Regarding adaptability in the digital business environment, the remaining independent variables imprinting it burden the Index with loads from 48% to 41%, and this observation is indicative of the up until now incomplete or piecemeal information of producers.

4.3.3 Factor 3: Consulting Services Index

Finally, an important element for PA adoption is its support by specialized agronomists in it, so the final third index (F3) aggregates the independent variables relating to Counseling Services. It is logical, in the context of adopting innovative techniques such as PA and SFTs, to invest in the co-operation or employment of

Table 29 Summary presentation of the factors comprising the F3 Indicator, regarding the data that crop producers consider that the PA should meet for the Consulting Services

Factor 3—Consulting Services Index	Factor loadings
V29_prefer_specialist_agriclt_advisor	0.880
V28_prefer_agriclt_advisor	0.797
V38_specialized_servises	0.751

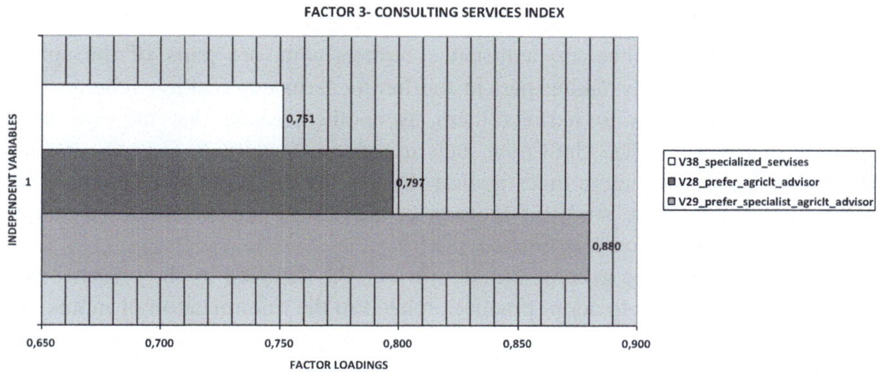

Fig. 14 Sections Bar-chart for loads carried by the independent variables that make up the F3—Consulting Services Index

skilled personnel in the overall Research and Development (R & D) framework. Here, the observations highlight the gravity of the counseling process by being valued as a separate index, with high scores (Table 29 and Fig. 14).

Producers, therefore, value the contribution of agronomists in the process of using PA techniques and SFTs to a very high score (>75%), with the highest of all, the presence of specialized advisors in agronomist-related issues (question 38).

5 Conclusions: Proposals

Although the percentage of producers who know the term PA appeared to be increased when compared to earlier studies (Mourtzinis et al. 2007), it did not exceed 16.3% in response rates. Far fewer, only 12.5%, were familiar with the practices of PA, mainly through information actions from companies and geotechnical practitioners that are involved in it. Even so, in a relevant question, people that are involved in crop production and in the research study by Michailidis et al. (2010) were found to be little interested in it, now foresee in a percentage of more than 60%, that the adoption of PA and SFTs will be a prerequisite for farmers in the future.

By examining the specific conditions that would help the crop producers to adopt the new cultivation techniques and technologies, the dimension of their inadequate information and their almost non-existent education emerged. That results as the vague documentation of all the manifestations of these practices in their daily occupation vocation with agriculture. This lack of education in a large extend is the reason that they cannot implement this technologies. Such a result is consistent with the result of the study carried out by Kountios (2014).

The profile of Greek producers refer to people aged up to 46 years old with a Middle (40.8%) or Higher (31.7%) education Level and a good familiarization (37.3%) with computing technology—necessary conditions for their training. They have the right profile to recognize the usefulness of the technology involved and the essence of the techniques towards sustainable management.

The existence of other 'on a distance farmers' that are parts of the spatial householdings but really taking part in the family farm exploitation schemes and influence the decisions to redirect them, is another reason that impinge upon adoption of PA and SFTs. Until now such an aspect—specific to Greece—has not been studied and needs more investigation. This factor could not be included in the four nation comparative study of Lawson et al. (2011) and constitutes a unique phenomenon observed only in Greece.

For these reasons, the general attitude towards PA depicted by the present study concerned the intention to adopt practices related to the minimization of inputs and monitoring the company's performance rather than using intelligence decision support systems. The adoption of a business scheme that involves managing their exploitation by intelligent decision making systems, efficiency measurement and innovative practices of distance counseling is not understood. The most of them don't comprehend that an adoption of that kind of schemes requires a new approach of entrepreneurship which is inevitable.

In general, however, the innovation of the whole system of SFTs is recognized by producers, mainly driven by the recognition of information technology and web applications as a new business scheme, albeit less in intensity than pure entrepreneurship and excluding the factor of the economic dimension. The willingness to invest, as well as the depreciation estimate, was slightly increased, despite the current economic situation of the holdings. This disconnection is in line with the findings of the most recent research of Kountios et al. (2017) about Young Farmers willing to build a competitive agricultural business.

Particularly strong was proven to be the relationship between geotechnical consultants and the producers' decision to adopt PA and SFTs. Indeed, their evaluation was that they would need more specialized agronomists.

Based on the data gathered on the crop producers' profile and given the very large number of observations, it is not wrong to assume that the composition of the rural population has changed in recent years and younger and more trained producers undertake to produce agricultural products in Greece.

Their intentions with regard to cultivation practice related to better management of inputs and were able to recognize entrepreneurship and counseling as a means for the future development of their holdings.

Through actions concerning their information and support of their businesses to adopt PA and SFTs, it is possible to restart agricultural primary production in the sense of redirecting it to more environmentally friendly and more economical practices.

The new technologies presented here concern the maintenance and modernization of agricultural and livestock farm machinery and at the same time the reorganization of inputs and the rational management of natural agricultural resources and hence a fundamental restructuring of everyday agricultural practice. The target population, through its responses, recognized the added value of new techniques and technologies, although many actions will still be needed in order for them to be adopted. But, more than anything, the target population comprehended the role they will play in the future.

The precision relates, in substance, of exercise of cultivation practices using the same method and the same result, under the condition of using greater reliability machines and minimal requirement calculations and corrections, thus requiring the continuous collection and disposal of data and metadata and their management by scientifically trained staff.

Therefore, with the purpose of adopting SFTs and PA for the modernization of agricultural holdings of crop production in the Region of Central Greece, the following actions are recommended:

- training of producers of large-scale plantations who are up to 50 years of age, or growers of over 150 acres in new techniques and technologies,
- updating and training of producers with high-yield crops such as trees in PA and SFTs,
- education of agronomists and their continuous training in the computerized applications of GIS and Computer Information Management Systems, as well as the techniques of Auto-steering, Recording Technologies, mapping of spatial variability and Variable Dose Applications,
- developing internet applications regarding combinational data and metadata to find data for the production process,
- eligibility of expenditure for the 'main occupation' farmers for computing systems, sensors and automation of procedures,
- an effort to detect the possibility of diversifying the products produced with PA and then marking them as products of sustainable agricultural practice,
- enhancing the entrepreneurship of family holdings adopting PA and SFTs,
- support for the development of group actions at local level to create large-scale data exchange networks and finally
- funding research for the specialization of techniques and SFTs and their evolution.

The main reason for suggesting the above policies is because PA and SFTs will play a key role in the economic development of agricultural holdings in Greece, in the future, in the context of their alignment with the general trend in the Member States of the European Union (European Parliament Department B Structural and Cohesion Policies 2014). Even more so, because they will be a prerequisite in the future, since they harmonize the elements of space, time, data and doses of inputs into the agricultural holding units, within an evolving integrated cultivation system, aiming at the sustainable conservation of natural agricultural resources and the ever-increasing demand for agricultural-food products (Image 3).

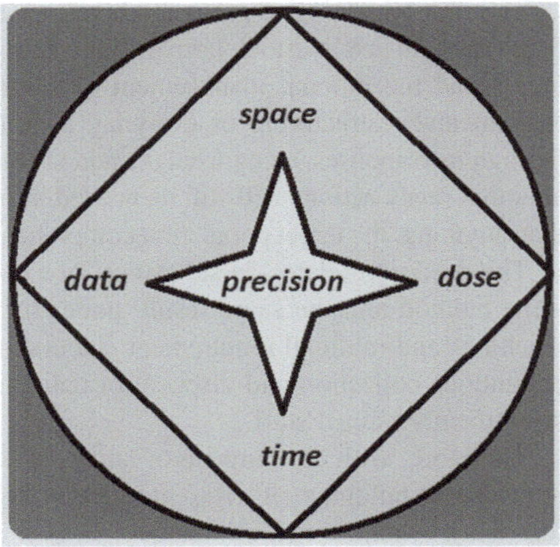

Image 3 A logo depicting the key elements that are consistent with the PA's farming practices for sustainability of agricultural resources. Source: Own processing

References

Anthopoulou, T., & Goussios, D. (2007). Rural geography. In T. Terkenlis, T. Iosifidis, I. Chorianopoulos, & I. Χωριανόπουλος (Eds.), *Anthropography* (pp. 234–274). Athens: Kritiki AE.

Auernhammer, H. (2001). Precision farming—the environmental challenge. *Computing Electronic Agriculture, 31*, 43–30.

Chalkos, G. (2013). *Economy and environment, methods of valuation and management*. Athens: Publications Liberal Books.

Chen, C., Pan, J., & Lam, S. K. (2014). A review of precision fertilization research. *Environmental Earth Sciences, 71*, 4073–4080. Berlin: Springer.

Cronbach, L. J., & Shavelson, R. J. (2004). My current thoughts on coefficient alpha and successor procedures. *Educational and Psychological Measurement, 64*(3), 391–418. https://doi.org/10.1177/0013164404266386.

Duquenne, M. N. (2016). Course notes spatial analysis of Master's studies program in "Spatial Analysis and Environmental management".

European Parliament Department B Structural and Cohesion Policies. (2014). *Study: "Precision Agriculture An Opportunity For EU Farmers- Potential Support with the CAP 2014–2020"*. Brussels: European Parliament. Retrieved from http://www.europarl.europa.eu/studies

EUROSTAT. (2014). *Rural Development Report 2014*. http://ec.europa.eu/atwork/synthesis/aar/doc/agri_aar_2014.pdf

FAO. (2010, November 3–5). Report of International Scientific Symposium on Biodiversity and Sustainable Diets, Rome.

Fountas, S., & Gemtos, T. (2015). *Precision agriculture*. Athens: Publications Hellenic Academic EBooks SEAB.

Gebbers, R., & Adamchuk, V. I. (2010). Precision agriculture and food security. *Science, 327*(5967), 828–831. https://doi.org/10.1126/science.1183899.Chemical Sciences. ISSN: 0975-8585.

Geiger, F., Bengtsson, J., Berendese, F., et al. (2010). Persistent negative effects of pesticides on biodiversity and biological control potential on European farmland. *Basic and Applied Ecology, 11*(2), 97–105.

Gerakis, P. A., Veresoglou, D. S., & Kalmpourtzi, K. L. (2008). *Sustainable development of agricultural resources*. Thessaloniki: Modern Education Publications.

Goussios, D., & Duquenne, M. N. (2003). L'exploitation agricole a` distance en Grece: mobilite, pluriactivite et ruralisation (Note). In Mediterranee, tome 100, 1-2-2003. Recherches recentes en geographie aixoise. pp. 45–48.

Headley, C. (2014). The role of precision agriculture for improved nutrient management on farms. *Journal of the Science of Food and Agriculture, 95*, 12–19.

Hellenic Statistical Authority (ELSTAT). (2014). *Census of Agriculture and Livestock 2009 – Structure survey of agricultural and livestock farms of Greece – Single format of metadata-SIMS2014.* Retrieved from http://www.statistics.gr/documents/20181/fea9f1cd-1a70-4c09-b135-95a42fe5841c

Hogan, T. P., & Cannon, B. (2007). *Psychological testing: A practical introduction* (2nd ed.). Hoboken, NJ: Wiley. ISBN 978-0-471-73807-7.

International Assessment of Agricultural Knowledge, Science and Technology for Development (IAASTD). (2009). Agriculture at a crossroads. In B. McIntyre, H. R. Herren, J. Wakhungu, & R. T. Watson (Eds.), *Synthesis report*. Washington, DC: Island Press.

Katter, T., Tiemann, S., Siebert, R., & Fountas, S. (2009). The role of communication and co-operation in the adoption of precision farming. *Precision Agriculture*. https://doi.org/10.1007/s11119-009-9150-0

Kountios, G. (2014). *Precision agriculture and information and communication technologies: Research of young farmers educational needs in Central Macedonia*. PhD Thesis, Aristotle University of Thessaloniki. Retrieved from www.didaktorika.gr/eadd/handle/10442/40246

Kountios, G., Ragkos, A., Bournaris, T., Papadavid, G., & Michailidis, A. (2017). Educational needs and perceptions of the sustainability of Precision Agriculture: Survey evidence from Greece. *Precision Agriculture*. https://doi.org/10.1007/s11119-017-9537-2

Kurth, T., Gocke, A., Wagner, K., & Corsini, L. (2015). *Crop Farming 2030. The reinvention of the sector*. USA: Boston Consulting Group.

Lawson, L. G., Pedersen, S. M., Sorensen, C. G., Pesonen, L., Fountas, S., Werner, A., Oudshoorn, F. W., Herold, L., Chatzinikos, T., Kirketerp, I. M., & Blackmore, S. (2011). A four nation survey of farm information management and advanced farming systems: A descriptive analysis of survey responses. *Computers and Electronics in Agriculture, 77*, 7–20.

Lei, P.-W., & Wu, Q. (2007). CTTITEM: SAS macro and SPSS syntax for classical item analysis. *Behaviour Research Methods, 39*(3), 527–530. https://doi.org/10.3758/BF03193021. PMID 17958163.

Michailidis, A., Samathrakis, B., Xatzitheodoridis, F., & Loizou, E. (2010). Adoption-diffusion of precision agriculture: Comparative analysis among the Greek regions. In M. Arambatzis, et al. (Eds.), *Innovative applications of information technology in the agricultural sector and the environment* (pp. 123–138). 3rd volume of scientific papers of the Hellenic Association for Information and Communication Technologies in Agriculture Food and Environment (HAICTA), Branch of Northern and Central Greece, Thessaloniki.

Mourtzinis, S., Fountas, S., & Gemtos, T. (2007). *Perspective of Greek farmers for precision agriculture* (Vol. 185, pp. 850–857). Proceedings of the 5th National Congress of Agricultural Engineering.

National Research Council. (1997). *Precision agriculture in the 21st century*. Washington, DC: National Academy Press.

Oliver, M. (2013). An overview of precision agriculture. In M. Oliver, T. Bishop, & B. Marchant (Eds.), *Precision Agriculture for sustainability and environmental protection* (pp. 3–12). Abingdon: Routledge.

Pannell, D. J., Marshall, G. R., Barr, N., Curtis, A., Vanclay, F., & Wilkinson, R. (2006). Understanding and promoting adoption of conservation practices by rural landholders. *Australian Journal of Experimental Agriculture, 46*, 1407–1424.

Pearson, K. (1900). X. On the criterion that a given system of deviations from the probable in the case of a correlated system of variables is such that it can be reasonably supposed to have arisen from random sampling. *Philosophcal Magazine Series 5, 50*(302), 157–175. https://doi.org/10.10801/14786440009463897. Retrieved from http://www.tandfonline.com.

Pison, G., Rousseeuw, J. P., Filzmozer, P., & Croux, C. (2003). Robust factor analysis. *Journal of Multivariate Analysis, 84*, 145–172.
Rogers, E. (2003). *Diffusion of innovations* (5th ed.). Simon and Schuster. ISBN 978-0-7432-5823-4.
Rogerson, P. (2001). *Statistical methods for geography*. London: Sage.
World Bank. (2007). *World Development Report 2008 "Agriculture for Development"*. Washington DC: World Bank/Oxford University Press.

Internet References

http://www.statistics.gr/el/statistics/-/publication/SAM04/
http://www.opekepe.gr
http://ec.europa.eu/agriculture/rural-development-2014-2020/country-files/el/factsheet-greece_en.pdf
http://ec.europa.eu/eurostat/ramon/nomenclatures
http://ec.europa.eu/atwork/synthesis/aar/index_en.htm
http://ec.europa.eu/agriculture/direct-support/greening/index_en.htm
http://ec.europa.eu/agriculture/cap-history/index_en.htm
http://www.europarl.europa.eu/studies
https://iris.gov.gr/webcenter/portal/irisgov/
http://sphweb.bumc.bu.edu/otlt/MPH-Modules/SB/BehavioralChangeTheories/BehavioralChangeTheories4.html
http://www.europarl.europa.eu/RegData/etudes/note/join/2014/529049/IPOL_AGRI_NT(2014)529049_EN.pdf

Multi-criteria Optimization Methods Applied in Agricultural Touring

Kyvele Constantina Diareme and Theodore Tsiligiridis

Abstract Agricultural tourism is considered a means of providing motor for growth in rural areas and year-round tourism flow, promoting local products and SMEs, encouraging the diversification of economic activity and in the long run a way of improving the quality of life in rural areas. For more than three decades now this concept goes hand to hand with avoiding/preventing the social and economic collapse of rural areas, and with multi-functionality. Increasing interest in tourism, tourism in rural areas and thematic tourism as well as tourists seeking fast and accurate information has led to the proposal of personalized (team) tours, rather than generic ones, with the use of recommender and geo-informatic systems alongside with smart applications, web services, context and location based services. To provide maximum functionality to users and engage them into using a service it is needed to represent/model the real daily multi-dimensional activities of a tourist during a trip with more than one objective function. Crucial to such a service are the formulations that model tourist trip problems, and the algorithms that generate and optimize the proposed tours. Therefore, this comprehensive review explores the multi-objective nature of agricultural touring, and focuses on multi-objective formulations that arose in the literature so far in touring, especially concerning tourism, and in regard of them being applied under agri-touristical scenarios. Under the scope of multi-objective optimization, we focus also in the related Orienteering and Team Orienteering Problem (OP/TOP). We consider selected non-dynamic/dynamic multi-objective route planning problems/methods, variants of the OP and TOP for route planning and scheduling problems, as well as for tourism and agricultural tourism, and the algorithms proposed and/or tested for the latter.

Keywords Agri-tourism · Agricultural touring · Orienteering Problem · Team Orienteering Problem · Multi-objective · Recommender systems · Tourist tour planner

---

K. C. Diareme (✉) · T. Tsiligiridis
Informatics Laboratory, Department of Agricultural Economics and Rural Development, School of Food, Biotechnology and Development, Agricultural University of Athens, Athens, Greece
e-mail: kkdiareme@aua.gr

J. Berbel et al. (eds.), *Multicriteria Analysis in Agriculture*, Multiple Criteria Decision Making, https://doi.org/10.1007/978-3-319-76929-5_11

List of Abbreviations

BOOP	Bi-objective orienteering problem
BOP	Bi-orienteering problem
EA	Evolutionary algorithm
EMO	Evolutionary multi-objective
EU	European Union
GRASP	Greedy randomized adaptive search procedure
MDLS	Multi directional local search
MOOP	Multi objective orienteering problem
MOTDOP	Multi-objective time-dependent orienteering problem
NSGA-II	Non-dominated sorting genetic algorithm II
OP	Orienteering problem
P-ACO	Pareto ant colony optimization
POI	Point of interest
PR	Path relinking
P-VNS	Pareto variable neighborhood search
SME	Small and medium-sized enterprises
TOP	Team orienteering problem
TTDP	Tourist trip design problem

1 Introduction

Agricultural tourism, often referred to as agri-tourism or tourism in rural areas, has been long considered a means of boosting rural economies, providing motor for growth and year-round tourism flow, encouraging the diversification of economic activity, promoting local products and SMEs, and ultimately improving the quality of life in rural areas and avoiding depopulation. This is not a new notion as for more than three decades now this concept is closely connected with avoiding/preventing the social and economic collapse of rural areas, pluriactivity and multi-functionality, especially in the European Union (EU). It should be noted that 44% of the EU—28 member states' territory is classified as rural, 88% accounts for rural and intermediate areas (European Commission 2016), and that since the 1990s the diversification of economic activities in rural areas and the development of agricutlural tourism were actively promoted through the Common Agricultural Policy (CAP) (Koutsouris et al. 2014, Campón-Cerro et al. 2017, Eusébio et al. 2017). As an extend agricultural tourism was promoted as beneficial for rural and less favored areas, for the aforementioned reasons; and in cases regarded as more intimate and less detrimental to culture and environment than mass tourism (Daugstad and Kirchengast 2013).

However there has been criticism to whether or not agricultural tourism has had any true positive effects to rural areas, although as Gao and Wu (2017) state "it should not be understood merely as a type of tourism, but also as a tool for the conservation and regeneration of rural society and rural culture". According to Campón-Cerro et al. (2017) rural tourism is often introduced into settings where residents are unfamiliar with providing services, much less regular tourism activities, which poses challenges to its professionalization. In other cases, it has been found that the improvement of, at least, the local farming families' incomes espoused by rural/agri-tourism development policies is not attained to a satisfactory degree, and in addition cash flows from agriculture have been higher than those from tourism (Koutsouris et al. 2014).

Nowadays, we see an even more increasing interest in tourism in rural areas and thematic tourism (with the use of technology) in general; main reasons are the trends of tourism diversification and of capitalizing on the signs of authenticity that tourists recognize in their travels, technological development (hardware, telecommunications and theoretical models), as well as changes in the needs of tourists/consumers. In addition recent scientific literature strengthens the interest towards agricultural tourism. Hüller et al. (2017) argue that establishing linkages of agricultural food production and rural tourism might contribute to the economic development of a district, the study of Eusébio et al. (2017) reveals the importance of offering different rural tourism products to the different population groups, and to Zasada et al. (2017) natural and landscape amenities present important regional environmental assets for rural development, competitiveness and social welfare. It should be noted that the literature indicates lack of destination management and destination marketing organizations specifically established for rural tourism destinations (Adeyinka-Ojo et al. 2014); personalized tour planning systems, and applications could help bridge this gap.

The above along with the fact that tourists seek fast information, use smartphones and tablets, and with the rapidly increasing volume of information, that often do not help them easily find what they are looking for, has diversified the need of services provided (Kramer et al. 2006; Souffriau and Vansteenwegen 2010). Tourists' needs have shifted from the typical situation where the visitor being in a destination would search for information in the Local Tourist Organization where they would be provided with a genetic tour, to the need of accurate, up to date personalized (team) tours. Personalized (team) tours are being proposed with the use of recommender systems, geo-informatic systems alongside with smart applications, web services as well as context and location based services, so as to provide the best possible information, with validity and reliability in the least time possible (Souffriau and Vansteenwegen 2010; Diareme and Tsiligiridis 2013). In this context there is a need to deal with real-life problems and not simplifications of them (to the extend that it is possible); to our view this will also provide maximum functionality to users and engage them into using a service. As such focus has been placed, also, in taking under consideration the real (or more realistic) Tourist Trip Design Problem (TTDP); meaning that effort is done in representing/modeling the real daily activities of a tourist during a trip, e.g. need for leisure time. Although in the simplest form touring problems to be solved consist of one objective function and the basic constrain, some form of cost referring to the cost of moving from Point of Interest (POI) to POI that most commonly denotes travel time, it is usually the case to see problems with more than one objective function. For instance, in order for a tourist to select a route, many criteria needs to be considered including cost, speed, availability, comfort level, reliability, etc. There is no optimal solution for this selection problem, e.g. one might want to choose a fast but inexpensive vehicle, while others might want a comfortable and reliable one. In fact, real-world problems are multi-objective (Tricoire 2012), and the same can be said for real-world touring problems. So, to provide solutions that could be applied under real life scenarios one has to deal with Multi Objective Optimization Problems (MOOPs). Those are problems that have multiple objectives or criteria, often conflicting with each other, that need to be optimized; in cases also mentioned as many-objective [for problems having more than three objectives (Chand and Wagner 2015)]. The conflict of these objectives or criteria arises as improvement in one can only be made to the detriment of one or more of the others (Odu and Charles-Owaba 2013). Therefore, using a multi-criteria decision analysis, a migrated system with an analyzing technique a decision can be made according to the Decision Maker's (DM's) preference so to choose the most desirable and satisfactory alternative under uncertain situations. To take things one step further, under cases, one can also deal with Dynamic Multi Objective Optimization Problems (DMOPs).

However, limited such examples of applications of methodologies or even problem specific methodologies for tourism in rural areas exist to our knowledge, and many of them refer only to bi-objective optimization. On the other hand, in the general tourism approach various problem definitions and optimization methodologies can be found, that could provide the basis of rural tourism specific problems taking also into account socioeconomic, sectorial, and territorial characteristics. Excessive problems have been studied so far depending on the specificities of different situations, e.g. Travelling Salesman Problem (TSP), Orienteering Problem (OP), and variations of them, the most interesting of which are the Travelling Salesperson Problem with Hotel Selection (TSPHS) (Vansteenwegen et al. 2012), the Time Dependent Team Orienteering Problem with Time Windows (TDTOPTW) (Garcia et al. 2013), the Time-Dependent Orienteering Problem with Stochastic Weigh and Time Windows (TD-OPSWTW) (Verbeeck et al. 2014). However, due to their computational complexity (NP-hard) the various algorithms proposed so far including computational intelligence provide good solutions if not optimal, and the problem remains open.

This research focuses on multi-objective formulations that arose in the literature so far in touring, especially concerning tourism, and in regard of them being applied under agricultural touring scenarios. Under the scope of Multi Objective Optimization (MOO) we focus also in the OP and TOP, which can provide a powerful basis for problem specific variants. It should be noted that extensions of the OP have been successfully applied to model the TTDP (Gavalas et al. 2014a). In addition the tourist planning problem with restrictions to travel distance and time basically coincides with the Team Orienteering Problem with Time Windows (TOPTW) constraints (Diareme and Tsiligiridis 2013) which is NP-hard problem, and research on time-dependent travel times, multi-constraints and multi-objectives of the OP variants is expected to help capture more realistic scenarios (Gunawan et al. 2016).

Specifically, we considered the following:

- Selected non-dynamic/dynamic multi-objective route planning problems/ methods.
- Cases of multi-objective variants of the OP and TOP for route planning and scheduling problems, as well as for tourism and agricultural tourism, and the algorithms proposed and/or tested for the latter.
- The relevant approximation methods used, neighborhood definition, and next POI insertion methodology during construction/initialization step for the algorithms proposed and/or tested for the known multi-objective variants of the OP and TOP for agricultural touring. This part is analysed in detail.

In this research we present an comprehensive review of the above, and explore the multi-objective nature of agricultural touring with an ultimate goal to locate multi-objective formulations to be applied or extended for use in touring in agricultural, rural areas. To make this research self-contained some fundamentals on the current state of route planning in the tourism sector, personalized touring, and multi-objective optimization are also presented, as well as the specificities of agricultural touring.

2 Personalized (Team) Touring

Routing refers on creating a path using a set of POIs, each one of which is assigned with a score (or cost). In its simplest form the problem is to maximize (minimize) the sum of scores (costs) collected, by visiting all POIs only once while taking under consideration other constraints of the problem. In this sense routing is the process of selecting best paths in a network. Personalized routing refers on planning an optimal route based on the recommendations and restrictions of an individual but also facilitating each user by personalizing their recommendations. In its simple form it is looking for a set of POIs to be visited once, so that the total score, benefit or profit is maximized subject to a constraint on the total travel cost or time. A score or profit is associated to each POI, and for each pair of POIs a travel cost is specified. Personalized routing can be outdoor or indoor with indoor typical use being finding optimal routes for impaired persons within confined spaces or finding the safest route or the route with the least obstacles; whereas personalized outdoor routing is about finding the shortest route within constraints, or the optimal one based on the user's preferences (in cases of conflicting objectives). The process is carried out through Decision Support Systems (DSSs) or web services and applications that can be handled by the user from a computer, smart phone or mobile guide. The input data are usually information about the POIs available (geographic location (latitude and longitude), scoring system) and the restrictions of the user, in our case a tourist (or group of tourists) who is in a given area. The output is the generated route. We consider that personalized outdoor routing can be divided in three phases: recommendation which refers to the information provided by the service to the user, route generation and route customization. In general, the main problems to be addressed involves selecting the techniques used to obtain data, evaluate data and scoring method, deciding how the user will communicate with the system, the methods used to generate and customize the optimal route, and mapping the results/proposed tour. We focus our work on route generation and route customization; by using the OP and TOP as a model.

Suppose there is a web service that generates personalized routes. The user connects with the service and the system lets the user know of the POIs he could visit, provided the system knows the location of the user. For each POI the geographical coordinates are known as well as the score profit. The scoring system reflects the "attractiveness" of a POI, and in a simple form it could be a measure of the tourists arriving in a city and the number of tourists visiting a certain place. Distances between POIs are known. In more complex models the data could contain even traffic or weather information, opening and closing hours of a venue, ticket price, minimum duration needed for the visit to be considered complete, etc.. The next step is to provide the user with a route to follow based on his constraints. In the simplest form the only constraint is the time available by the user (say, Tmax). For the route generation we distinguish three cases, routes where start and finish points are the same (tours), with different start and finish points (paths) and routes generated for more than one tourist or groups of tourists (team). In the generalized version of the problem more than one tourists, or groups of tourists, are in one place. Each one of them wants to follow a tour and then return to the starting point and meet with the others. Due to the limited time they have available they cannot visit all places, hence they must follow the route that provides them with the highest profit possible in the time available. This problem can be modeled as TOP, in the case of a single visitor the problem is similar with the single-competitor OP; both problems are presented in a following section. Scoring system is defined/reflects the satisfaction a user gains by visiting a control point.

As explained in their simplest form the problems to be solved consist of one objective function and the basic constrain; some form of cost referring to the cost of travelling from POI to POI that most commonly denotes travel time. In addition, specific problems are being formulated to provide solutions for problems that complement tourism in rural regions, such as cycle tourists without preferences or with preference, e.g. culture oriented or gastronomic fan. One example, although not a multi-objective one, is the multi-commodity orienteering problem with network design (Malucelli et al. 2015). Although traditionally focus has been placed, to our knowledge, to single objective optimisation and bi-objective optimisation, recently focus has been placed, also, in taking under consideration the real (or more realistic) Tourist Trip Design Problem (TTDP); meaning that effort is done in representing/modelling the real daily activities of a tourist during a trip.

2.1 Current State in Route Planning for Tourism

One of the early attempts to create a mobile tourist guide was Cyberguide (1995–1997), of the Georgia Institute of Technology, which provided information based on the position and orientation of the user and focused on how portable computers could assist in exploring physical spaces and cyberspaces. Since then, other systems have been developed such as GUIDE (Cheverst et al. 2000) and Gulliver's Genie (O'Hare and O'Grady 2003). All of the above take into account the position of the user, provide information about the POIs located around him and help him choose which points he wants to visit, hence for the generation of a personalized route the user must select by himself the points that interest him (Souffriau et al. 2008). During the last decade systems and applications have been developed that, instead of recommending pre-packaged tours or sorting POIs by estimated interest value as recommender systems do, they determine the combination of POIs that maximize the joint interest (Souffriau and Vansteenwegen 2010), e.g. the Dynamic Tour Guide (Hagen et al. 2005) that calculates personal tours on the fly, the system of Lee et al. (2009) that allows planning personalized travel routes to Tainan City, China. In the past few years even more Personalized Electronic Tourist Guides (PETGs) relying on mobile computing have been developed (De Falco et al. 2015) as well as tour planner systems, expert systems and travel assistants. Recent examples include the CityTrip planner (Vansteenwegen et al. 2011a) and the eCOMPASS multimodal tourist tour planner (Gavalas et al. 2015). For a more thorough review on recommender systems and applications to provide personalized tours, multi-criteria recommender systems as well as algorithmic approaches the interested reader is referred to Kabassi (2010), Anacleto et al. (2014), Borràs et al. (2014), Gavalas et al. (2014a, b), and Nilashi et al. (2017). It should be noted that some latest works take also under consideration data from social networks, e.g. TRIPBUILDER (Brilhante et al. 2015), PLANTOUR (Cenamor et al. 2017), and the "Filter-first, tour-second" framework for Personalized multi-period tour recommendations (Kotiloglu et al. 2017). In future we expect that the adoption of crowd-source route planning in tourism and the use of Volunteered Geographic Information (VGI) will create a new era for e-tourism and personalized tour recommendation.

2.2 The Orienteering and the Team Orienteering Problems

The well-known OP was first introduced by Tsiligirides (1984), and is based on the sport of orienteering (orienteering event and score orienteering event). Orienteering is a mixture of cross-country running and navigation through a forest using a map and a compass where players have to visit all control points, or where control points have scores and players have to visit only some of them so that they maximize the profit collected in a specified amount of time. It becomes apparent that orienteering is an optimization problem. The OP can be defined using a complete graph where each vertex is associated with a score that denotes the profit from visiting the corresponding POI or its attractiveness. The score of each vertex can be collected at most once, travel time between each pair of vertices is known, and each player has a predefined starting and ending POI. Those two POIs (start and end) can be the same physical location and thus the term tour is used. The goal of the OP is to determine a path or tour that maximizes the sum of scores collected from the visited vertices subject to a given path or tour length restriction; this means that the sequence of the vertices visited is under consideration also.

Besides the Single Competitor OP another case is the TOP where a team of M members has to determine at most M paths, with respect to a fixed time constraint, from the start POI to the end one through a subset of locations to maximize the total score collected. In this case the score of POI is attributed only to the first player that will visit the specific POI. As one can easily deduct the TTDP is closely related to the OP (Lin and Yu 2015) and TOP, where scoring system is defined by the satisfaction a user gains by visiting a control point (Diareme and Tsiligiridis 2013) or the attractiveness of a POI based on historical data, other tourists preferences and not only. It should be noted that OP and its variants, as well as TTDP problems, are of high complexity; even the most basic version of the OP is NP-hard (Golden et al. 1987). Although there have been a number of exact methods and heuristics proposed no efficient solution techniques for the problem at hand exist, especially when considering large scale problems with multiple objectives, multiple objectives and travel time dependency (Mei et al. 2016) thus a number of heuristics, meta/hyper-heuristics, and evolutionary algorithms have been proposed in the literature depending on the OP/TOP variant.

The two problems have been used to model a variety of problem situations and thus a wide range of variants have been proposed e.g. the Team Orienteering Problem with Time Windows, the arc OP, the TOP with Decreasing Profits (DP-TOP) (Murat Afsar and Labadie 2013). For a detailed survey, the interested reader is referred to Vansteenwegen et al. (2011b) and Gunawan et al. (2016). However, despite the general interest in the OP and TOP, the fact that both are characterized by an inherent conflict between the profit collected and the distance traveled (Matl et al. 2017), and the fact that multi-objectives of the OP variants are expected to help capture more realistic scenarios (Gunawan et al. 2016) multi-objective formulations of the OP and TOP have received attention only during the last 10 years.

2.3 Multi-criteria Analysis and Multi-objective Optimization

In single-objective optimization one single solution exists that is better than all others, and given two or more solutions those can be compared so that the best one is found. When considering problems with several objectives (multi-objective problems) it is difficult to compare solutions and decide the best. There are many ways to approach a multi-objective problem with the most widely-used to be the weighted sum method or utility functions, namely the one—which combines the individual objectives functions into a single composite one to use single-objective methods, keeping the objectives separate from each other in a model and approximating the Pareto set (Zitzler et al. 2004; Jozefowiez et al. 2008; Schilde et al. 2009). Scalar methods require setting the weights and having a priori knowledge of preferences, whereas the quality of the approximations can be limited (Jozefowiez et al. 2008). The drawback of this approach lies in the need for correct selection of the weights or utility functions to characterize the Decision Maker's preferences. In reality, the problem of precisely and accurately selecting these weights is unsolved. Small perturbations in the weights can lead to very different solutions and for this reason Decision Makers often prefer a set of promising solutions given the multiple objectives (Konak et al. 2006).

An alternative approach is to determine an entire Pareto optimal solution set or a representative subset, namely, a set of solutions that are non-dominated with respect to each other. In other words, since it can be difficult to compare solutions in multi-objective optimization problems, rather than having a single optimum, there can be a set of alternative trade-offs, the Pareto-optimal solutions. These solutions are optimal in the wider sense that no other solutions in the search space are superior to them when all objectives are considered (Chand and Wagner 2015; Cui et al. 2017). Note that such solutions cannot be improved in any objective without causing a degradation in at least one other; when moving between Pareto solutions there is always a certain amount of lose in one objective to achieve a certain amount of gain in the other. In practice, this approach is often preferred to the former since the final solution of the Decision Maker is always a trade-off between crucial parameters, however, the size of the Pareto set increases with the increase in the number of objectives.

To define a multi-objective decision problem we consider an n-dimensional decision variable vector $\boldsymbol{x} = \{x_1, x_2, \ldots, x_n\}$ in the solution space $\boldsymbol{X}$. Then the problem is to find a vector x* that minimizes a given set of k objective functions $f(\boldsymbol{x}^*) = \{f_1(\boldsymbol{x}^*),\ f_2(\boldsymbol{x}^*), \ldots, f_k(x^*)\}$ subject to a number of constraints of the form $g_j(\boldsymbol{x}^*) = b_j$; $j = 1, \ldots, m$ and bound on the decision variables. As we have noted, objectives under consideration conflict with each other, optimizing x with respect to a single objective often presents not acceptable results with respect to the other objectives. Since a perfect multi-objective solution that simultaneously optimizes each objective function cannot be achieved, an acceptable solution is to investigate a set of solutions, each of which satisfies the objectives at an appropriate level without being dominated by any other solution.

Assuming all objective functions are for minimization, a feasible solution $\boldsymbol{x}$ is said to dominate another feasible solution $\boldsymbol{y}$ ($\boldsymbol{x} \succ \boldsymbol{y}$), if and only if, $f_i(\boldsymbol{x}) \leq f_i(\boldsymbol{y})$ for $i = 1, \ldots, k$ and $f_j(\boldsymbol{x}) > f_j(\boldsymbol{y})$ for at least one objective function j. A solution is said to be *Pareto optimal* if it is not dominated by any other solution in the solution space. A Pareto optimal solution can't be improved with respect to any objective without deteriorating at least one other objective. The set of all feasible non-dominated solutions in X is referred to as the *Pareto optimal set*, whereas for a given Pareto optimal set, the corresponding objective function values in the objective space is called the *Pareto front*.

Under the above considerations, the aim of a multi-objective optimization algorithm is to identify solutions in the Pareto optimal set. For many multi-objective problems to identify the entire Pareto optimal set cannot be achieved due to its enormous size. Hence, a practical approach is to investigate a set of solutions that represent, as much as possible, the Pareto optimal one. This is usually referred as the *best-known Pareto set*. So one could say that the aim of multi-objective optimization is not only to find the optimal solution but the set of Pareto-optimal solutions. In the above sense, multi-objective optimization algorithms are aimed to obtain the Pareto set, covering the whole Pareto front, i.e. the representation of the optimal set of solutions (Mora et al. 2013). Due to complexity generating the whole Pareto set can be computationally expensive or infeasible, so good approximations are sought after rather than optimal solutions. Approximating the Pareto set is itself multi-objective.

The second approach is that of Multi-Objective Evolutionary Algorithms (MOEAs). Evolutionary algorithms operate in a set of candidate solutions on the basis of selection and variation, can handle large search spaces, can generate multiple alternative trade-offs in a single optimization run, and exploit similarities of solutions by recombination. It should be noted that although some of the approaches prevail, based on the literature, the superiority of one single approach is under consideration. The selection of the method depends, amongst others, on the available technology, on the type of information provided in the problem, designer's preferences and solution requirements.

However, in the absence of preferences or algorithms that assess the quality of the fronts, it is very difficult to decide which trade-off is better. In addition, one has again to consider also the role of the Decision Maker; consider at which point the decisions are made, if the articulation of preferences is a priori, a posteriori (as in the case of Genetic Algorithms), during the optimization process [as in the case of interactive Evolutionary Multi-Objective Optimization algorithms (interactive EMO)] or none at all, whether the decision space is continuous or discrete, decision criteria, and hierarchy. It should be mentioned that in the event of solutions that cannot be discriminated in terms of Pareto dominance, cone dominance-sort can be used where the user can drive the search process towards a preferred part of the search space (Purevsuren et al. 2015). Overview of the Multi-Criteria Decision Making (MCDM) and multi-objective optimization methods can be found in Odu and Charles-Owaba (2013), Bandyopadhyay and Saha (2013), and Giagkiozis and Fleming (2015).

The touring problem in its simplest form can be considered as a multi-objective problem (bi-objective problem) since a tourist needs to gain the maximum profit from visiting POIs while trying to minimize commute time or money spend; or trying to keep those under a specified level. In order to deal with real life situations, as said, the problems have several objectives; often or at least some of them conflicting or equally important. Let's also consider that still in a relatively simple case maximizing the profit and minimizing money spend (for commute time or ticket for a POI) can be conflicting (prestigious/popular venues might have higher priced ticket) and thus not optimized simultaneously. Similarly, in rural areas, there are typical situations were POIs are located far from another or are not easily accessible; the above can lead to high commute time from one POI to the next. Other conflicting objectives could be maximizing the number of POIs visited (for tourists who want to visit as many places as possible regardless to the satisfaction they gain from each one), and maximizing leisure time.

Generating a personalized tour for a rural region is in general a problem with small or medium sets i.e. number of POIs. However it should be noted that although there is a rise in the interest towards multi criteria analysis and multi-objective optimization not a wide range of models/algorithms that deal with high complexity problems (large scale problems) can be located; contrary to what the situation is for single-objective optimization.

3 Selected Cases of Multi-objective Problems

One basic aim of this work is to locate problem formulations and methodologies that could be used in agricultural tourism for the real (or more realistic) TTDP. Moreover, as already mentioned, in order to provide the reader with the ability to gain a thorough understanding of the explored topic we have also included selected cases of non-dynamic/dynamic multi-objective route planning problems/methods. In order that we facilitate the search, there was a need to answer the following twofold question:

1. "Which are the cases of multi-objective problems applied for route planning (including tour planning for tourism)?"
2. "Which are the cases of multi-objective optimization (including bi-objective and dynamic problems) applied in agricultural tourism, or tourism in general, for the proposal of tours, and which of them are variants of the OP or the TOP?"

Although this is not a systematic review, relevant publications were selected using appropriate search keywords from three libraries, comprehensive databases, namely Science Direct, Scopus and IEEE Xplore. Publications have been selected based on the language (only English publications were explored) and the availability of the full text. In practice, no publication year restriction was applied, although for the selection of cases with a more general application publications of the last 5 years were preferred. Moreover, publications were not excluded from the review based on their type so as to mitigate the possibility of bias (grey literature and unpublished studies were not included to avoid potential reduction of the validity of the research). The selected 20 publications are presented below. For a review of previous years on multi-objective vehicle routing problems the reader is referred to Jozefowiez et al. (2008), Cornu et al. (2017) for a non-exhaustive review of meta-heuristics

Table 1 Multi-objective problems for various routing, tour planning applications

No.	Authors	Problem proposed	Objectives
1	Domuța et al. (2012)	Intermodal and multi-objective time-dependent shortest path problem (IMTDSPP)	2
2	Changdar et al. (2014)	Multi-objective solid travelling salesman problem (TSP)	2
3	Molina et al. (2014)	A multi-objective eco-efficiency model for the Fixed Heterogeneous Fleet VRP with Time Windows (HVRPTW)	3
4	Kovacs et al. (2015)	The multi-objective generalized consistent vehicle routing problem (MOGenConVRP)	3
5	Cui et al. (2016)	Multi-objective 4PL routing problem on time varying networks	2
6	Lei et al. (2017)	Multi-objective Bus Route Plan Model	2
7	Braekers et al. (2016)	Bi-Objective Home Care Routing and Scheduling Problem (BI–HCRSP)	2
8	Toro et al. (2017)	Green Capacitated Location-Routing Problem (G-CLRP)	2
9	Guo et al. (2017)	Robust Dynamic Multi-objective Vehicle Routing Optimization Method	2
10	Nedjati et al. (2017)	Covering tour Location Routing Problem with Replenishment at intermediate depots (CLRPR)	2

approaches on Multi-Objective Travelling Salesman Problem (MOTSP), Cui et al. (2017) for a review of mutli-objective optimisation methods and applications in energy saving.

The problems presented in Table 1 are cases of non-dynamic/dynamic multi-objective route planning problems. The problems presented in Table 2 are multi-objective problems and algorithms for tour planning in tourism and agritourism, and multi-objective variants of the OP or TOP. The known variants of OP and TOP, as presented in Table 2, that have been applied/tested for tourism or agricultural tourism scenarios are further analyzed. It should be noted that the concept behind most of the problems and/or the formulations (of both tables) refers to many objectives e.g. three or five. However the application/test of the proposed methodologies (evaluation of algorithms proposed) is being done using the number of objectives denoted in the corresponding column (column "objectives").

3.1 Multi-objective Variants of the (Team) Orienteering Problem for Tourism and Agricultural Tourism

Below are presented the known multi-objective variants of the OP and TOP that have been applied or tested for tourism and/or agricultural tourism scenarios. For the presented problems, the known algorithms proposed and/or tested are also presented. For those algorithms approximation methods used as well as neighborhood

Table 2 Multi-objective problems and algorithms for (1) tour planning in tourism and agri-tourism (2) based on OP, TOP

No.	Authors	Title	Problem based on	Objectives	MOO approach
11	Hasuike et al. (2013)	Interactive multi-objective route planning for sight-seeing on Time-Expanded Networks under various conditions	Time-Expanded Network (TEN)	2	Single objective
12	Chen et al. (2015)	Multi-objective Orienteering Problem with Time Windows: An Ant Colony Optimization Algorithm	Variant of the OP with time windows (OPTW)	2	Decomposition into single objective, Weighted
13	Matl et al. (2017)	Bi-objective orienteering for personal activity scheduling	OP with time windows (OPTW)	2	Set of Pareto solutions
14	Schilde et al. (2009)	Metaheuristics for the bi-objective orienteering problem	Variant of the OP	2	Set of Pareto solutions
15	Tricoire (2012)	Multi-directional local search	Application on the BOP	2	Set of Pareto solutions
16	Purevsuren et al. (2015)	Evolutionary Multi-Objective Optimization Algorithms with Path Relinking for Bi-Orienteering Problem (BOP)	Application on the BOP	2	Set of Pareto solutions
17	Marti et al. (2015)	Multi-objective GRASP with Path Relinking	Application on the BOP	2	Weighted, depending on PR variant
18	Rezki and Aghezzaf (2017)	The bi-objective orienteering problem with budget constraint: GRASP_ILS	Variant of the BOP	2	Weighted
19	De Falco et al. (2015)	A Multi-objective Evolutionary Algorithm for Personalized Tours in Street Networks	Variant of the TOP with time windows (TOPTW)	5	Set of Pareto solutions
20	Mei et al. (2016)	Efficient meta-heuristics for the Multi-Objective Time-Dependent Orienteering Problem	Variant of the OP	2	Set of Pareto solutions

definition, and next POI insertion methodology during construction/initialization step are analyzed; where possible.

Schilde et al. (2009) introduced the Multi Objective Orienteering Problem (MOOP) for planning individual tourist routs in cities and rural areas. The MOOP is a multi-objective extension of the OP where different categories of POIs exists (e.g. culture, leisure etc.) and each POI provides different benefits for each category; focus is placed on tours, start and end node are the same physical positions, in order to find all Pareto efficient ones and let the user (tourist) select the one to follow.

To determine the candidate solutions the authors applied two algorithms, both combined with a Path Relinking (PR) procedure, for problems with two objectives values (bi-objective case); hence each POI has two scores associated with. The problem is therefore referred to in the literature as the Bi-Objective Orienteering Problem (BOOP), or Bi-Orienteering Problem (BOP). The algorithms are an adaptation of the Pareto Ant Colony Optimization (P-ACO) metaheuristic of Doerner et al. (2004), and a multi-objective extension of the Variable Neighborhood Search (Pareto VNS, P-VNS) of Mladenović and Hansen (1997). P-ACO uses a population of artificial ants, each one of them assigned an initial tour. During the construction phase POIs are inserted in the tours based on a priority measure (named heuristic information) that corresponds to the ratio between benefit and cost for each candidate POI, the pheromone information and local pheromone update, and a decision rule for selecting the vertex to be added in a tour. After the construction of the initial tours iterative improvement consisting of three operators is applied to each tour, efficiency of all solutions is checked (efficient ones are stored in an external memory), global pheromone update is performed, and a new colony of ants is used for each next iteration. In P-VNS storage for efficient solutions is used as well. The initial tour is created by means of a greedy algorithm, using a random weight vector (initially set to an extreme point) solutions are evaluated in the next phases of shaking and iterative improvement. The latter is similar to P-ACO. PR is applied in the end to improve results of both P-ACO and P-VNS.

Tricoire (2012) proposed a multi-directional local search (MDLS) that was also applied for the BOOP and according to the authors provided better results for, at the time, within comparable computational times. The stochastic metaheuristic for multi-objective optimization bears some similarities with the Pareto Local Search (PLS) of Paquete et al. (2004). According to the author the core idea of MDLS, i.e. selecting a solution, searching around it in each direction and then updating the archive, is new. It consists of iteratively improving a given non-dominated set of solutions by exploring neighborhoods using single-objective local search; based on the key idea of using different local searches, each of them working on a single objective. Although any single-objective local search method can be used for directional search in MDLS, large neighborhood search (LNS) is used, neighbourhood is selected randomly (each one is equiprobable). For the BOOP the list of non-dominated sets is ordered in decreasing values of the first objective, and the dominance rule is that if two tours visit the same set of control points, the one with the shorter duration dominates the other one.

Purevsuren et al. (2015) presented two hybridization schemes of interactive Evolutionary multi-objective (EMO) algorithms applied for the BOP in order for Decision Makers to interact and modify preference information during the optimisation process. Specifically, those are two hybridization schemes of the Cone dominance-based interactive EMO algorithm (CDEMO) (Purevsuren et al. 2015), and a modified version of the Non-dominated Sorting Genetic Algorithm II (NSGA-II) of Deb et al. (2002) with PR to promote convergence of the interactive algorithm. Two shorting criteria are used to order solutions of the population i.e. Pareto dominance and cone-dominance. In the first scheme (iEMO-PR) PR is launched in every iteration, applied to each pair in the elite set, and the population is updated with the resulting solutions of these PR. Whereas the second scheme (iEMO-EvoPR) evolves the elite set with PR and intensification procedure with PR is launched every given number of iterations. A local search technique, namely the Hill Climbing algorithm, is applied to the best solutions of the generated paths to find the local optima. According to the authors "The experimental results show that the proposed hybrid approaches can converge to the preferred solution faster and require less user feedback", with the second scheme giving the best results.

Marti et al. (2015) have identified and classified the ways in which the Greedy Randomized Adaptive Search Procedure (GRASP) (Feo and Resende 1995) with PR can be applied to MOPs, and proposed an adaptation of GRASP with PR that was also applied for the BOP. Each GRASP iteration consists in constructing a trial solution using some greedy randomized construction procedure, then local search is applied from the constructed solution in order to find the best local optimum over all local searches. The authors have tested more than one method for the applications of the proposed algorithm. For the BOP three improvement methods are proposed, namely pure, sequential, and combined, matched with a corresponding local search. Author's experimental results show that pure-ordered, random-sequential-combined, and random weighted methods are the ones with the better performance. Three GRASP methods are obtained for the BOP with the associated three PR implementation, Pure-PR (pure construction method, pure local search), Seq-PR (sequential combined construction method, sequential combined local search), and Weight-PR (using the combined weight methods).

Rezki and Aghezzaf (2017) introduced a new variant of the BOOP, namely the bi-objective orienteering problem with budget constraint (BOOPBC), where POIs have also an entrance fee (apart from the two scores) and additional constraint is imposed on the tourist's budget. To find an approximation of the set of Pareto optimal solutions, authors developed a hybrid metaheuristic based on GRASP which uses a weighted sum of the objective functions approach, greedy randomised construction phase, a Restricted Candidate List (RCL) and an external archive of sets of non-dominated solutions, and Iterated Local Search (ILS) that guides the improvement phase; replacing the local search of GRASP.

De Falco et al. (2015) presented an optimizer to plan multiple–day walking itineraries in old city centers, tailored to tourists' personal interests. This personalized multiple–day walking tour takes into account both hard constrains and soft temporary ones; namely a set of POIs with scores, waiting and visiting times, daily opening hours, a variety of tourist's trip constraints and environmental context in a 3-hour basis weather forecasting. In addition, distances between POIs are represented by a real–valued triangular matrix to give the actual structure of the streets in the area of the POIs. The problem is, quoting the authors, "An extension of the Team Orienteering Problem with Time Windows where the number of team members is replaced by the number of days available for the tourist to stay." This practically means that the output tour (in vector format) can be used in more than one days. Every day the examination of the solution restarts from the leftmost position in the vector as for the previous day and all the POIs already visited are skipped. Start and end POIs can be different locations and can change from day to day. In addition, the optimizer uses previous tours to avoid sending the tourist to the same place again unless he/she explicitly requires so. The five objectives (some contrasting) that the optimizer has to optimize, to our view, help tackle a problem that is very similar to a real-life situation. In order that near optimal personalized multiple-day tours, based on the concept of Pareto optimal set, to be found, a multi-objective Evolutionary Algorithm (EA) is employed. The components of the EA are standard (apart from mutation), this means that the generation of the initial population is random. The algorithm has been tested successfully for a set of 20 POIs of the old city center of Naples, Italy and for a 2-day tour.

Mei et al. (2016) proposed a Multi-Objective Time-Dependent Orienteering Problem (MOTDOP) to handle multiple preferences and time-dependent cost (travel time) simultaneously and proposed a set of non-dominated solutions (tours). Each POI has a score list to indicate the degrees of desirability for various preferences, e.g. categories that the POI belongs to. Two metaheuristics which solve the MOTDOP are employed, namely the Multi-Objective Memetic Algorithm (MOMA) and the Multi-Objective Ant Colony System (MACS) which is largely based on P-ACO (previously discussed) where the 2-opt step has been replaced with a local search procedure to take time dependent travel time into account. Concerning the next POI insertion during the construction (initialization) stage in the first metaheuristic apart from the starting and ending POI all intermediate POIs are inserted one by one after a random shuffle step takes place. The metaheuristics are evaluated using two generated sets of instances with two objectives (bi-objective case).

4 Specificities of Agricultural Touring

Agricultural tourism seems to be in need of a *being-in-the world* approach and entrepreneurial and technological training of rural population in order to achieve its full potential; if possible. One has to consider, amongst others, the characteristics of local entrepreneurs and the contribution of their businesses to the local economy, as well as the double role of the farmer and tourism services/facilities provided and their motivation in doing so, previously established linkages, local products production and cultural profile. From a technological point of view, and bearing in mind that the efficient proposal of tours that will meet the needs of tourists in rural areas effectively will add value to any agricultural tourism effort, we briefly explore some key element to be taken under consideration when modeling rural tourist activities.

The typical situation in a general tourism routing scenario is that all POIs are connected with a known road, and that we can compute travel time. Of course, the score profit assigned to each POI is known or established using some methodology. Amongst others, POIs can have time windows that could vary depending the day and time of year, fee, and perhaps waiting time and time needed for the visit to be considered successful. However, traveling in and urban area can be very different that traveling to a rural area. Even more taking a trip that includes visiting places in more than one rural areas of a wider place, i.e. island, or region can pose some challenges. Thus, agricultural touring has some additional specificities, presented below, that should be taken under consideration when working on extending multi-objective problems for agricultural touring and problem specific formulations.

- Availability of POIs differs depending the time period of the year.
- There are cases where all POIs cannot be represented in a complete graph.
- Accessibility and commute type from one POI to another can differ depending the time period of the year.
- Since under agricultural touring scenarios the touring areas can be remote, secluded there are other considerations such as to know how safe is the proposed tour; this extends to being able to take into consideration the safety of a commute path linking two POIs and the commute method.

In addition to the above, one could consider how to ensure the inclusion of certain local small businesses and SMEs in a tour so as to promote local development policies but without biasing the results of the proposed tour and of the general objective of personalized (team) touring.

5 Discussion

Strengthening rural areas can be a cross disciplinary effort in order to create a vital ecosystem where local communities can grow and prosper. As already explained, agricultural tourism can provide a variety of beneficial effects. The main goal of this work was to present multi-objective formulations and algorithms that could be applied or extended for touring in agricultural areas. Following the methodology presented we located 20 publications. In total seven of the publications located refer to formulations that have been applied to agricultural touring or tested in rural areas and have used OP/TOP or a variation of them. Those seven formulations amount in a total of ten proposed algorithms that were presented; performance comparison or comparison of computational time will be a subject for future work.

As a general remark we could say that no formulation was located that took into consideration all basic specificities of agricultural touring as presented in Sect. 4. In addition the concept behind most of the problems and/or the formulations refers to many objectives, e.g. three or five objectives. However out of the cases presented only in two algorithms were applied/tested using more than two objectives; a situation that is not limited in tourism and agritourism but appears in other multi-objective touring applications as well. It is worth mentioning that, the five objectives (some contrasting) of De Falco et al. (2015) that the optimizer has to optimize, to our view, help tackle a problem that is close to a real-life situation.

It is apparent that more methods that deal effectively with more than two objectives are needed as in future multi-objective rural specific formulations, along with the application of new Information and Communication Technologies (ICTs) and Geographic Information Science trends (e.g. Volunteered Geographic Information) could help propose tours of high added value to tourists and thus increase year-round tourism flow, and in the long run help local economies enjoy the extended benefits of agricultural tourism.

6 Conclusion

In this research, the multi-objective nature of tours for agricultural tourism has been explored. The known concept of proposing personalized/team tours has been presented as well as the reasons for why in agritourism research should focus, also, in multi-objective optimization. Focus was placed on the multi-objective nature of agricultural touring, formulations that arose in the literature so far in touring, especially concerning tourism, and in regard of them being applied under agritouristical scenarios. Under the scope of multi-objective optimization focus was placed also on the related Orienteering Problem (OP). Selected cases of dynamic multi-objective, multi-objective (and bi-objective) route planning problems/methods, cases of multi-objective (and bi-objective) variants of the OP and TOP mainly for route planning and scheduling problems, and the known multi-objective variants of the OP and TOP for tourism, agricultural tourism and the algorithms proposed and/or tested were presented. The formulations and algorithms located, based on the methodology described, can be applied or even extended. The nature of agricultural touring and its complexity was explored, and insight was given to elements that should be taken under consideration in an interdisciplinary effort to model the real agricultural touring problem in future.

Acknowledgments We would like to thank the two anonymous reviewers for their useful comments and feedforward.

References

Adeyinka-Ojo, S. F., Khoo-Lattimore, C., & Nair, V. (2014). A framework for rural tourism destination management and marketing organisations. *Procedia – Social and Behavioral Sciences, 144*, 151–163. https://doi.org/10.1016/j.sbspro.2014.07.284

Anacleto, R., Figueiredo, L., Almeida, A., & Novais, P. (2014). Mobile application to provide personalized sightseeing tours. *Journal of Network and Computer Applications, 41*, 56–64. https://doi.org/10.1016/j.jnca.2013.10.005

Bandyopadhyay, S., & Saha, S. (2013). *Some single-and multiobjective optimization techniques*. Berlin: Springer.

Borràs, J., Moreno, A., & Valls, A. (2014). Intelligent tourism recommender systems: A survey. *Expert Systems with Applications, 41*, 7370–7389. https://doi.org/10.1016/j.eswa.2014.06.007

Braekers, K., Hartl, R. F., Parragh, S. N., & Tricoire, F. (2016). A bi-objective home care scheduling problem: Analyzing the trade-off between costs and client inconvenience.

European Journal of Operational Research, 248, 428–443. https://doi.org/10.1016/j.ejor.2015.07.028

Brilhante, I. R., Macedo, J. A., Nardini, F. M., et al. (2015). On planning sightseeing tours with TripBuilder. *Information Processing and Management, 51*, 1–15. https://doi.org/10.1016/j.ipm.2014.10.003

Campón-Cerro, A. M., Hernández-Mogollón, J. M., & Alves, H. (2017). Sustainable improvement of competitiveness in rural tourism destinations: The quest for tourist loyalty in Spain. *Journal of Destination Marketing and Management, 6*, 252–266. https://doi.org/10.1016/j.jdmm.2016.04.005

Cenamor, I., De Rosa, T., Núñez, S., & Borrajo, D. (2017). Planning for tourism routes using social networks. *Expert Systems with Applications, 69*, 1–9. https://doi.org/10.1016/j.eswa.2016.10.030

Chand, S., & Wagner, M. (2015). Evolutionary many-objective optimization: A quick-start guide. *Surveys in Operations Research and Management Science, 20*, 35–42. https://doi.org/10.1016/j.sorms.2015.08.001

Changdar, C., Mahapatra, G. S., & Kumar Pal, R. (2014). An efficient genetic algorithm for multi-objective solid travelling salesman problem under fuzziness. *Swarm and Evolutionary Computation, 15*, 27–37. https://doi.org/10.1016/j.swevo.2013.11.001

Chen, Y., Sun, W.-J., & Chiang, T.-C. (2015). Multiobjective orienteering problem with time windows: An ant colony optimization algorithm. In IEEE (Ed.), *Technologies and Applications of Artificial Intelligence (TAAI)* (pp. 128–135). Tainan: IEEE.

Cheverst, K., Davies, N., Mitchell, K., & Friday, A. (2000). *Experiences of developing and deploying a context-aware tourist guide: The GUIDE project* (pp. 20–31). Proceedings of the 6th Annual International Conference on Mobile Computing and Networking. https://doi.org/10.1145/345910.345916

Cornu, M., Cazenave, T., & Vanderpooten, D. (2017). Perturbed decomposition algorithm applied to the multi-objective traveling salesman problem. *Computers and Operations Research, 79*, 314–330. https://doi.org/10.1016/j.cor.2016.04.025

Cui, Y., Huang, M., & Liu, J. (2016). *Multi-objective 4PL routing problem on time varying networks* (pp. 271–275). 2016 12th International Conference on Natural Computation, Fuzzy Systems and Knowledge Discovery (ICNC-FSKD). https://doi.org/10.1109/FSKD.2016.7603185

Cui, Y., Geng, Z., Zhu, Q., & Han, Y. (2017). Review: Multi-objective optimization methods and application in energy saving. *Energy, 125*, 681–704. https://doi.org/10.1016/j.energy.2017.02.174

Daugstad, K., & Kirchengast, C. (2013). Authenticity and the pseudo-backstage of agri-tourism. *Annals of Tourism Research, 43*, 170–191. https://doi.org/10.1016/j.annals.2013.04.004

De Falco, I., Scafuri, U., & Tarantino, E. (2015). *A multiobjective evolutionary algorithm for personalized tours in street networks. EvoApplications. Lecture Notes in Computer Science* (Vol. 9028, pp. 115–127). Springer. https://doi.org/10.1007/978-3-319-16549-3

Deb, K., Pratap, A., Agarwal, S., et al. (2002). A fast and elitist multiobjective genetic algorithm: NSGA-II. *IEEE Transactions on Evolutionary Computation, 6*, 182–197. https://doi.org/10.1109/4235.996017

Diareme, K. C., & Tsiligiridis, T. (2013). *Personalized outdoor routing: Route planning techniques and algorithms* (pp. 246–251). 2nd International Symposium and 24th National Conference on Operational Research, Athens.

Doerner, K., Gutjahr, W. J., Hartl, R. F., et al. (2004). Pareto ant colony optimization: A metaheuristic approach to multiobjective portfolio selection. *Annals of Operations Research, 131*, 79–99. https://doi.org/10.1023/B:ANOR.0000039513.99038.c6

Domuţa, C., Tipi, N. S., Radu, D., et al. (2012). *Finding intermodal and multi-objective time-dependent best routes with exact algorithm* (pp. 531–535). 2012 I.E. International Conference on Automation, Quality and Testing, Robotics. https://doi.org/10.1109/AQTR.2012.6237768

European Commission. (2016). CAP CONTEXT INDICATORS 2014–2020, 2016 update. Brussels.

Eusébio, C., Carneiro, M. J., Kastenholz, E., et al. (2017). Who is consuming the countryside? An activity-based segmentation analysis of the domestic rural tourism market in Portugal. *Journal of Hospitality and Tourism Management, 31*, 197–210. https://doi.org/10.1016/j.jhtm.2016.12.006

Feo, T. A., & Resende, M. G. C. (1995). Greedy randomized adaptive search procedures. *Journal of Global Optimization, 6*, 109–133. https://doi.org/10.1007/BF01096763

Gao, J., & Wu, B. (2017). Revitalizing traditional villages through rural tourism: A case study of Yuanjia Village, Shaanxi Province, China. *Tourism Management, 63*, 223–233. https://doi.org/10.1016/j.tourman.2017.04.003

Garcia, A., Vansteenwegen, P., Arbelaitz, O., et al. (2013). Integrating public transportation in personalised electronic tourist guides. *Computers and Operations Research, 40*, 758–774. https://doi.org/10.1016/j.cor.2011.03.020

Gavalas, D., Konstantopoulos, C., Mastakas, K., & Pantziou, G. (2014a). Mobile recommender systems in tourism. *Journal of Network and Computer Applications, 39*, 319–333. https://doi.org/10.1016/j.jnca.2013.04.006

Gavalas, D., Konstantopoulos, C., Mastakas, K., & Pantziou, G. (2014b). A survey on algorithmic approaches for solving tourist trip design problems. *Journal of Heuristics, 20*, 291–328. https://doi.org/10.1007/s10732-014-9242-5

Gavalas, D., Kasapakis, V., Konstantopoulos, C., et al. (2015). The eCOMPASS multimodal tourist tour planner. *Expert Systems with Applications, 42*, 7303–7316. https://doi.org/10.1016/j.eswa.2015.05.046

Giagkiozis, I., & Fleming, P. J. (2015). Methods for multi-objective optimization: An analysis. *Information Sciences(NY), 293*, 338–350. https://doi.org/10.1016/j.ins.2014.08.071

Golden, B. L., Levy, L., & Vohra, R. (1987). The orienteering problem. *Naval Research Logistics, 34*, 307–318. https://doi.org/10.1002/1520-6750(198706)34:3<307::AID-NAV3220340302>3.0.CO;2-D

Gunawan, A., Lau, H. C., & Vansteenwegen, P. (2016). Orienteering Problem: A survey of recent variants, solution approaches and applications. *European Journal of Operational Research, 255*, 315–332. https://doi.org/10.1016/j.ejor.2016.04.059

Guo, Y., Cheng, J., Luo, S., & Gong, D. (2017). Robust dynamic multi-objective vehicle routing optimization method. *IEEE/ACM Transactions on Computational Biology and Bioinformatics*. https://doi.org/10.1109/TCBB.2017.2685320

Hasuike, T., Katagiri, H., Tsubaki, H., & Tsuda, H. (2013). Interactive multi-objective route planning for sightseeing on time-expanded networks under various conditions. *Procedia Computer Science, 22*, 221–230. https://doi.org/10.1016/j.procs.2013.09.098

Hüller, S., Heiny, J., & Leonhäuser, I.-U. (2017). Linking agricultural food production and rural tourism in the Kazbegi district – A qualitative study. *Annals of Agrarian Science, 15*, 40–48. https://doi.org/10.1016/j.aasci.2017.02.004

Jozefowiez, N., Semet, F., & Talbi, E. (2008). Multi-objective vehicle routing problems. *European Journal of Operational Research, 189*, 293–309. https://doi.org/10.1016/j.ejor.2007.05.055

Kabassi, K. (2010). Personalizing recommendations for tourists. *Telematics and Informatics, 27*, 51–66. https://doi.org/10.1016/j.tele.2009.05.003

Konak, A., Coit, D. W., & Smith, A. E. (2006). Multi-objective optimization using genetic algorithms: A tutorial. *Reliability Engineering and System Safety, 91*, 992–1007. https://doi.org/10.1016/j.ress.2005.11.018

Kotiloglu, S., Lappas, T., Pelechrinis, K., & Repoussis, P. P. (2017). Personalized multi-period tour recommendations. *Tourism Management, 62*, 76–88. https://doi.org/10.1016/j.tourman.2017.03.005

Koutsouris, A., Gidarakou, I., Grava, F., & Michailidis, A. (2014). The phantom of (agri)tourism and agriculture symbiosis? A Greek case study. *Tourism Management Perspectives, 12*, 94–103. https://doi.org/10.1016/j.tmp.2014.09.001

Kovacs, A. A., Parragh, S. N., & Hartl, R. F. (2015). The multi-objective generalized consistent vehicle routing problem. *European Journal of Operational Research, 247*, 441–458. https://doi.org/10.1016/j.ejor.2015.06.030

Kramer, R., Modsching, M., Hagen, K., et al. (2006). A city guide agent creating and adapting individual sightseeing tours. *International Journal of Computational Intelligence Research, 2*, 191–206.

Lee, C. S., Chang, Y. C., & Wang, M. H. (2009). Ontological recommendation multi-agent for Tainan City travel. *Expert Systems with Applications, 36*, 6740–6753. https://doi.org/10.1016/j.eswa.2008.08.016

Lei, S., Li, Z., Wu, B., & Wang, H. (2017). *Research on multi-objective bus route planning model based on taxi GPS data* (pp. 249–255). 2012 International Conference on Cyber-Enabled Distributed Computing and Knowledge Discovery (CyberC). https://doi.org/10.1109/CyberC.2016.56

Lin, S. W., & Yu, V. F. (2015). A simulated annealing heuristic for the team orienteering problem with time windows. *Applied Soft Computing, 37*, 632–642. https://doi.org/10.1016/j.ejor.2011.08.024

Malucelli, F., Giovannini, A., & Nonato, M. (2015). Designing single origin-destination itineraries for several classes of cycle-tourists. *Transportation Research Procedia, 10*, 413–422. https://doi.org/10.1016/j.trpro.2015.09.091

Marti, R., Campos, V., Resende, M. G. C., & Duarte, A. (2015). Multiobjective GRASP with path relinking. *European Journal of Operational Research, 240*, 54–71. https://doi.org/10.1016/j.ejor.2014.06.042

Matl, P., Nolz, P., Ritzinger, U., et al. (2017). Bi-objective orienteering for personal activity scheduling. *Computers and Operations Research, 82*, 69–82. https://doi.org/10.1016/j.cor.2017.01.009

Mei, Y., Salim, F. D., & Li, X. (2016). Efficient meta-heuristics for the multi-objective time-dependent orienteering problem. *European Journal of Operational Research, 254*, 443–457. https://doi.org/10.1016/j.ejor.2016.03.053

Mladenović, N., & Hansen, P. (1997). Variable neighborhood search. *Computers and Operations Research, 24*, 1097–1100. https://doi.org/10.1016/S0305-0548(97)00031-2

Molina, J. C., Eguia, I., Racero, J., et al. (2014). Multi-objective vehicle routing problem with cost and emission functions. *Procedia – Social and Behavioral Sciences, 160*, 254–263. https://doi.org/10.1016/j.sbspro.2014.12.137

Mora, A. M., García-Sánchez, P., Merelo, J. J., & Castillo, P. A. (2013). Pareto-based multi-colony multi-objective ant colony optimization algorithms: An island model proposal. *Soft Computing, 17*, 1175–1207. https://doi.org/10.1007/s00500-013-0993-y

Murat Afsar, H., & Labadie, N. (2013). Team orienteering problem with decreasing profits. *Electronic Notes in Discrete Mathematics, 41*, 285–293. https://doi.org/10.1016/j.endm.2013.05.104

Nedjati, A., Izbirak, G., & Arkat, J. (2017). Bi-objective covering tour location routing problem with replenishment at intermediate depots: Formulation and meta-heuristics. *Computers and Industrial Engineering, 110*, 191–206. https://doi.org/10.1016/j.cie.2017.06.004

Nilashi, M., Bagherifard, K., Rahmani, M., & Rafe, V. (2017). A recommender system for tourism industry using cluster ensemble and prediction machine learning techniques. *Computers and Industrial Engineering, 109*, 357–368. https://doi.org/10.1016/j.cie.2017.05.016

O'Hare, G. M. P., & O'Grady, M. J. (2003). Gulliver's Genie: A multi-agent system for ubiquitous and intelligent content delivery. *Computer Communications, 26*, 1177–1187. https://doi.org/10.1016/S0140-3664(02)00252-9

Odu, G. O., & Charles-Owaba, O. E. (2013). Review of multi-criteria optimization methods – theory and applications. *IOSR Journal of Engineering, 3*, 01–14. https://doi.org/10.9790/3021-031020114

Paquete, L., Chiarandini, M., & Stützle, T. (2004). Pareto local optimum sets in the biobjective traveling salesman problem: An experimental study. *Metaheuristics for Multiobjective Optimization, 535*, 177–199. https://doi.org/10.1007/978-3-642-17144-4_7

Purevsuren, D., Cui, G., & Ur Rehman, S. (2015, November). *Evolutionary multiobjective optimization algorithms with path relinking for Bi-Orienteering problem* (pp. 132–135). Proceedings of the IEEE International Conference on Software Engineering and Service Science (ICSESS). https://doi.org/10.1109/ICSESS.2015.7339021

Rezki, H., & Aghezzaf, B. (2017). *The bi-objective orienteering problem with budget constraint: GRASP_ILS* (pp. 25–30). IEEE 2017 International Colloquium on Logistics and Supply Chain Management (LOGISTIQUA).

Schilde, M., Doerner, K. F., Hartl, R. F., & Kiechle, G. (2009). Metaheuristics for the bi-objective orienteering problem. *Swarm Intelligence, 3*, 179–201. https://doi.org/10.1007/s11721-009-0029-5

Souffriau, W., & Vansteenwegen, P. (2010). *Tourist trip planning functionalities: State-of-the-art and future. Lecture Notes in Computer Science (including Subser Lect Notes Artif Intell Lect Notes Bioinformatics)* (Vol. 6385, pp. 474–485). https://doi.org/10.1007/978-3-642-16985-4_46

Souffriau, W., Vansteenwegen, P., Vanden Berghe, G., & Van Oudheusden, D. (2008). A greedy randomised adaptive search procedure for the team orienteering problem. In *EU/MEeting* (pp. 1–9). France: Troyes.

Ten Hagen, K., Kramer, R., Hermkes, M., et al. (2005). Semantic matching and heuristic search for a dynamic tour guide. *Information and Communication Technology in Tourism*, 149–159.

Toro, E. M., Franco, J. F., Echeverri, M. G., & Guimarães, F. G. (2017). A multi-objective model for the green capacitated location-routing problem considering environmental impact. *Computers and Industrial Engineering, 110*, 114–125. https://doi.org/10.1016/j.cie.2017.05.013

Tricoire, F. (2012). Multi-directional local search. *Computers and Operations Research, 39*, 3089–3101. https://doi.org/10.1016/j.cor.2012.03.010

Tsiligirides, T. (1984). Heuristic methods applied to orienteering. *The Journal of the Operational Research Society, 35*, 797–809. https://doi.org/10.1057/jors.1984.162

Vansteenwegen, P., Souffriau, W., & Van Oudheusden, D. (2011a). The orienteering problem: A survey. *European Journal of Operational Research, 209*, 1–10. https://doi.org/10.1016/j.ejor.2010.03.045

Vansteenwegen, P., Souffriau, W., Vanden Berghe, G., & Van Oudheusden, D. (2011b). The city trip planner: An expert system for tourists. *Expert Systems with Applications, 38*, 6540–6546. https://doi.org/10.1016/j.eswa.2010.11.085

Vansteenwegen, P., Souffriau, W., & So, K. (2012). The travelling salesperson problem with hotel selection. 207–217. https://doi.org/10.1057/jors.2011.18

Verbeeck, C., Sörensen, K., Aghezzaf, E.-H., & Vansteenwegen, P. (2014). A fast solution method for the time-dependent orienteering problem. *European Journal of Operational Research, 236*, 419–432. https://doi.org/10.1016/j.ejor.2013.11.038

Zasada, I., Häfner, K., Schaller, L., et al. (2017). A conceptual model to integrate the regional context in landscape policy, management and contribution to rural development: Literature review and European case study evidence. *Geoforum, 82*, 1–12. https://doi.org/10.1016/j.geoforum.2017.03.012

Zitzler, E., Laumanns, M., & Bleuler, S. (2004). A tutorial on evolutionary multiobjective optimization. *Metaheuristics for Multiobjective Optimization, 535*, 3–37. https://doi.org/10.1007/978-3-642-17144-4_1

Life Cycle Assessment and Multi-criteria Analysis in Agriculture: Synergies and Insights

Emmanouil Tziolas, Thomas Bournaris, Basil Manos, and Stefanos Nastis

Abstract The simultaneous and increasing needs for safe and quality food products, along with the environmental and socio-economic sustainability, develop a multi-level problem with controversies and arbitrary assumptions for farmers and policy makers. In order to assess the aspect of sustainability in agricultural production, different impact assessment tools could be implemented. Although LCA gives the potential to develop alternative scenarios in order to achieve the optimal environmental performance, in the context of sustainability, at the same time subjective measures are developed which are difficult to quantify. Multi-criteria analysis (MCA) is the key to solve the current weakness, since it takes into account multiple criteria in a wide assortment of aspects and thus it could integrate sustainability elements. The purpose of this study is to outline the integration of LCA and MCA methodologies and develop a complete literature review regarding the sustainability of the agricultural sector through the above mentioned methodological merge. In this review we analyze scientific papers integrating LCA and different multi-criteria methodologies in agriculture. Through this analysis, we determine the connection between the methodologies through a variety of aspects regarding (a) the number and nature of multi-criteria methods integrated with LCA, (b) the way of integration between the methods in a technical perspective and (c) the benefits developed through the integration as well as the final conclusions which could only be elicited through this complex process. Studies which implemented LCA and MCA simultaneously illustrated positive economic and environmental results, since LCA focused on environmental sustainability and the multi-criteria modeling dealt with the subjective measures of LCA.

E. Tziolas (✉) · T. Bournaris · B. Manos · S. Nastis
Department of Agricultural Economics, Aristotle University of Thessaloniki, Thessaloniki, Greece
e-mail: etziolas@agro.auth.gr

J. Berbel et al. (eds.), *Multicriteria Analysis in Agriculture*, Multiple Criteria Decision Making, https://doi.org/10.1007/978-3-319-76929-5_12

1 Introduction

Environmental pollution and agricultural emissions are strongly connected with farming practices and other factors such as temperature and rainfall (Sabiha et al. 2016). Regarding human activities in agriculture, about 47% of CH_4 and 58% of N_2O emissions are produced by anthropogenic practices (Smith et al. 2007), mainly from the implementation of nitrogen fertilizers and manure production (FAO 2003). Although organic agriculture is a step towards sustainable agricultural development with less environmental impacts (Lorenz and Lal 2016), a conclusive strategy for agriculture as a whole has not yet been conducted.

The European Commission, in the framework of sustainable production patterns and development, promotes a cradle-to-gate approach for agricultural production systems through LCA. The International Reference Life Cycle Data System (IRLCD) handbook is a guide regarding the necessity of an integrated environmental impact analysis in the European region (EC 2010, 2011), developed by the Institute for Environment and Sustainability in the European Commission Joint Research Centre (JRC). The increasing awareness from the European perspective is reasonable and justifiable as 20–60% of environmental impacts are correlated with the food sector (Tukker et al. 2006).

The assessment of environmental impacts in agricultural production has always been a difficult part in studies, as future decisions of farmers cannot be predicted. New directives regarding the impacts of climate change in agriculture resulted to a decrease of emissions between 1990 and 2004 (Bessou et al. 2011), though the ambitious targets of the European Union for the year 2020 necessitate more effort by the EU members (EC 2016). Sustainability management was always a difficult task for policy makers, as it has two different approaches based on numerical data or societal evidence (bottom-up approach, stakeholder involvement etc.) which could result to conflicting interests (De Luca et al. 2017).

The decisions of farmers are affected by climate change and other factors such as technology, agricultural policy, demand and supply as well. In the decision-making process, these factors play a greater role for farmers in comparison with the environmental impacts of climate change (Chiotti and Johnston 1995; Hermans et al. 2010). Thus, the optimization of several conflicting criteria develops a complex and multi-level problem regarding completely different aspects in agriculture.

Multi-criteria analysis (MCA) takes into account all the conflicting options, through a modular procedure, in order to determine overall preferences and support the decision-making process (Belton and Stewart 2002; Kurka and Blackwood 2013). MCA has been implemented successfully in various agricultural systems (Bournaris et al. 2014; Dace and Blumberga 2016) optimizing different criteria throughout the European region. Consequently, the integration between MCA and LCA methodologies could be a thorough methodological tool for the assessment of rural development plans.

LCA is implemented in agricultural systems more than 15 years (Brentrup et al. 2004) and the potential development of alternative scenarios to conclude to the optimum environmental efficiency (Beccali et al. 2010) conforms to the methodological philosophy of MCA. Since MCA quantifies different terms to produce credible results, all the subjective terms (such as biodiversity or landform) of LCA should be quantified to reduce uncertainty based on a numerical approach. Thus, the combination of MCA tools and LCA enlarges upon the significant advantage of eliminating the subjective assumptions of LCA (Miettinen and Hamalainen 1997; Gaudreault et al. 2009).

The aim of this study is to highlight the integration of LCA and MCA in agricultural production systems through a literature review. Furthermore, the merge of these two tools will be analyzed in a greater context, assessing advantages and potential disadvantages, the amount of different methodological approaches and the limits of the methodological merge on a technical perspective. Finally, based on the abovementioned we will elicit conclusions for the integrated methodological framework as a decision support system in agriculture.

2 The Analysis of LCA

2.1 A Brief History of LCA

Concerns about resource and energy depletion as well as future prospects for sustainable development had their first reflection on the academic community on the 1960s decade (EPA 2006). These studies could be considered as partial LCAs as they focus basically on energy efficiency and pollution control, but the dawn of LCA as we know it today happened on 1974 for beverage container alternatives by the Environmental Protection Agency and for the load of different packaging products by Basler and Hofman. Between 1970 and 1990, the absence of a common application protocol of LCA resulted to entirely divergent methodological strategies, terminology and outcomes. The lack of international scientific discussion and the implementation of different methodological platforms were the main causes for the inability of rational LCA approach (Guinee et al. 2011). Though at start LCA was focusing on the environmental aspect, the social and economic aspects of a three-dimension sustainability framework were introduced. Jorgesen et al. (2008) illustrates the methodologies for social LCA and Hunkeler et al. (2008) the economic trajectory with environmental Life Cycle Costing (LCC).

The increasing interest became more prominent, as the standardization of methods and procedures from the International Organization for Standardization (ISO) was published in 1997. Nowadays, the two standards related to LCA are:

i. ISO 14040: 2006 Environmental management—Life cycle assessment—Principles and framework
ii. ISO 14044: 2006 Environmental management—Life cycle assessment—Requirements and guidelines

Each standard is reviewed every 5 years and the last confirmation for the above standards was in 2016. Although the ISO standards describe principles and provide guidelines for LCA, there is not any thorough information on how to conduct an LCA study. Guinee et al. (2011) state that LCA methods are not standardized in detail by ISO, thus an assortment of approaches depending on the addressed questions is developed. The Co-ordination Action for Innovation in Life Cycle Analysis for Sustainability (CALCAS) was commissioned by the European Commission to define the research frontiers, to identify potential gaps and to bring closer the scientific community about LCA (CALCAS 2009).

The framework of Life Cycle Sustainability Analysis (LCSA), which is a broader LCA, contains environmental, economic and social LCA broadening the analysis objective related to the goal and scope definition, the modeling and the interpretation phase. LCSA is the final step of the LCA, proposed by the CALCAS research, taking into account sustainability not only as an environmental aspect, but also as a complete procedure to preserve prosperity in the planet. It is not an automated procedure, but an interdisciplinary process utilizing multiple methodological aspects based on the addressed question. Nevertheless, it could be defined as the evolution of impact assessment which mixes manifold methodological principles in order to completely evaluate a problem (van der Sluijs 2002; Blind and Refsgaard 2007; Henriksen et al. 2007).

2.2 *The Framework of LCA*

Life cycle assessment (LCA) is a methodological framework related to the life cycle of a product or a process, in order to estimate and assess the environmental impacts such as climate change, land use, the depletion of resources, water use and others (Rebitzer et al. 2004). Another definition of LCA describes it as a systematic set of procedures for compiling and examining the inputs and outputs of materials and energy and the associated environmental impacts directly attributable to the functioning of a product or service system throughout its life cycle (ISO 2006). Therefore, LCA is a comprehensive tool which investigates the life cycle of an activity or a procedure without shifting potential problems to other regions, environments or products (Finnveden et al. 2009).

The life cycle of a product or an activity varies based on the perspective of the interested stakeholder, thus specific borders should be established. In this context, multiple approaches are implemented to fulfill a specific module in LCA (ILCD 2010). In Fig. 1 the variations on LCA borders are depicted and cradle to grave is referring to a life cycle from the extraction of the resource until the end of its life. Cradle to gate is a partial assessment from the resource extraction to the gate of the industry, before the formulation of the final product or process. Gate to gate is also a partial LCA which focuses solely on the production chain, explaining the added value process. The cradle to cradle variation is a recycling procedure, in which the

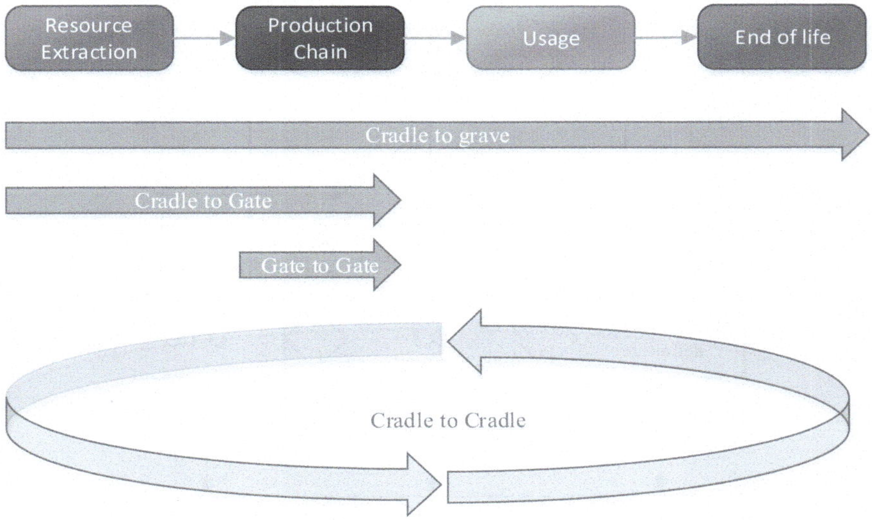

Fig. 1 Variations on LCA borders

sustainable process continues to rejuvenate its disposals after the end of their cycle (e.g. glass bottles).

LCA is a four-step procedure which contains the goal and scope definition, the Life Cycle Inventory (LCI) analysis, the Life Cycle Impact Assessment (LCIA) and the interpretation (Finnveden et al. 2009; Matthews et al. 2014). A brief framework of LCA steps is illustrated in Fig. 2.

- Goal and scope definition: In this phase technical details are included to determine the description of a product system, such as functional units, system boundaries, assumptions, impact categories and limitations (Rebitzer et al. 2004; Peters 2016). Therefore, a complete depiction and justification of the targets is illustrated through the current phase. In addition, the type of LCA should be decided based on the choices made in this phase. Attributional and Consequential LCA play a significant role in the product system modeling in the later phases (Rebitzer et al. 2004).
- Life Cycle Inventory: This analysis involves the development of a quantified and accumulated flow with all the system's inputs and outputs. This phase could be developed with three different methods: process LCA, economic input output LCA and hybrid approach based on the technique of calculation, the relative advantages and the potential limitations for the research (Islam et al. 2016).
- Life Cycle Impact Assessment: In the phase of LCIA potential impacts are identified and evaluated, based on the LCI flow results (Laurin and Dhaliwal 2017). In addition, these impacts are categorized and assigned to multiple classes. Finally, there is an optional step of weighting and normalization in which the available data is elaborated in order to facilitate the procedure for decision-makers (Verones et al. 2017).

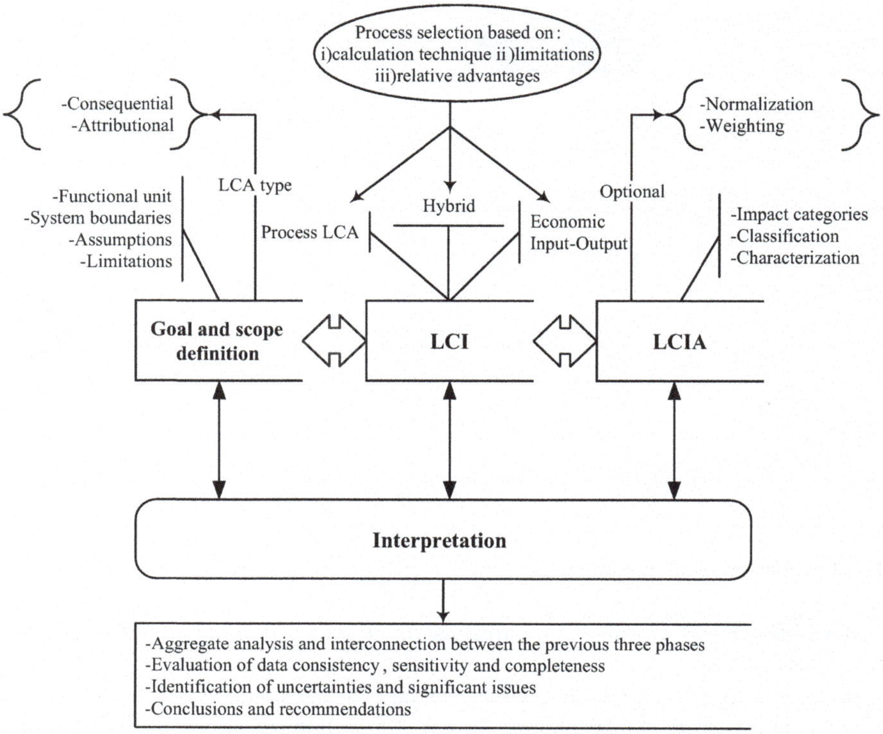

Fig. 2 LCA framework

- Interpretation: The interpretation phase consists of an aggregate analysis and interconnection between the previous three phases to elicit conclusions and recommendations (Castellani et al. 2017). Furthermore, all the uncertainties and significant issues related with the LCA study are mentioned in the interpretation phase (Finnveden et al. 2009).

This thorough assessment with multiple stages is an advantageous tool for scientists, as LCA affiliates simultaneously with environmental, economic, social and scarce resources trade-offs, providing rational insights (EC 2016). Furthermore, due to new environmental business policies, industries and small—medium enterprises are benefited by performing LCA, since it takes into account the complete life cycle of products (Hunkeler et al. 2004) focusing on reduced expenditures, better façade for local and regional communities and recycled raw materials. Policy makers are satisfied as well, ensuring prosperity to society and promoting sustainability options to industries. Therefore, LCA has been implemented in the agricultural sector several times to reduce future uncertainty and to promote sustainability.

2.3 LCA in Agriculture

The challenges for a long-term environmental strategy designed to specifically address all the different threats from farming practices have not been tackled yet by the European Community, as there is not specific legislation on a very important aspect of agriculture such as soil management (EC 2014). In this context, sustainability in agriculture could only be investigated by scholars implementing different impact assessment tools in the primary sector.

Holman et al. (2017) state that the European Agricultural Policy focuses inaccurately to a continuous increase of productivity rather than to a development of a multifunctional body which will promote sustainable production. The implementation of multiple integrated impact assessment methodologies has been suggested by several studies to enhance the management of complex agricultural systems (Parson 1995; Harris 2002; Parker et al. 2002). The implementation of LCA in agriculture could be the most challenging approximation in the evaluation of environmental sustainability, since the methodological framework for agricultural products is complex and data demanding in comparison with typologies of other products (Notarnicola et al. 2015).

Multiple studies successfully implementing LCA on food products, dairy and meat products, waste management, land and water usage in agriculture have been reported (Roy et al. 2009). Emphasizing on the optimum scenarios, Roy et al. (2009) exalt the usage of LCA in the agricultural sector citing manifold case studies and concluding that the current methodology evaluates food security as well as environmental impacts. Environmental impacts have been assessed for the rice processing chain in Italy (Blengini and Busto 2009) and Japan (Breiling et al. 1999), for tomato production (Munoz et al. 2004; Hayashi 2006), for organic and conventional apple production (Keyes et al. 2015) and for vinification (Vázquez-Rowe 2012). Furthermore, grassland farming was assessed in Germany to identify potential impacts from farming intensity (Haas et al. 2001), as well as pesticide impact on human health and the environment (Margni et al. 2002).

The last decade shows an important increase of implementing LCA methodology in biomass production (Tziolas et al. 2017a), which is highlighted by the European Parliament (2009) to evaluate the sustainability of different bio-energy pathways utilizing LCA (Sastre et al. 2016). Biomass production and its evaluation as an extraction resource (Parajuli et al. 2017; Razza et al. 2016; Eranki and Dale 2011) is related to different types of energy feedstock with their respective methods of conversion to multiple end uses such as bio-diesel (Togarcheti et al. 2017), bio-gas (Sundaram et al. 2017), bio-fuel (Seghetta et al. 2016; Budsberg et al. 2012; Yu and Tao 2009), residue exploitation (Hiloidhari et al. 2017; Gaudreault et al. 2016; Schaubroeck et al. 2013; Cherubini and Ulgiati 2010) and power (heat and electricity) generation (Yongmei et al. 2016; Murphy et al. 2016; Arteaga-Pérez et al. 2015; Xiao et al. 2009).

Territorial management is another aspect that concerns policy-makers, stakeholders and farmers, as the uncertainty of agricultural production and environmental and market constraints are usually unpredicted obstacles (Dogliotti et al. 2014). Farm management of a whole region is a major concern for the involved parties and environmental constraints could be evaluated through LCA. Average data for the assessment of arable crops in France were elaborated through LCA in order to analyze agricultural production models (Corrado et al. 2017). Capitanescu et al. (2017) implemented LCA to encompass the entire production chain in Luxembourg and Tasca et al. (2016) evaluated two agri-food supply chains of vegetables in northern Italy. Consequently, LCA is a core methodological tool for assessing whole agricultural systems and whole regions as well, in order to balance conflicting goals and directions.

LCA in agriculture is implemented more than 15 years (e.g. primary production, geographical production systems, optimization of environmental efficiency) (Brentrup et al. 2004). The alternative scenarios developed by the LCA methodology in order to result to the optimum option of environmental efficiency are a huge advantage (Beccali et al. 2010). The outbreak of LCA implementation in the scientific community is justified because it is a credible and thorough impact assessment tool, which could be used for the whole agricultural production scale: from simply commodity products to whole agricultural systems, as mentioned above. Moreover, LCA illustrates a versatile character as it could be integrated with other methodological tools in order to reduce uncertainty and elicit credible results.

3 LCA and MCA

Although LCA is still evolving as a quantitative tool with great potential, implemented in the agricultural sector towards sustainable agricultural systems and welfare of rural areas (Hayashi et al. 2005), there are specific issues. One weakness of LCA is the approach of subjective terms which sometimes could be difficult to quantify such as biodiversity or land formation. Another weakness is the uncertainty which may arise from LCA usage and the difficulty to interpret the given information (Boufateh et al. 2011). Furthermore, the environmental score achieved through LCA increases the uncertainty, as it could be interpreted in numerous ways (Rowley and Peters 2009). In this context, these disadvantages could be minimized or even bypassed based on a numerical and deterministic approach in order to elicit concurrent conclusions.

Multi-Criteria Analysis (MCA) is a decision-making and planning tool with multiple aspects, which involves multiple (sometimes conflicting) criteria. MCA in the agricultural sector is a common tool that strategically organizes and supports an assortment of challenging decisions of farmers, policy-makers and various stakeholders. Numerous studies have been published integrating MCA to support decision-making related with water management (Xevi and Khan 2005; Bournaris et al. 2015; Banihabib and

Shabestari 2017), disparities in rural areas (Popescu and Bara 2015) biomass production (Tziolas et al. 2017b; Caprara and Martelli 2016; Kylili et al. 2016), economic aspects (Tiwari et al. 1999), input management (Gómez-Limón et al. 2004), policy impacts and scenarios (Manos et al. 2006, 2013; Riesgo and Gómez-Limón 2006; Bournaris and Manos 2012; Bournaris et al. 2014), management of whole agricultural regions (Bournaris et al. 2009; Manos et al. 2010) etc. It is obvious that MCA contributed to the radical reform of the agricultural production sector in various aspects, integrating multiple methodological MCA aspects within the scientific community. Thus, MCA plays a significant role for the primary sector and a formulation of an integrated LCA and MCA methodology could prevent environmental hazards, based on the current regulations, and optimize multiple conflicting criteria with the highest standards of credibility.

The evaluation phase is the most important phase in impact assessment, as the results of the methodological framework should elicit conclusions and support rationally the decision-making process. Thus, MCA could "borrow" tools to the LCA in order to help the endeavor to optimize the environmental efficiency and the evaluation stage (Rowley and Shields 2011; Dorini et al. 2011). Furthermore, the holistic focus of MCA to the decision-making process could assist to the management of impact assessment scales, the indicators selection, weighting and the aggregation of indicators (Finkbeiner et al. 2010).

3.1 MCA and LCA Integration in Agriculture

In the last few years impact assessment studies implement multiple methodological tools simultaneously rather than single assessments based on one methodological aspect (Falcone et al. 2016). Although the integration of various methodological aspects could be confusing and sometimes difficult to interpret, the academic community converges to the integration of methodologies in order to produce more credible results. The usage of MCA in LCA is widespread for different scientific branches (Hermann et al. 2007; Rabl and Holland 2008; Myllyviita et al. 2012; Cai et al. 2017; Maia Angelo et al. 2017) highlighting the practicality and utility of the methodology integration.

Various MCA methods have been implemented with LCA in order to assess environmental, economic and social sustainability in agricultural production. The distinction between MCA categories is based on the single criterion approach and the outranking approach since interactive methods (trial and error approximation) are not applicable in LCA (Benoit and Rousseaux 2003). Multi-Attribute Utility Theory (MAUT) (Lipuscek et al. 2010; Myllyviita et al. 2012), Outranking methods (PROMETHEE and ELECTRE) (Kralisch et al. 2013; Castellini et al. 2012), Analytic Hierarchy Process (AHP) (Dinh et al. 2009; De Luca et al. 2015a), Simple Multi-Attribute Rating Technique (SMART) (Myllyviita et al. 2014), VIKOR (Visekriterijumska Optimizacija IKompromisno Resenje) (Falcone et al. 2016), TOPSIS (Technique of Order Preference Similarity to Ideal Solution) (Karklina

Table 1 MCA tools applicable with LCA

Value or utility-based models (single criterion)	Outranking methods	Other methods	Interactive methods
Multi-Attribute Utility Theory (MAUT)	PROMETHEE (I–II)	NAIADE	Not applicable with LCA
Analytical Hierarchy Process (AHP)	ELECTRE (I–IV)		
Simple Multi-Attribute Rating Technique (SMART)	EXPROM (I, II)		
Simple Additive Weighting (SAW)	MELCHIOR	Stochastic Multi-objective Acceptability Analysis (SMAA)	
TOPSIS	QUALIFLEX		
VIKOR	REGIME		
	ORESTE		

et al. 2015) are some of the most used methodological frameworks in agriculture along with LCA. The conjunction between LCA and MCA techniques does not have to be exclusive and other tools, as well as multiple methodological aspects, could integrate to the study. In fact, other participation methods such as interviews, focus groups, questionnaires etc. are implemented to various LCA and MCA studies (De Luca et al. 2017) (Table 1).

3.2 MCA Framework in LCA

The classification of MCA methods in LCA studies is a complex procedure and the ways of integration vary from complete fusion to independent methodological direction. MCA methods could be implemented in all four phases of LCA, but the most frequent occasion is as a supplementary tool to add more assessment information and converge to more realistic evaluations (De Luca et al. 2017). The complete opposite procedure is also a possibility, in which LCA constitutes a small part in a MCA framework elaborating data from the LCA as indicators. Finally, there is the fully merged condition where the methodologies are applied simultaneously, eliciting conclusions from the hybrid methodology and not from each one separately.

The implementation and the phase where each methodological aspect interferes are depending on the way of integration between MCA and LCA. At first, we will focus on the assumption that MCA has a complementary role in LCA which is depicted in Fig. 3 and in continuation we will examine the exact opposite.

The goal and scope definition phase of LCA contains technical details such as limitations, functional unit, borders etc. that should be identified. MCA defines the objectives and the scenarios under assessment and promotes an objective measure to the subjective criteria of LCA (Guitouni and Martel 1998; Benoit and Rousseaux 2003). When the problem and objective definition of MCA is determined, the uncertainty of the goal and scope definition phase is limited or even eliminated.

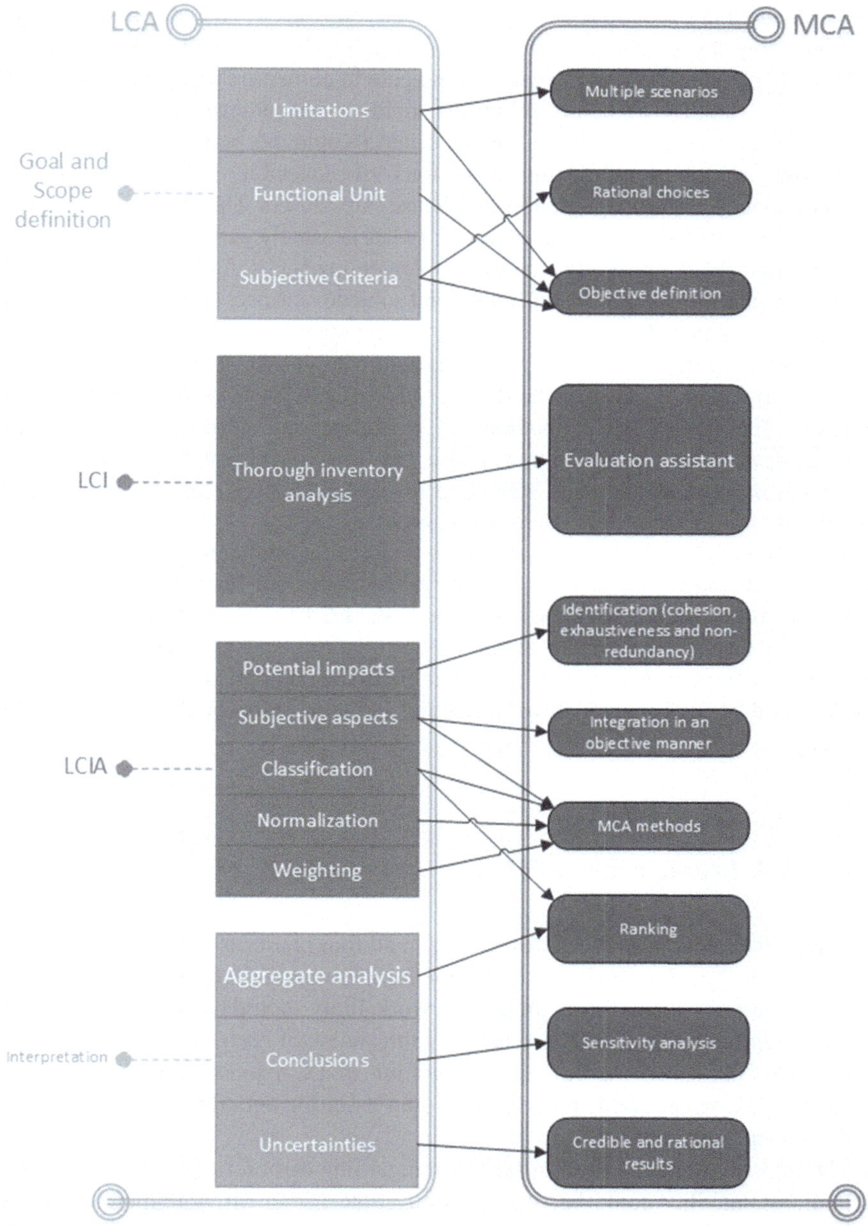

Fig. 3 Synergies of MCA tools with LCA

The LCI phase is data demanding because multiple inputs and outputs should be quantified in order to proceed to the next phase. In this phase MCA has little to none role to play, as other methods (e.g. focus groups, interviews etc.) are more significant to the broadening of the inventory. In some occasions MCA is a criteria evaluation assistance, but it is essential to perform a thorough inventory analysis in advance.

LCIA is the phase that MCA could participate in numerous ways and in multiple phases (Dias et al. 2016). Potential impacts are identified and evaluated in LCIA and the formulation of a coherent group of criteria could be performed with MCA techniques to comply with the conditions of cohesion, exhaustiveness and non-redundancy effectively (Benoit and Rousseaux 2003). The categorization and the assignment to multiple classes phase of LCIA is usually performed with ranking MCA techniques. Another advantage of implementing MCA in LCIA is that it could integrate economic, environmental and social subjective aspects to the assessment in an objective manner (De Luca et al. 2017; Myllyviita et al. 2014). Finally, normalization and weighting could be implemented with various MCA methods as well and the most frequent applied methods are AHP, ELECTRE, PROMETHEE and TOPSIS. The use of MCA methods in weighting and normalization provides transparency to the interpretation phase, as the implementation without the sub-stages of normalization and weighting reveals problems to the aggregation and the comparison of the results (Bengtsson and Steen 2000).

The interpretation is the most complex phase as it connects to the other three phases and the aggregate analysis and the elicitation of conclusions requires attention. Sensitivity analysis could be integrated to the LCA through MCA techniques in order to investigate the impact of the input variations and eliminate uncertainty. Furthermore, the conclusions are more credible as the comparison between different results could be implemented through ranking and the results could be interpreted in an easier way.

3.3 LCA Framework in MCA

Naturally, the integration of MCA methods in all four phases is not an easy task and the choice of the appropriate method must fulfill some requirements based on the nature of the assessment. Practicability and workability should be major principles to the selection method, as well as sensitivity thresholds of each method (Benoit and Rousseaux 2003). Nevertheless, the integration of LCA tools to a Multi-Criteria Decision Making (MCDM) process is also an asset for agriculturists to provide a well-rounded assessment.

Through LCA a plethora of information regarding carbon footprint and environmental sustainability could become a useful guide for an MCA framework. LCA aims to the environmental aspects and in the problem definition stage of an MCA approach it could be a great asset for the objectives choice and the potential problems that MCA could not recognize (Fig. 4). LCA is a complete procedure, depicting from cradle to grave the lifetime of a product or a procedure and their impact to the

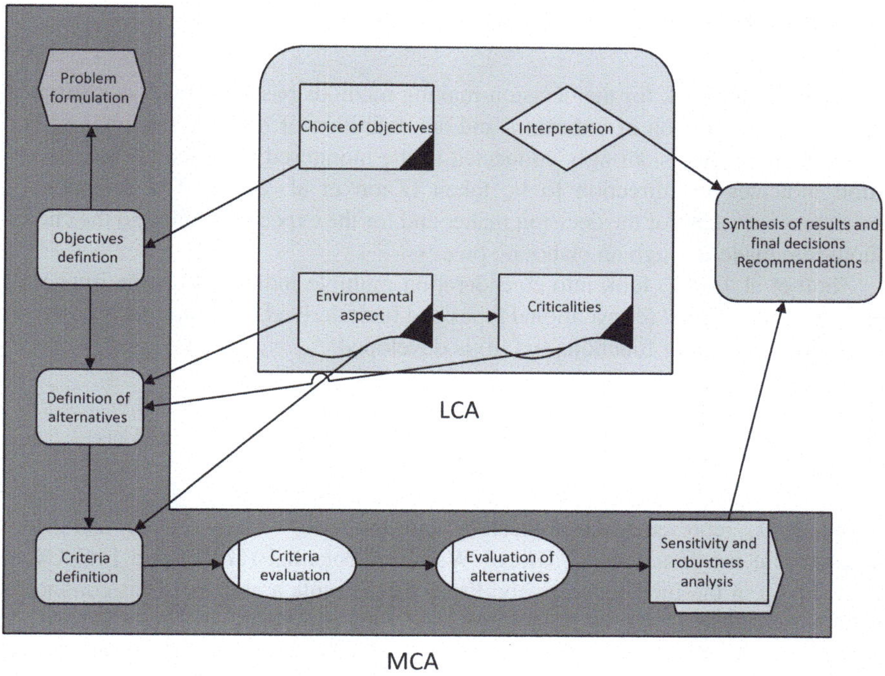

Fig. 4 Synergies of LCA in MCA

environment, illustrating the necessity to be performed for agricultural products. Moreover, LCA is implemented in order to set a system of boundaries (environmental and economic), considering different variations even on the same agricultural system (De Luca et al. 2015b). Specifically, in a biomass production system, different approaches to the lifetime of a product (cradle to grave, cradle to gate etc.) could be considered and provide better interpretation to the results. Through LCA carbon footprints are identified and the results could be used in a techno-economic analysis for a whole region (Cucek et al. 2012). It is obvious that insights from LCA could provide additional data and develop multiple alternatives, as well as highlight criticalities to the MCA system.

Based on the abovementioned, MCA tools have plenty applications in the LCA trajectory and in various stages of the LCA. Nevertheless, the choice of the appropriate MCA model to the integration with LCA plays a significant role to the whole assessment, to the corresponding weights and to the criteria approach (Seppala et al. 2002). On the other hand, LCA in an MCDM system could intervene to crucial stages of the MCA, providing additional data and working mainly supplementary. In the following chapter, multiple integrations of MCA methods to the LCA in agriculture are illustrated, as well as the phase and the way of integrations.

4 Implementing MCA Tools to LCA

A general framework for the decision-making methods based on LCA is illustrated through the evaluation of indicators and the definition of multiple parameters. The usage of indicators is strongly connected to the monitored system and their fluctuation indicates the direction to be taken (Zhou et al. 2007). The selection of indicators is crucial for the decision maker and for the expected results, so the choice should be made through an elaborate process.

Zhou et al. (2007) took into consideration multiple indicators for the impact of different types of fuel (conventional types and fuels derived from biomass), based on LC tools. A group of functions ($q_i(x_i)$) is developed:

$$q_i(x_i) = \begin{cases} 0 & if \quad x_i \quad \leq \quad MIN(i) \\ \left(\frac{x_i - MIN_{(i)}}{MAX_{(i)} - MIN_{(i)}}\right)^{\lambda}, & if \quad MIN(i) \quad < x_i \leq \quad MAX(i) \\ 1 & if \quad x_i \quad > \quad MAX(i) \end{cases}$$

Where x_i represents the indicators and the two values *MIN*(i) and *MAX* (i) represent the functions of Global Warming Potential (GWP) and Life Cycle Cost (LCC) of the fuels respectively. From this formula a new matrix is conducted with various values for each indicator and fuel combination. These values are normalized sustainability indicators which could be used in order to elicit safe conclusions. Hayashi et al. (2014) aggregated two types of indicators (positive and negative) and the normalization equations are:

$$r_i^+ = \left(r_{Ai}^+ - T_i\right)\left(r_{\max i}^+ - T_i\right)^{-1} \; if \; r_{Ai}^+ > T_i$$
$$r_i^+ = \left(r_{Ai}^+ - T_i\right)T_i^{-1} \; if \; r_{Ai}^+ \leq T_i$$
$$\text{(AND)}$$
$$r_j^- = -\left(r_{Aj}^- - T_j\right)\left(r_{\max j}^- - T_j\right)^{-1} \; if \; r_{Aj}^- > T_j$$
$$r_j^- = \left(r_{Aj}^- - T_j\right)T_j^{-1} \; if \; r_{Aj}^- \leq T_j$$

Where A_i or A_j depicts the indicator's value before the normalization and T_i or T_j depicts the threshold subscript. Additionally to the normalization procedure, an aggregative function is conducted in order to allocate weights to the indicator of interest and this is the main insight of the additive aggregation methods (Dias et al. 2016):

$$S(q; w) = \sum_{i=1}^{n} w_i r_i$$
$$w_1, \ldots, w_n \geq 0 \; and \; \sum_{j}^{n} w_j = 1$$

Where w_i represents the weight aspect of vector w and r_i represents the aggregated indicators, based on the wanted criteria. Thus multiple conclusions could be elicited, through the elaboration with various weights depending on the decision-maker. This simple method applies the principles of MCA in the normalization and weighting phases of LCA in order to facilitate the assessment of rating between different options.

4.1 *Stochastic Multi-attribute Analysis (SMAA) and LCA*

A methodology implemented with LCA is not necessarily integrated to the core of the analysis, but it could be used alongside to the LCA. Usually, MCA tools are financial evaluators to the feasibility of an LCA project. Reeb et al. (2016) implemented SMAA in parallel with LCA to develop a distribution function biased to environmental preference and then assess four additional criteria related to the financial feasibility of a biomass feedstock system with MCA. Furthermore, Dias et al. (2016) integrated SMAA in LCIA for the aggregation of various impact categories for comparing alternative biodiesel chains.

SMAA is implemented in order to interpret the LCIA results and to add safety to the feasibility analysis, because the financial analysis is incorporated, enabling stakeholders to make rational decisions. In order to identify and evaluate the environmental preference score, multiple environmental impacts (global warming, acidification, ozone depletion etc.) were depicted as coefficients of variation for each feedstock scenario model. The use of SMAA in LCIA has developed seven different ranks and not numerical data, allocating weights biased to the environmental preference. Thus, weights for each impact category were generated from the lower value (− −) to the higher (+ +). Rogers and Seager (2009) describe the SMAA-LCIA as an approach following the principles of PROMETHEE outranking to help the normalization process of the inventory.

Then the LCIA results are depicted as an aggregated single score, weighted and referred as the environmental preference score for each feedstock which could be easily ranked and understood. The complete assessment comprises an assortment of criteria for the feasibility evaluation of technical and financial aspects. Multiple costs, yields and transport distances are integrated to an MCA model along with the single score of environmental preference for each feedstock, so that the interpretation of LCA is complete and rational. Three different MCA methods were employed from Reeb et al. (2016) to develop an overall rank for each feedstock depending on the financial, technical and environmental criteria.

The unweighted ranking method in which all the criteria have the same weights for alternative j and the X_{ij} is the rank for each criterion i for alternative j. The equation for N criteria and S_j score of alternative is depicted as:

$$S_j = \sum_{i=1}^{N} X_{i,j}$$

The same principles are applied to the next equation, where the weight Y_i for each criterion i is incorporated to the equation and the sum of all fractional weights is equal to 1:

$$S_j = \sum_{i=1}^{N} X_{i,j} Y_i$$

In that way, each weight could be represented as a percentage and it could be applied to each criterion based on the decision-making of the stakeholder. Furthermore, the results are easily understood and logically derived from the decision-making process. The final method comprises the addition of more weights to the cumulative rank based on the magnitude difference between alternatives which is an endeavor to normalize raw data. Through iterative, constrained randomization multiple ranges were developed and with the implementation of the above equations, the final rank order was created. This methodology implements MCA aspects to the LCA results in order to interpret them in a secure perspective and evaluate the LCIA results with substantial credibility, since the decisions of stakeholders could be integrated.

4.2 Analytic Hierarchy Process (AHP) and LCA

AHP is a simpler form of MAUT making comparisons between alternatives to develop the criteria ranking. AHP is a common methodological approach in LCA, as it has been characterized as the best technique for sustainable development (Narayanan et al. 2007). In the decision-making process of LCA, the definition of the optimal alternative solution is a difficult procedure and many times it could be unattainable. The necessity for a compromise is the core of AHP by implementing subjective judgments and ranking the alternatives through weighting (von Doderer and Kleynhans 2014).

The process of AHP in LCA could be implemented in various phases in order to assess the alternatives or to rank the final decisions. First of all, the construction of a decision tree for the depiction of the goal, the alternatives and the criteria should be formulated. Then, priorities between the elements are defined through rational judgments based on two-way comparisons; the scoring procedure. Thus, weights which will determine the final decision are developed. The comparison between weighted alternatives develops a group of overall priorities and the final ranking is established.

Cristobal (2011) implemented AHP in an LCA project to determine the weights among multiple renewable energy exploitation industries in Spain. The alternatives were compared based on the criteria of the generated power, the operating hours, the carbon dioxide per year and the costs. The alternatives were the different types of industries which exploited various forms of renewable energy. Though the VIKOR method was used for the final consideration, the decision maker's weights (W_n) were determined with AHP, based on the criteria. A scale of verbal judgments (2 = minimal importance to 9 = absolute importance and 1 = comparison between the same) developed a matrix with a numerical depiction of relative importance for each criterion:

$$\begin{bmatrix} 1 & \frac{W_1}{W_2} & \cdots & \frac{W_1}{W_n} \\ \frac{W_2}{W_1} & 1 & \cdots & \frac{W_2}{W_n} \\ \vdots & \vdots & \ddots 1 & \vdots \\ \frac{W_n}{W_1} & \frac{W_n}{W_2} & \cdots & 1 \end{bmatrix}$$

Narayanan et al. (2007) implemented AHP and LCA in order to assess the sustainable development of indicators, as well as to compare the alternatives for bio-diesel exploitation from specific crops. Von Doderer and Kleynhans (2014) integrate AHP to the interpretation phase of LCA to determine the performance data and evaluate them as well. The LCA results are transformed and normalized into multiple scores for each Lignocellulosic Bio-energy Systems (LBS) for the Worcester Biomass Procurement Area (WBPA) and then weighted for the final assessment.

4.3 VIKOR and LCA

VIKOR is also a method that seeks for a compromise solution, in a problem with conflicting criteria, rather than the optimal solution (Falcone et al. 2016) of other methods. Ren et al. (2015) implemented AHP for the definition of weights and VIKOR method for the determination of the most sustainable sequence scenario for the bio-ethanol production in China. This study (Ren et al. 2015) included environmental, economic and social criteria through Life Cycle methodologies and the cultivation scenarios were based on wheat, corn and cassava crops. The VIKOR method was integrated to the core of the LC methodologies determining the compromise solution.

Falcone et al. (2016) assessed vine-growing sustainability in South Italy by implementing LC tools for the environmental and economic sustainability assessment, while the VIKOR method was used to rank the scenarios determined by the sustainability indices. The environmental and economic indices synthesized a composite index of sustainability and the scenarios took values between 0 and 1. The four scenarios referred to agricultural practices (e.g. conventional, organic etc.) related to vine-growing parameters.

The state-of-art for the VIKOR method is thoroughly illustrated from Yazdani and Graeml (2014). One of the first steps for implementing VIKOR in LCA is the determination of the best (x_i^+) and worst (x_i^-) values of all the criteria (i) for each alternative (x_i) according to the following equations:

$$x_i^+ = max(x_{ij}), where\ j = 1, 2, \ldots, n$$
$$x_i^- = min(x_{ij}), where\ j = 1, 2, \ldots, n$$

The next step for the VIKOR method is the calculation of the range between the alternative x_j and the positive ideal solution (S_j), while the R_j represents the range between the alternative x_j and the negative ideal solution:

$$S_j = \sum_{i=1}^{n} w_i \frac{x_i^+ - x_{ij}}{x_i^+ - x_i^-}$$
$$R_j = max\left[w_i\left(\frac{x_i^+ - x_{ij}}{x_i^- - x_{ij}}\right)\right]$$

The above equations include weights (w) for each criterion (i), but it is not necessary to elicit them through the VIKOR method. As described in a previous chapter, (Ren et al. 2015) implemented AHP for the definition of criteria, while Falcone et al. (2016) assumes that the weights are equal. Another occasion is that the potential interested parties could determine the weights based on the preferences of farmers, policy makers, stakeholders etc. The calculation of the value Q_j will create an index of different scenarios, which will take values from 0 to 1, with values trending to 0 being the most ideal based on the equations:

$$S^* = min(S_j), S^- = max(S_j), j = 1, 2, \ldots, n \quad \text{(AND)}$$
$$R^* = min(R_j), R^- = max(R_j), j = 1, 2, \ldots, n$$
$$Q_j = v\left(\frac{S_j - S^*}{S^- - S^*}\right) + (1 - v)\left(\frac{R_j - R^*}{R^- - R^*}\right)$$

Where v represents the weight of the alternative with the major group utility and usually is equal to 0.5. When the v is over 0.5 then the created index will tend to indicate mainly positive attitude while the complete opposite will occur when v value is below 0.5. Thus, the created index is easy to understand, credible and filled with data from multiple perspectives (environmental, economic and social) and scenarios for agricultural purposes.

4.4 TOPSIS and LCA

The Technique of Order Preference by Similarity (TOPSIS) is an MCA method based on the concept that the optimal alternative should have high proximity to the ideal solution, while being at a distance from the negative ideal solution (Zyoud and Fuchs-Hanusch 2017). TOPSIS in LCA has been implemented mainly on the assessment for the performance of a system (Karklina et al. 2015; Zare et al. 2016). In agriculture, Karklina et al. (2015) assessed a bio-methane production and the distribution system in Latvia with social LCA and the TOPSIS method. The criteria or social performances were related to employment, welfare, environmental and resource security, while the alternatives included different types of biomass exploitation industries.

The TOPSIS method requires the construction of a decision making matrix in which the criteria are presented as x_j and the alternatives as A_i. The weights are represented as w_j and the normalized data as b_{ij}:

$$\begin{array}{c} \\ A_1 \\ A_2 \\ \vdots \\ A_i \\ \vdots \\ A_n \end{array} \begin{array}{c} \begin{matrix} w_1b_1 & w_2b_2 & \cdots & w_jb_j & \cdots & w_nb_n \end{matrix} \\ \begin{bmatrix} w_1b_{11}^k & w_2b_{12}^k & \cdots & w_jb_{1j}^k & \cdots & w_nb_{1n}^k \\ w_1b_{21}^k & w_2b_{22}^k & \cdots & w_jb_{2j}^k & \cdots & w_nb_{2n}^k \\ \vdots & \vdots & \cdots & \vdots & \cdots & \vdots \\ w_1b_{i1}^k & w_2b_{i2}^k & \cdots & w_jb_{ij}^k & \cdots & w_nb_{im}^k \\ \vdots & \vdots & \cdots & \vdots & \cdots & \vdots \\ w_1b_{n1}^k & w_2b_{n2}^k & \cdots & w_jb_{nj}^k & \cdots & w_nb_{nm}^k \end{bmatrix} \end{array}$$

The ideal solutions are determined through the following equations:

$$A^+ = max_iw_jb_{ij}\ (\text{AND})\ A^- = min_iw_jb_{ij}$$

The Euclidean distance between the ideal solution (Ed^+) and the least ideal solution (Ed^-) is illustrated by the equations:

$$Ed^+ = \sqrt{\sum_{j=1}^{n}\left(t_{ij} - t_{ij}^+\right)^2}, \text{ where } t_{ij} = b_{ij}w_i$$

$$Ed^- = \sqrt{\sum_{j=1}^{n}\left(t_{ij} - t_{ij}^-\right)^2}, \text{ where } t_{ij} = b_{ij}w_is$$

Finally the relative proximity to the ideal solution is generated by the comparison of the Euclidean distances through the equation:

$$P_i^* = \frac{Ed_i^+}{\left(Ed_i^+ - Ed_i^-\right)}$$

Determined by the TOPSIS method, Karklina et al. (2015) conducted a social LCA and through the results of the TOPSIS method, multiple ratings were developed for each biomass exploitation industry. Albeit the fact that all the weights were assumed as equal, TOPSIS method was integrated to the core of LCA, elaborating data in the LCIA and the Interpretation phase.

4.5 Simple Multi attribute Rating Technique (SMART) and LCA

SMART is a comprehensive model in order to justify decisions implementing qualitative and quantitative data (Risawandi and Rahim 2016), based on a linear additive model. The method's mindset is strongly affected by the ranking MCA fundamentals, but also takes into account the human perspective as inputs (Kasie 2013). It is obvious that SMART has many similarities with other ranking MCA methods and the differences hinge to the model's inputs, which sometimes could be detail demanding. Furthermore, the simplified character of SMART is tending to rank the top alternatives very similarly (Hobbs and Meier 2000).

Myllyviita et al. (2012) assessed through LCA the environmental impact of two biomass production chains and implemented the SMART method to calculate the weights for the environmental impacts. The study embodied the participation of environmental specialists through questionnaires in order to weight and identify impact categories. The panelists identified new environmental impacts (such as biodiversity, nutrient balance of soil etc.), not included in the LCA method, and the SMART method gave the opportunity to integrate them to the assessment. The simplicity and the versatile character of SMART method was the key for the selection, because it can easily be modified and the panelists had a manageable task for setting the weights, even if they did not understand the methodology. The overall impact score was derived by the following equation:

$$U_k = \sum\nolimits_i w_i b_{ik} c_i$$

Where w_i is the weight for each impact category i, b_{ik} is the impact assessment score for the production chain k and c_i is the normalization factor. The normalization factor (c_i) was defined in the LCIA phase through the LCA technique. The other way of internal normalization is calculated by comparing the non-normalized scores of the criteria for each impact category by:

$$Nor_i = \frac{b_{ik}}{\sum\nolimits_k b_{ik}}$$

Thus the overall impact score is described by the following equation:

$$U_k = \sum\nolimits_i w_i Nor_i$$

The implementation of the SMART method in the LCIA phase of LCA facilitated the procedure with the determination of weights and the elicitation of credible results. Furthermore, the insights of panelists were depicted, with the simplicity that characterizes the SMART method, through the questionnaire assessment and the determination of weights for each impact category. Finally, the addition of environmental impacts outside of the LCA grasp and the relatively easy integration of them to the assessment's core, developed a much better assessment of the four biomass production chains.

4.6 *PROMETHEE and LCA*

The PROMETHEE method was firstly implemented by Brans (1982) and it is an outranking method between conflicting criteria, which ranks the finite set of alternatives and selects the optimal solution (Behzadian et al. 2010). Outranking methods and the PROMETHEE method specifically are widely implemented for environmental problems, especially when there are numerous discrete alternatives to select (Herva and Roca 2013). The ideology of PROMETHEE is based on a pair-wise comparison of alternatives with the given criteria (Mohamadabadi et al. 2009), trying to enrich the dominance relationship among the multiple alternatives (Ghafghazi et al. 2010). The preference function between the two alternatives (*a* and *b*) takes values between 0 and 1 and is illustrated as follows:

$$p(a,b) = \begin{cases} 0 & if\, f(a) \leq f(b) \\ p[f(a), f(b)] & if\, f(a) > f(b) \end{cases}$$

The assumption is reasonable through the following equation in order to help the decision making process:

$$p[f(a), f(b)] = p[f(a) - f(b)]$$

The next step is the development of the weighted preference index in order to identify the overall preference of the alternative *a* over the alternative *b* and w_h is the relative weight considered by the decision maker for the criterion *h*:

$$\pi(a,b) = \frac{\sum_{h=1}^{k} w_h p_h(a,b)}{\sum_{h=1}^{k} w_h}$$

Based on the abovementioned, the PROMETHEE method introduces the three outranking flow measures (leaving, entering and net flow), which calculate the final ranking:

$$Leaving\ flow: \varphi^{+}(a) = \sum\nolimits_{x \in K} \pi(a, x)$$
$$Entering\ flow: \varphi^{-}(a) = \sum\nolimits_{x \in K} \pi(x, a)$$
$$Net\ flow: \varphi(a) = \varphi^{+}(\alpha) - \varphi^{-}(\alpha)$$

The net flow should be the highest in order to rank as first, because the leaving flow represents the outranking of the alternative *a* in comparison with the other alternatives (set of *K* alternatives) and the entering flow represents the complete opposite. Mohamadabadi et al. implemented the PROMETHEE method to rank the different types of fuels (ethanol, diesel, bio-diesel etc.), based on five criteria (GHG emissions, costs etc.). The life cycle of all the alternatives was assessed and the PROMETHEE method contributed to the final assessment with the ranking of the mentioned alternatives.

4.7 REGIME and LCA

The REGIME method focuses on the evaluation of the vector distances between the criteria, introducing the concept of a solution area (Nijkamp et al. 1993). The major advantage of this method is the simplicity and the ease of qualitative data integration, while the big disadvantage is the lack of critical information when the amount of quantitative data is quite too high.

Finco et al. (2012) utilize the REGIME method through LCA for four different biomass types, namely sunflower, soy, palm and rape seed. The criteria were the green house gas emissions, the land-use change and the energy balance and the assessment was implemented for different scenarios (economic and environmental). The ideology of the REGIME method is based on the simple qualitative prioritization of criteria and the simple qualitative prioritization of the alternative performances for each criterion (Nijkamp et al. 1993). For every criterion, a pair of alternatives is compared and develops a specific vector for the comparison which takes three values (−1, 0 +1).

Therefore, an index is conducted with all the vector values based on the comparisons of all the alternatives. The logic symbols (+ and −) represent the degree of dominance of one option over another. The performance score is an aggregate probability measure which is illustrated as follows:

$$p_i = \frac{1}{I-1} \sum\nolimits_{j \neq i} p_{ij}$$

Where *I* is the amount of the alternatives and p_{ij} is the preference of one option over another. Although the REGIME method can use multiple types of information, it is not the best evaluation method for LCA as some information may be lost and the environmental assessment could be directed to false conclusions.

5 Conclusions

Agricultural systems are usually complex schemes with multiple interests and various parties involved, while uncertainty plays a significant role in agricultural production. The necessity for tools to evaluate the sustainability of agricultural production, to identify and include the opinions of stakeholders, to elaborate with vast amounts of information and to facilitate this convoluted situation is an aspiration for the scientific community and the involved parties (Buchholz et al. 2009). Renewable energy systems based on agriculture demonstrate an increasing trend and MCA tools are applied in various projects to converge on sustainable solutions and to guide the potential stakeholders (Giampietro et al. 2006).

Nowadays, the most pressing environmental problem is the raise of the greenhouse gas emissions (Rebolledo-Leiva et al. 2017), which is integrated to an assortment of agricultural practices and to the transportation of primary goods. Although MCA tools are credible and easy to understand, the implementation of each method could create different results (Teshome et al. 2014) and usually MCA methods are perceived as a technocratic approach (Geneletti 2013). Furthermore, as an impact assessment tool in agriculture, MCA should be formulated in order to clearly definite the nature and the complete framework of the problem.

On the other hand, LCA is an environmental impact assessment tool, which takes into consideration all the inputs and outputs of a procedure or a product, from the start of their existence to the final disposal or to the desired level of production. Nowadays, the exploitation of renewable resources and the optimal farm management practices develop a compound problem with induced environmental problems and farm income fluctuations.

Policy makers, farmers and stakeholders, in the search of the golden dawn for environmental sustainability and socio-economic development, are in need of credible policy strategies. The integration of LCA and MCA methods develops a powerful tool for the assessment of agricultural production in multiple stages. The LCA approach enhances the environmental perspective, taking into account all the potential environmental hazards and depicting the carbon footprint of each action taken. From the opposing point of view, LCA introduces measures which could be subjective and difficult to interpret. Thus, MCA tools are implemented to identify this dissonance, elaborate with objective measures and depict values in a numerical and easy to understand way. In agriculture, the two methods above have been implemented for several occasions, but mostly in biomass exploitation projects. Converge to biomass is caused by the renewable source character of biomass and the potential derivatives of fuels (bio-gas, bio-diesel etc.), which have an impact to the environment and the social welfare.

The LCA and MCA methods could be implemented in a fully merged framework in which they are implemented simultaneously with the same importance, as a hybrid (Cucek et al. 2011). Another approach is the one that MCA tools are integrated in various stages of the LCA, as an assisting process, to support the decision making process in a rational positivist way (Scott et al. 2016). The last way of integration is the implementation of LCA in the context of a wider framework of MCA, mainly for environmental purposes (Lipuscek et al. 2010).

MCA tools could be integrated to all the phases of LCA based on the purposes of the interested part. In the goal and scope definition and LCI phases, MCA tools are mostly integrated in order to define the main objective, thus being the two phases with the least synergies with MCA tools. In the LCIA phase, classification, weighting and normalization could be calculated by MCA tools, to add an extra layer of credibility to the results. Furthermore, in the interpretation phase the results could be analyzed with MCA tools for a better approximation of the decision making process. On the other hand, when LCA is used as a secondary assessment tool, it could enhance the objective definition with an environmental tone. Besides that, it could add the environmental aspect to the alternative solutions and indicate certain criticalities.

Regarding the agricultural sector, MCA methods implemented with LCA are usually utility-based models (single criterion) and outranking methods. Additive aggregation, VIKOR, TOPSIS and Simple Multi Attribute Rating Technique (SMART) utilize multiple equations in order to integrate weights in the LCIA phase and along with that, normalize the data to elicit coherent conclusions. Analytic Hierarchy Process (AHP) is a method for the development of weights through continuous comparisons and it could also be implemented for the final ranking of alternatives. Finally, outranking methods, such as PROMETHEE, ELECTRE and REGIME, depend on a pair-wise comparison between alternatives in order to evaluate the interpretation phase of LCA. They are the most implemented tools with LCA and they could be used in the other phases as well. The abundance of MCA tools offers a variety in the method choice, but also generates confusion for the right methodological approach. Nevertheless, the integration of LCA and MCA tools is a powerful tool for the complete assessment of agriculture, regarding the main sectors of human welfare. Environmental sustainability and socio-economic development are matters that concern the policy-makers and people in general. Thus, impact assessment tools should be credible and possible to rely on. The LCA and MCA integration develops an all-around tool that takes into account all the possible externalities and elicits solid and sustainable solutions in agriculture.

Acknowledgements The article is written as part of a PhD study at the Department of Agricultural Economics, School of Agriculture, Aristotle University of Thessaloniki (AUTh), Greece. The study is funded by General Secretariat for Research and Technology (GSRT) and the Hellenic Foundation for Research and Innovation (HFRI).

References

Arteaga-Pérez, L. E., Vega, M., Rodríguez, L. C., Flores, M., Zaror, C. A., & Ledón, Y. C. (2015). Life-Cycle Assessment of coal–biomass based electricity in Chile: Focus on using raw vs torrefied wood. *Energy for Sustainable Development, 29*, 81–90. ISSN 0973-0826, https://doi.org/10.1016/j.esd.2015.10.004

Banihabib, M. E., & Shabestari, M. H. (2017). Fuzzy hybrid MCDM model for ranking the agricultural water demand management strategies in arid areas. *Water Resources Management, 31*(1), 495–513.

Beccali, M., Cellura, M., Iudicello, M., & Mistretta, M. (2010). Life cycle assessment of Italian citrus-based products. Sensitivity analysis and improvement scenarios. *Journal of Environmental Management, 91*, 1415–1428.

Behzadian, M., Kazemzadeh, R. B., Albadvi, A., & Adhdasi, M. (2010). PROMETHEE: A comprehensive literature review on methodologies and applications. *European Journal of Operational Research, 200*, 198–215.

Belton, V., & Stewart, T. J. (2002). *Multiple criteria decision analysis: An integrated approach* (1st ed.). Norwell, MA: Kluwer Academic.

Bengtsson, M., & Steen, B. (2000). Weighting in LCA-approaches and applications. *Environmental Progress, 19*, 101–109.

Benoit, V., & Rousseaux, P. (2003). *International Journal of LCA, 8*, 74. https://doi.org/10.1007/BF0297843

Bessou, C., Ferchaud, F., Gabrielle, B., et al. (2011). Biofuels, greenhouse gases and climate change: A review. *Agronomy for Sustainable Development, 31*(1), 1–79. https://doi.org/10.1051/agro/2009039.

Blengini, G., & Busto, M. (2009). The life cycle of rice: LCA of alternative agrifood chain management systems in Vercelli (Italy). *Journal of Environmental Management, 90*, 1512–1522.

Blind, M. W., & Refsgaard, J. C. (2007). Operationalising uncertainty in data and models for integrated water resources management. *Water Science and Technology, 56*(9), 1–12. https://doi.org/10.2166/wst.2007.593

Boufateh, I., Perwuelz, A., & Rabenasolo, B. (2011). Multiple criteria decision-making for environmental impacts optimization. *International Journal Business Performance and Supply Chain Modeling, 3*(1), 28–42.

Bournaris, T., & Manos, B. (2012). European union agricultural policy scenarios' impacts on social sustainability of agricultural holdings. *International Journal of Sustainable Development & World Ecology, 19*(5), 426–432. https://doi.org/10.1080/13504509.2012.670670.

Bournaris, T., Papathanasiou, J., Moulogianni, C., & Manos, B. (2009). A fuzzy multicriteria mathematical programming model for planning agricultural regions. *New Medit, 8*(4), 22–27.

Bournaris, T., Moulogianni, C., & Manos, B. (2014). A multicriteria model for the assessment of rural development plans in Greece. *Land Use Policy, 38*, 1–8.

Bournaris, T., Papathanasiou, J., Manos, B., Kazakis, N., & Voudouris, K. (2015). Support of irrigation water use and eco-friendly decision process in agricultural production planning. *Operational Research, 15*(2), 289–306.

Brans, J. P. (1982). Lingenierie de la decision. Elaboration dinstruments daide a la decision. Methode PROMETHEE. In R. Nadeau & M. Landry (Eds.), *Laide a la Decision: Nature, Instruments et Perspectives Davenir* (pp. 183–214). Quebec: Presses de Universite Laval.

Breiling, M., Tatsuo, H., & Matsuhashi, R. (1999). *Contribution of rice production to Japanese greenhouse gas emissions applying life cycle assessment as a methodology*. University of Tokyo, Japan. Accessed July 12, 2017, from http://www.breiling.org/publ/lcaricejap-en.pdf

Brentrup, F., Kusters, J., Kuhlmann, H., & Lammel, J. (2004). Environmental impact assessment of agricultural production systems using the life cycle assessment methodology. I. Theoretical concept of a LCA method tailored to crop production. *European Journal of Agronomy, 20*, 247–264.

Buchholz, T., Rametsteiner, E., Volk, T. A., & Luzadis, V. A. (2009). Multi criteria analysis for bioenergy systems assessments. *Energy Policy, 37*(2), 484–495. ISSN 0301-4215. https://doi.org/10.1016/j.enpol.2008.09.054

Budsberg, E., Rastogi, M., Puettmann, M. E., & Johnson, L. (2012). *Life-cycle assessment for the production of bioethanol from willow biomass crops via biochemical conversion*. Accessed July 30, 2017, from https://www.researchgate.net/publication/259910766_Life-Cycle_Assessment_for_the_Production_of_Bioethanol_from_Willow_Biomass_Crops_via_Biochemical_Conversion. https://doi.org/10.13073/FPJ-D-12-00022.1

Cai, Y., Applegate, S., Yue, W., Cai, J., Wang, X., Liu, G., & Li, C. (2017). A hybrid life cycle and multi-criteria decision analysis approach for identifying sustainable development strategies of Beijing's taxi fleet. *Energy Policy, 100*, 314–325. ISSN 0301-4215, https://doi.org/10.1016/j.enpol.2016.09.047

CALCAS (Co-ordination Action for innovation in Life-Cycle Analysis for Sustainability). (2009). *D20 Blue Paper on Life Cycle Sustainability Analysis Revision 1 after the open consultation*. Final version of Deliverable D20 of Work package 7, revised with the comments received during the Open Consultation.

Capitanescu, F., Marvuglia, A., Navarrete Gutiérrez, T., & Benetto, E. (2017, March). Multi-stage farm management optimization under environmental and crop rotation constraints. *Journal of Cleaner Production, 147*(20), 197–205. ISSN 0959-6526. https://doi.org/10.1016/j.jclepro.2017.01.076

Caprara, C., & Martelli, R. (2016). Multi-criteria analysis of suitability for energy crops under structural and environmental constraints: A case study in a northeastern Italian region. *Transactions of the ASABE, 59*(4), 815–828.

Castellini, C., Boggia, A., Cortina, C., Dal Bosco, A., Paolotti, L., Novelli, E., & Mugnai, C. (2012). A multicriteria approach for measuring the sustainability of different poultry production systems. *Journal of Cleaner Production, 37*, 192–201.

Castellani, V., Sala, S., & Benini, L. (2017, January 1). Hotspots analysis and critical interpretation of food life cycle assessment studies for selecting eco-innovation options and for policy support. *Journal of Cleaner Production, 140*(Part 2), 556–568. ISSN 0959-6526. https://doi.org/10.1016/j.jclepro.2016.05.078

Cherubini, F., & Ulgiati, S. (2010). Crop residues as raw materials for biorefinery systems – A LCA case study. *Applied Energy, 87*(1), 47–57.

Chiotti, Q. P., & Johnston, T. (1995). Extending the boundaries of climate change research: A discussion on agriculture. *Journal of Rural Studies, 11*, 335–350.

Corrado, S., Castellani, V., Zampori, L., & Sala, S. (2017). Systematic analysis of secondary life cycle inventories when modelling agricultural production: A case study for arable crops. *Journal of Cleaner Production*. https://doi.org/10.1016/j.jclepro.2017.03.179.

Cristóbal, J. R. S. (2011). Multi-criteria decision-making in the selection of a renewable energy project in spain: The Vikor method. *Renewable Energy, 36*(2), 498–502. ISSN 0960-1481. https://doi.org/10.1016/j.renene.2010.07.031

Čuček, L., Klemeš, J. J., Varbanov, P., & Kravanja, Z. (2011). Life cycle assessment and multi-criteria optimization of regional biomass and bioenergy supply chains. *Chemical Engineering Transactions, 25*, 575–580. https://doi.org/10.3303/CET1125096.

Čuček, L., Varbanov, P. S., Klemeš, J. J., & Kravanja, Z. (2012). Total footprints-based multi-criteria optimisation of regional biomass energy supply chains. *Energy, 44*(1), 135–145. ISSN 0360-5442. https://doi.org/10.1016/j.energy.2012.01.040

Dace, E., & Blumberga, D. (2016). How do 28 European Union Member States perform in agricultural greenhouse gas emissions? It depends on what we look at: Application of the multi-criteria analysis. *Ecological Indicators, 71*, 352–358. ISSN 1470-160X. https://doi.org/10.1016/j.ecolind.2016.07.016

De Luca, A. I., Iofrida, N., Strano, A., Falcone, G., & Gulisano, G. (2015a). Social life cycle assessment and participatory approaches: A methodological proposal applied to citrus farming in Southern Italy. *Integrated Environmental Assessment and Management, 11*(3), 383–396. https://doi.org/10.1002/ieam.1611

De Luca, A. I., Molari, G., Seddaiu, G., Toscano, A., Bombino, G., Ledda, L., Milani, M., & Vittuari, M. (2015b). Multidisciplinary and innovative methodologies for sustainable management in agricultural systems. *Environmental Engineering and Management Journal, 14*(7), 1–11. Accessed July 13, 2017, from http://www.eemj.icpm.tuiasi.ro/pdfs/vol14/no7/Full/11_1052_De_Luca_14.pdf

De Luca, A. I., Iofrida, N., Leskinen, P., Stillitano, T., Falcone, G., Strano, A., & Gulisano, G. (2017). Life cycle tools combined with multi-criteria and participatory methods for agricultural sustainability: Insights from a systematic and critical review. *Science of the Total Environment, 595*(1), 352–370. ISSN 0048-9697, https://doi.org/10.1016/j.scitotenv.2017.03.284

Dias, L. C., Passeira, C., & Malça, J. F. F. (2016). Integrating life-cycle assessment and multi-criteria decision analysis to compare alternative biodiesel chains. *Annals of Operations Research*. https://doi.org/10.1007/s10479-016-2329-7

Dinh, L. T. T., Guo, Y., & Mannan, M. S. (2009). Sustainability evaluation of biodiesel production using multicriteria decision-making. *Environmental Progress & Sustainable Energy, 28*(1), 38–46.

Dogliotti, S., Rodríguez, D., López-Ridaura, S., Tittonell, P., & Rossing, W. (2014). Designing sustainable agricultural production systems for a changing world: Methods and applications. *Agricultural Systems, 126*, 1–2.

Dorini, G., Kapelan, Z., & Azapagic, A. (2011). Managing uncertainty in multiple-criteria decision making related to sustainability assessment. *Clean Technology Environmental Policy, 13*, 133–139.

EC EU Energy in Figures, Statistical Pocketbook. (2016). Accessed June 28, 2017, from https://ec.europa.eu/energy/sites/ener/files/documents/pocketbook_energy-2016_web-final_final.pdf

Eranki, P. L., & Dale, B. E. (2011). Comparative life cycle assessment of centralized and distributed biomass processing systems combined with mixed feedstock landscapes. *GCB Bioenergy, 3*, 427–438. https://doi.org/10.1111/j.1757-1707.2011.01096.x

European Commission. (2016, June 8). *European platform on life cycle assessment (LCA)*. Accessed July 15, 2017, from http://ec.europa.eu/environment/ipp/lca.htm

European Commission – Directorate-General for Communication (European Commission). (2014). *Environment A healthy and sustainable environment for present and future generations*. Luxembourg: Publications Office of the European Union. ISBN 978-92-79-42633-9. https://doi.org/10.2775/90841

European Commission – Joint Research Centre – Institute for Environment and Sustainability: International Reference Life Cycle Data System (ILCD). (2010, March). *Handbook – General guide for Life Cycle Assessment – Detailed guidance* (1st ed.). EUR 24708 EN. Luxembourg: Publications Office of the European Union.

European Commission – Joint Research Centre – Institute for Environment and Sustainability: International Reference Life Cycle Data System (ILCD). (2011, November). *Handbook – Recommendations for life cycle impact assessment in the European context* (1st ed.). EUR24571 EN. Luxemburg: Publications Office of the European Union.

European Parliament. (2009). Directive 2009/28/EC of the European parliament and of the council of 23 April 2009. *Official Journal of the European Union, 140*(16), 16–62. https://doi.org/10.3000/17252555.L_2009.140.eng.

Falcone, G., De Luca, A., Stillitano, T., Strano, A., Romeo, G., & Gulisano, G. (2016). Assessment of environmental and economic impacts of vine-growing combining life cycle assessment, life cycle costing and multicriterial analysis. *Sustainability, 8*(8), 793. https://doi.org/10.3390/su8080793

Finco, A., Bentivoglio, D., & Nijkamp, P. (2012). *International Journal of Foresight and Innovation Policy, 8*(2–3), 173–188.

Finkbeiner, M., Schau, E. M., Lehmann, A., & Traverso, M. (2010). Towards life cycle sustainability assessment. *Sustainability, 2*, 3309–3322.

Finnveden, G., Hauschild, M. Z., Ekvall, T., Guinée, J., Heijungs, R., Hellweg, S., Koehler, A., Pennington, D., & Suh, S. (2009). Recent developments in Life Cycle Assessment. *Journal of Environmental Management, 91*(1), 1–21. ISSN 0301-4797. https://doi.org/10.1016/j.jenvman.2009.06.018

Food and Agriculture Organization (FAO). (2003). *World agriculture: Towards 2015/2030 an FAO perspective*. ISBN: 92 5 104835 5 (FAO paperback).

Gaudreault, C., Samson, R., & Stuart, P. (2009). Implications of choices and interpretation in LCA for multi-criteria process design: De-inked pulp capacity and cogeneration at a paper mill case study. *Journal of Cleaner Production, 17*, 1535–1546.

Gaudreault, C., Bently Wigley, T., Margni, M., Verschuyl, J., & Titus, K. V. B. (2016). Addressing biodiversity impacts of land use in life cycle assessment of forest biomass harvesting. *WIREs Energy and Environment*. https://doi.org/10.1002/wene.211

Geneletti, D. (2013). *Multi-criteria analysis. LIAISE Toolbox*. Accessed July 31, 2017, from http://www.liaise-kit.eu/ia-method/multi-criteria-analysis

Ghafghazi, S., Sowlati, T., Sokhansanj, S., & Melin, S. (2010). A multicriteria approach to evaluate district heating system options. *Applied Energy, 87*, 1134–1140.

Giampietro, M., Mayumi, K., & Munda, G. (2006). Integrated assessment and energy analysis: Quality assurance in multi-criteria analysis of sustainability. *Energy, 31*, 59–86.

Gómez-Limón, J. A., Riesgo, L., & Arriaza, M. (2004). Multi-criteria analysis of input use in agriculture. *Journal of Agricultural Economics, 55*(3), 541–564.

Guinée, J. B., Heijungs, R., Huppes, G., Zamagni, A., Masoni, P., Buonamici, R., Ekvall, T., & Rydberg, T. (2011). Life cycle assessment: Past, present, and future. *Environmental Science & Technology, 45*(1), 90–96. https://doi.org/10.1021/es101316v

Guitouni, A., & Martel, J.-M. (1998). Tentative guidelines to help choosing an appropriate MCDA method. *European Journal of Operational Research, 109*(2), 501–521. ISSN 0377-2217, https://doi.org/10.1016/S0377-2217(98)00073-3

Haas, G., Wetterich, F., & Köpke, U. (2001). Comparing intensive, extensified and organic grassland farming in southern Germany by process life cycle assessment. *Agriculture, Ecosystems and Environment, 83*(1–2), 43–53.

Harris, G. (2002). Integrated assessment and modeling – science for sustainability. In R. Costanza & S. E. Joergensen (Eds.), *Understanding and solving environmental problems in the 21st century* (pp. 5–17). London: Elsevier.

Hayashi, K. (2006). Environmental indicators for agricultural management: Integration and decision making. *International Journal of Materials & Structural Reliability, 4*(2), 115–127.

Hayashi, K., Gaillard, G., Nemecek, T., & Reckenholz, A. F. (2005). *Life cycle assessment of agricultural production systems: Current issues and future perspectives* (pp. 25–26). Proceedings of the International Seminar on Technology Development for Good Agriculture Practice in Asia and Oceania, Epochal Tsukuba.

Hayashi, T., van Ierland, E. C., & Zhu, X. (2014). A holistic sustainability assessment tool for bioenergy using the Global Bioenergy Partnership (GBEP) sustainability indicators. *Biomass and Bioenergy, 66*, 70–80. ISSN 0961-9534, https://doi.org/10.1016/j.biombioe.2014.01.040

Henriksen, H. J., Kjær, J., Brüsh, W., Jacobsen, L.-B., Jensen, J. D., Grinderslev, D., & Andersen, P. (2007). Environmental benefits and social cost – An example of combining Bayesian networks and economic models for analysing pesticide management instruments. *Nordic Hydrology, 38*(4–5), 351–371. https://doi.org/10.2166/nh.2007.017

Hermann, B. G., Kroeze, C., & Jawjit, W. (2007). Assessing environmental performance by combining life cycle assessment, multi-criteria analysis and environmental performance indicators. *Journal of Cleaner Production, 15*(18), 1787–1796. ISSN 0959-6526. https://doi.org/10.1016/j.jclepro.2006.04.004

Hermans, C. M. L., Geijzendorffer, I. R., Ewert, F., Metzger, M. J., Vereijken, P. H., Woltjer, G. B., & Verhagen, A. (2010). Exploring the future of European crop production in a liberalized market, with specific consideration of climate change and the regional competitiveness. *Ecological Modelling, 221*, 2177–2187.

Herva, M., & Roca, E. (2013). Review of combined approaches and multi-criteria analysis for corporate environmental evaluation. *Journal of Cleaner Production, 39*, 355–371. ISSN 0959-6526. https://doi.org/10.1016/j.jclepro.2012.07.058

Hiloidhari, M., Baruah, D. C., Singh, A., Kataki, S., Medhi, K., Kumari, S., Ramachandra, T. V., Jenkins, B. M., & Thakur, I. S. (2017). Emerging role of Geographical Information System (GIS), Life Cycle Assessment (LCA) and spatial LCA (GIS-LCA) in sustainable bioenergy planning. *Bioresource Technology*. ISSN 0960-8524. Accessed March 15, 2017, from https://doi.org/10.1016/j.biortech.2017.03.079

Hobbs, B., & Meier, P. (2000). *Energy decisions and the environment: A guide to the use of multicriteria methods*. Dordrecht: Kluwer Academic Publishers.

Holman, I. P., Brown, C., Janes, V., & Sandars, D. (2017). Can we be certain about future land use change in Europe? A multi-scenario, integrated-assessment analysis. *Agricultural Systems, 151*, 126–135.

Hunkeler, D., Saur, K., Rebitzer, G., Schmidt, W., Jensen, A., Stranddorf, H., & Christiansen, K. (2004). *Life cycle management*. Pensacola, FL: SETAC.

Hunkeler, D., Lichtenvort, K., & Rebitzer, G. (Eds.). (2008). *Environmental life cycle costing*. Pensacola, FL: SETAC. ISBN 9781420054705 – CAT# 54708.

Islam, S., Ponnambalam, S. G., & Lam, H. L. (2016). Review on life cycle inventory: Methods, examples and applications. *Journal of Cleaner Production, 136*, 266–278. ISSN 0959-6526. https://doi.org/10.1016/j.jclepro.2016.05.144

ISO 14040. (2006). *International Standard Environmental management – Life cycle assessment – Principles and framework*. Geneva: International Organization for Standardization.

ISO 14044. (2006). *International Standard Environmental management – Life cycle assessment – Requirements and guidelines*. Geneva: International Organization for Standardization.

Jørgensen, A., Le Bocq, A., & Hauschild, M. Z. (2008). Methodologies for social Life Cycle Assessment – A review. *International Journal of Life Cycle Assessment, 13*, 96–103.

Karklina, K., Slisane, D., Romagnoli, F., & Blumberga, D. (2015). *Social life cycle assessment of biomethane production and distribution in Latvia* (Vol. II, pp. 128–132). Environment. Technology. Resources. Proceedings of the 10th International Scientific and Practical Conference, Rezekne, Latvia. https://doi.org/10.17770/etr2015vol2.628

Kasie, F. M. (2013). Combining simple multiple attribute rating technique and analytical hierarchy process for designing multi-criteria performance measurement framework. *Global Journal of Researches in Engineering Industrial Engineering, 13*(1), 15–30.

Keyes, S., Tyedmers, P., & Beazley, K. (2015). Evaluating the environmental impacts of conventional and organic apple production in Nova Scotia, Canada, through life cycle assessment. *Journal of Cleaner Production, 104*, 40–54.

Kralisch, D., Staffel, C., Ott, D., Bensaid, S., Saracco, G., Bellantoni, P., & Loeb, P. (2013). Process design accompanying life cycle management and risk analysis as a decision support tool for sustainable biodiesel production. *Green Chemistry, 15*(2), 463.

Kurka, T., & Blackwood, D. (2013). Selection of MCA methods to support decision making for renewable energy developments. *Renewable and Sustainable Energy Reviews, 27*, 225–233. ISSN 1364-0321. https://doi.org/10.1016/j.rser.2013.07.001

Kylili, A., Christoforou, E., Fokaides, P. A., & Polycarpou, P. (2016). Multicriteria analysis for the selection of the most appropriate energy crops: The case of Cyprus. *International Journal of Sustainable Energy, 35*(1), 47–58.

Laurin, L., & Dhaliwal, H. (2017). Life cycle environmental impact assessment. In M. A. Abraham (Ed.), *Encyclopedia of sustainable technologies* (pp. 225–232). Oxford: Elsevier. https://doi.org/10.1016/B978-0-12-409548-9.10060-0.

Lipuscek, I., Bohanec, M., Oblak, L., & Zadnik Stirn, L. (2010). A multi-criteria decision-making model for classifying wood products with respect to their impact on environment. *International Journal of Life Cycle Assessment, 15*(4), 359–367.

Lorenz, K., & Lal, R. (2016). Environmental impact of organic agriculture. *Advances in Agronomy, 139*, 99–152. ISSN 0065-2113. https://doi.org/10.1016/bs.agron.2016.05.003
Maia Angelo, A. C., Saraiva, A. B., Clímaco, J. C. N., Infante, C. E., & Valle, R. (2017). Life cycle assessment and multi-criteria decision analysis: Selection of a strategy for domestic food waste management in Rio de Janeiro. *Journal of Cleaner Production, 143*(1), 744–756. ISSN 0959-6526. https://doi.org/10.1016/j.jclepro.2016.12.049
Manos, B., Bournaris, T., Kamruzzaman, M., Begum, M., Anjuman, A., & Papathanasiou, J. (2006). Regional impact of irrigation water pricing in Greece under alternative scenarios of European policy: A multicriteria analysis. *Regional Studies, 40*(9), 1055–1068.
Manos, B., Papathanasiou, J., Bournaris, T., & Voudouris, K. (2010). A multicriteria model for planning agricultural regions within a context of groundwater rational management. *Journal of Environmental Management, 91*(7), 1593–1600.
Manos, B., Bournaris, T., Chatzinikolaou, P., Berbel, J., & Nikolov, D. (2013). Effects of CAP policy on farm household behaviour and social sustainability. *Land Use Policy, 31*, 166–181.
Margni, M., Rossier, D., Crettaz, P., & Jolliet, O. (2002). Life cycle assessment of pesticides on human health and ecosystems. *Agriculture, Ecosystem and Environment, 93*(1–3), 379–392.
Matthews, J., Parr, C., Araoye, O., & McManus, M. (2014, July). Environmental auditing of a packaging system for redesign: A case study exploration. *Journal of Clean Energy Technologies, 2*(3), 267.
Miettinen, P., & Hamalainen, R. P. (1997). How to benefit from decision analysis in environmental life cycle assessment (LCA). *European Journal of Operational Research, 102*, 279–294.
Mohamadabadi, H. S., Tichkowsky, G., & Kumar, A. Ã. (2009). Development of a multi-criteria assessment model for ranking of renewable and non-renewable transportation fuel vehicles. *Energy, 34*, 112–125. https://doi.org/10.1016/j.energy.2008.09.004
Muñoz, P., Antón, A., Montero, J. I., & Castells, F. (2004). *Using LCA for the improvement of waste management in greenhouse tomato production*. Proceedings of the Fourth International Conference on Life Cycle Assessment in the Agri-Food Sector, Bygholm, Denmark.
Murphy, F., Sosa, A., McDonnell, K., & Devlin, G. (2016). Life cycle assessment of biomass-to-energy systems in Ireland modelled with biomass supply chain optimisation based on greenhouse gas emission reduction. *Energy, 109*, 1040–1055. ISSN 0360-5442. https://doi.org/10.1016/j.energy.2016.04.125
Myllyviita, T., Holma, A., Antikainen, R., Lähtinen, K., & Leskinen, P. (2012). Assessing environmental impacts of biomass production chains – Application of life cycle assessment (LCA) and multi-criteria decision analysis (MCDA). *Journal of Cleaner Production, 29*, 238–245. ISSN 0959-6526. https://doi.org/10.1016/j.jclepro.2012.01.019
Myllyviita, T., Leskinen, P., & Seppälä, J. (2014). Impact of normalization, elicitation technique and background information on panel weighting results in life cycle assessment. *International Journal of Life Cycle Assessment, 19*, 377–386.
Narayanan, D., Zhang, Y., & Mannan, M. S. (2007). Engineering for sustainable development (ESD) in bio-diesel production. *Process Safety and Environmental Protection Trans IChemE, Part B, 85*(B5), 349–359. https://doi.org/10.1205/psep07016
Nijkamp, P., Van Wissen, L., & Rima, A. (1993). A household life cycle model for residential relocation behaviour. *Socio-Economic Planning Sciences, 27*(1), 35–53. https://doi.org/10.1016/0038-0121(93)90027-G.
Notarnicola, B., Tassielli, G., Renzulli, P. A., & Lo Giudice, A. (2015). Life Cycle Assessment in the agri-food sector: An overview of its key aspects, international initiatives, certification, labeling schemes and methodological issues. In B. Notarnicola, R. Salomone, L. Petti, P. A. Renzulli, R. Roma, & A. K. Cerutti (Eds.), *Life cycle assessment in the agri-food sector case studies, methodological issues and best practices*. Switzerland: Springer.
Parajuli, R., Knudsen, M. T., Djomo, S. N., Corona, A., Birkved, M., & Dalgaard, T. (2017, May). Environmental life cycle assessment of producing willow, alfalfa and straw from spring barley as feedstocks for bioenergy or biorefinery systems. *Science of the Total Environment, 586*(15), 226–240. ISSN 0048-9697. https://doi.org/10.1016/j.scitotenv.2017.01.207

Parker, P., et al. (2002). Progress in integrated assessment and modeling. *Environmental Modelling and Software, 17*, 209–217.

Parson, E. A. (1995). Integrated assessment and environmental policy making. *Energy Policy, 23*, 463–475.

Peters, K. (2016). Methodological issues in life cycle assessment for remanufactured products: A critical review of existing studies and an illustrative case study. *Journal of Cleaner Production, 126*, 21–37. ISSN 0959-6526. https://doi.org/10.1016/j.jclepro.2016.03.050

Popescu, G., & Bara, S. (2015). Regional development disparities in Romanian agriculture and rural development: A multi-criteria approach. In G. Popescu & A. Jean-Vasile (Eds.), *Agricultural management strategies in a changing economy* (pp. 1–29). Hershey, PA: IGI Global. https://doi.org/10.4018/978-1-4666-7521-6.ch001.

Rabl, A., & Holland, M. (2008). Environmental assessment framework for policy applications: Life cycle assessment, external costs and multi-criteria analysis. *Journal of Environmental Planning and Management, 51*(1), 81–105. https://doi.org/10.1080/09640560701712275

Razza, F., Sollima, L., Falce, M., Costa, R. M. S., Toscano, V., Novelli, A., Ciancolini, A., & Raccuia, S. A. (2016). Life cycle assessment of cardoon production system in different areas of Italy. *Acta Horticulturae, 1147*, 329–334. https://doi.org/10.17660/ActaHortic.2016.1147.46

Rebitzer, G., Ekvall, T., Frischknecht, R., Hunkeler, D., Norris, G., Rydberg, T., Schmidt, W.-P., Suh, S., Weidema, B. P., & Pennington, D. W. (2004, July). Life cycle assessment: Part 1: Framework, goal and scope definition, inventory analysis, and applications. *Environment International, 30*(5), 701–720. ISSN 0160-4120

Rebolledo-Leiva, R., Angulo-Meza, L., Iriarte, A., & González-Araya, M. C. (2017). Joint carbon footprint assessment and data envelopment analysis for the reduction of greenhouse gas emissions in agriculture production. *Science of the Total Environment, 593*, 36–46. ISSN 0048-9697. https://doi.org/10.1016/j.scitotenv.2017.03.147

Reeb, C., Venditti, R., Gonzalez, R., & Kelley, S. (2016, May 11). Environmental LCA and financial analysis to evaluate the feasibility of bio-based sugar feedstock biomass supply globally: Part 2. Application of multi-criteria decision-making analysis as a method for biomass feedstock comparisons. *BioResources*. Accessed July 18, 2017, from http://ojs.cnr.ncsu.edu/index.php/BioRes/article/view/BioRes_11_3_6062_Reeb_Environmental_LCA_Financial_Analysis_Sugar_Feedstock

Ren, J., Manzardo, A., Mazzi, A., Zuliani, F., & Scipioni, A. (2015). Prioritization of bioethanol production pathways in China based on life cycle sustainability assessment and multicriteria decision-making. *International Journal of Life Cycle Assessment, 20*(6), 842–853.

Riesgo, L., & Gómez-Limón, J. A. (2006). Multi-criteria policy scenario analysis for public regulation of irrigated agriculture. *Agricultural Systems, 91*(1–2), 1–28.

Risawandi, & Rahim, R. (2016). Study of the simple multi-attribute rating technique for decision support. *IJSRST, 2*(6). Print ISSN: 2395-6011; Online ISSN: 2395-602X.

Rogers, K., & Seager, T. P. (2009). Environmental decision-making using life cycle impact assessment and stochastic multiattribute decision analysis: A case study on alternative transportation fuels. *Environmental Science and Technology, 43*(6), 1718–1723. https://doi.org/10.1021/es801123h

Rowley, H. V., & Peters, G. (2009). *Multi-criteria methods for the aggregation of life cycle impacts*. Proceedings of Sixth Australian Conference on Life Cycle Assessment, Australian Life Cycle Assessment Society, Australia.

Rowley, H. V., & Shiels, S. (2011). *Valuation in LCA: Towards a best-practice approach*. Proceedings of Seventh Australian Conference on Life Cycle Assessment, Australian Life Cycle Assessment Society, Australia.

Roy, P., Nei, D., Orikasa, T., Xu, Q., Okadome, H., Nakamura, N., & Shiina, T. (2009). A review of life cycle assessment (LCA) on some food products. *Journal of Food Engineering, 90*(1), 1–10. ISSN 0260-8774. https://doi.org/10.1016/j.jfoodeng.2008.06.016

Sabiha, N.-E., Salim, R., Rahman, S., & Rola-Rubzen, M. F. (2016). Measuring environmental sustainability in agriculture: A composite environmental impact index approach. *Journal of*

Environmental Management, 166, 84–93. ISSN 0301-4797. https://doi.org/10.1016/j.jenvman.2015.10.003

Sastre, C. M., Carrasco, J., Barro, R., González-Arechavala, Y., Maletta, E., Santos, A. M., & Ciria, P. (2016, October). Improving bioenergy sustainability evaluations by using soil nitrogen balance coupled with life cycle assessment: A case study for electricity generated from rye biomass. *Applied Energy, 179*(1), 847–863. ISSN 0306-2619

Schaubroeck, T., Alvarenga, R. A. F., Verheyen, K., Muys, B., & Dewulf, J. (2013). Quantifying the environmental impact of an integrated human/industrial-natural system using life cycle assessment; A case study on a forest and wood processing chain. *Environmental Science and Technology, 47*(23), 13578–13586. https://doi.org/10.1021/es4046633

Scott, R. P., Cullen, A. C., Fox-Lent, C., & Linkov, I. (2016). Can carbon nanomaterials improve CZTS photovoltaic devices? Evaluation of performance and impacts using integrated life-cycle assessment and decision analysis. *Risk Analysis, 36*, 1916–1935.

Seghetta, M., Hou, X., Bastianoni, S., Bjerre, A.-B., & Thomsen, M. (2016). Life cycle assessment of macroalgal biorefinery for the production of ethanol, proteins and fertilizers – A step towards a regenerative bioeconomy. *Journal of Cleaner Production, 137*, 1158–1169. ISSN 0959-6526, https://doi.org/10.1016/j.jclepro.2016.07.195

Seppälä, J., Basson, L., & Norris, G. A. (2002). Decision analysis frameworks for life-cycle impact assessment. *Journal of Industrial Ecology, 5*, 45–68.

Smith, P., Martino, D., Cai, Z., Gwary, D., Janzen, H., Kumar, P., McCarl, B., Ogle, S., O'Mara, F., Rice, C., Scholes, B., & Sirotenko, O. (2007). Agriculture. In B. Metz, O. R. Davidson, P. R. Bosch, R. Dave, & L. A. Meyer (Eds.), *Climate Change 2007: Mitigation*. Contribution of Working Group III to the Fourth Assessment Report of the Intergovernmental Panel on Climate Change. Cambridge: Cambridge University Press.

Sundaram, S., Kolb, G., Hessel, V., & Wang, Q. (2017). Energy-efficient routes for the production of gasoline from biogas and pyrolysis oil process design and life-cycle assessment. *Industrial and Engineering Chemistry Research, 56*(12), 3373–3387. https://doi.org/10.1021/acs.iecr.6b04611

Tasca, A. L., Nessi, S., & Rigamonti, L. (2016). Environmental sustainability of agri-food supply chains: An LCA comparison between two alternative forms of production and distribution of endive in northern Italy. *Journal of Cleaner Production*. https://doi.org/10.1016/j.jclepro.2016.06.170.

Teshome, A., de Graaff, J., & Stroosnijder, L. (2014). Evaluation of soil and water conservation practices in the north-western Ethiopian highlands using multi-criteria analysis. *Frontiers in Environmental Science, 2*, 60. https://doi.org/10.3389/fenvs.2014.00060

Tiwari, D. N., Loof, R., & Paudyal, G. N. (1999). Environmental-economic decision-making in lowland irrigated agriculture using multi-criteria analysis techniques. *Agricultural Systems, 60* (2), 99–112.

Togarcheti, S. C., Mediboyina, M. K., Chauhan, V. S., Mukherji, S., Ravi, S., & Mudliar, S. N. (2017, July). Life cycle assessment of microalgae based biodiesel production to evaluate the impact of biomass productivity and energy source. *Resources, Conservation and Recycling, 122*, 286–294. ISSN 0921-3449. https://doi.org/10.1016/j.resconrec.2017.01.008

Tukker, A., Huppes, G., Guinée, J., Heijungs, R., de Koning, A., van Oers, L., & Suh, S. (2006). *Environmental Impact of Products (EIPRO) analysis of the life cycle environmental impacts related to the final consumption of the EU-25*. EUR 22284 EN. Catalogue number: ISBN-10. European Communities.

Tziolas, E., Bournaris, T., Nastis, S., & Manos, B. (2017a, June 8–10). *A taxonomy survey of Life Cycle Assessment used for the evaluation of biomass production*. 6th International Symposium and 28th National Conference on Operational Research, Thessaloniki, Greece.

Tziolas, E., Manos, B., & Bournaris, T. (2017b). Planning of agro-energy districts for optimum farm income and biomass energy from crops residues. *Operational Research, 17*(2), 535–546.

US EPA – National Service Center for Environmental Publications (NSCEP). (2006, May). *Life cycle assessment: Principles and practice*. EPA/600/R-06/060.

van der Sluijs, J. P. (2002). Definition of integrated assessment. In M. K. Tolba (Ed.), *Encyclopedia of global environmental change. Responding to global environmental change* (Vol. 4). Chichester: Wiley.

Vázquez-Rowe, I., Villanueva-Rey, P., Iribarren, D., Teresa Moreira, M., & Feijoo, G. (2012). Joint life cycle assessment and data envelopment analysis of grape production for vinification in the Rias Baixas appellation (NW Spain). *Journal of Cleaner Production, 27*, 92–102. https://doi.org/10.1016/j.jclepro.2011.12.039.

Verones, F., Bare, J., Bulle, C., Frischknecht, R., Hauschild, M., Hellweg, S., Henderson, A., Jolliet, O., Laurent, A., Liao, X., Lindner, J. P., de Souza, D. M., Michelsen, O., Patouillard, L., Pfister, S., Posthuma, L., Prado, V., Ridoutt, B., Rosenbaum, R. K., Sala, S., Ugaya, C., Vieira, M., & Fantke, P. (2017, September 10). LCIA framework and cross-cutting issues guidance within the UNEP-SETAC Life Cycle Initiative. *Journal of Cleaner Production, 161*, 957–967. ISSN 0959-6526. https://doi.org/10.1016/j.jclepro.2017.05.206

von Doderer, C. C. C., & Kleynhans, T. E. (2014). Determining the most sustainable lignocellulosic bioenergy system following a case study approach. *Biomass and Bioenergy, 70*, 273–286. ISSN 0961-9534. https://doi.org/10.1016/j.biombioe.2014.08.014

Xevi, E., & Khan, S. (2005). A multi-objective optimisation approach to water management. *Journal of Environmental Management, 77*(4), 269–277.

Xiao, J., Shen, L., Zhang, Y., & Jiqing, G. (2009). Integrated analysis of energy, economic, and environmental performance of biomethanol from rice straw in China. *Industrial and Engineering Chemistry Research, 48*(22), 9999–10007. https://doi.org/10.1021/ie900680d

Yazdani, M., & Graeml, F. R. (2014). VIKOR and its applications: A state-of-the-art survey. *International Journal of Strategic Decision Sciences, 5*(2), 56–83.

Yongmei, W., Xingping, Z., Luo, K., & Yuan, L. (2016). *External performance of biomass power generation in china based on life cycle analysis*. Accessed July 15, 2017, from https://www.researchgate.net/publication/312203847_External_Performance_of_Biomass_Power_Generation_in_China_Based_on_Life_Cycle_Analysis. https://doi.org/10.1166/jbmb.2016.1612

Yu, S., & Tao, J. (2009). Economic, energy and environmental evaluations of biomass-based fuel ethanol projects based on life cycle assessment and simulation. *Applied Energy, 86*, S178–S188. ISSN 0306-2619, https://doi.org/10.1016/j.apenergy.2009.04.016

Zare, R., Nouri, J., Abdoli, M. A., & Atabi, F. (2016). Application integrated fuzzy TOPSIS based on LCA results and the nearest weighted approximation of FNs for industrial waste management-aluminum industry: Arak-Iran. *Indian Journal of Science and Technology, 9*(2).

Zhou, Z., Jiang, H., & Qin, L. (2007). Life cycle sustainability assessment of fuels. *Fuel, 86*(1), 256–263. ISSN 0016-2361. https://doi.org/10.1016/j.fuel.2006.06.004

Zyoud, S. H., & Fuchs-Hanusch, D. (2017). A bibliometric-based survey on AHP and TOPSIS techniques. *Expert Systems with Applications, 78*, 158–181. ISSN 0957-4174. https://doi.org/10.1016/j.eswa.2017.02.016

The manufacturer's authorised representative in the EU is Springer Nature Customer Service Centre GmbH, Europaplatz 3, 69115 Heidelberg, Germany. If you have any concerns regarding our products, please contact ProductSafety@springernature.com

Printed and bound by CPI Group (UK) Ltd, Croydon, CR0 4YY
05/10/2026
02228144-0009